Introduction to Drought Tolerance in Plants

INTRODUCTION TO DROUGHT TOLERANCE IN PLANTS

Edited by:

Shobhit Kumar Singh

Delve
PUBLISHING

www.delvepublishing.com

Introduction to Drought Tolerance in Plants

Shobhit Kumar Singh

Delve Publishing
224 Shoreacres Road
Burlington, ON L7L 2H2
Canada
www.delvepublishing.com
Email: orders@arclereducation.com

ISBN: 978-1-77469-417-6 (Hardcover)

ABOUT THE EDITOR

Dr. Shobhit Kumar Singh is working as Scientist C in the Department of Plant Physiology and Breeding, Tocklai Tea Research Institute, Tea Research Association, Jorhat, Assam, India. He completed his Ph.D. in Agriculture with special reference to the development of Bradyrhizobium japonicum strains for symbiotic nitrogen fixation in Soybean crop (Glycine max) in the year 2012. He has 08 years of research experience in the field of Crop Improvement with special reference to Conventional Plant Breeding, Biodiversity of plant, varietal development of soybean and tea plants. He has long term experience in tea varietal development program.

TABLE OF CONTENTS

List of Figures .... *xiii*

Preface .... *xvii*

Introduction **1**

Chapter 1 Introduction to Drought **3**

1.1 Introduction 4

1.2 Desertification 6

1.3 Drought 9

1.4 Breeding 10

1.5 Drought Tolerance 12

1.6 Plant Adaptations 14

1.7 Environmental Conditions 18

Chapter 2 Drought Stress Tolerance **21**

2.1 Introduction 22

2.2 Drought Stress 23

2.3 Plant Health 23

2.4 Drought and Salinity 25

2.5 Evolution of Plants 27

2.6 Breeding 28

2.7 Drought Sensing 30

2.8 Climate Change 31

2.9 Photorespiration 33

2.10 ABA 37

2.11 Gibberellins 37

2.12 Ethylene 38

2.13 Jasmonic Acid 38

Chapter 3 **Biochemical Aspects of Drought Tolerance** **43**
3.1 Introduction 44
3.2 Temperature 45
3.3 Agricultural Productivity 46
3.4 Anthropogenic Actions 49
3.5 Agriculture and Climate Change 50
3.6 Stages 52
3.7 Drought Resistance 54
3.8 Water Uptake 54
3.9 Abiotic Stressors 56
3.10 Plant Functions 60

Chapter 4 **Molecular Breeding** **61**
4.1 Introduction 62
4.2 Water Stress 63
4.3 Plant Adaptations 64
4.4 Flowering 69
4.5 Abiotic Stimuli 70
4.6 Soil-Water Deficiency 72
4.7 Phytoremediation 74
4.8 Heavy Metals 75
4.9 Microbiome 77

Chapter 5 **Plant Responses to Drought Stress** **83**
5.1 Introduction 84
5.2 Drought Adaptation 85
5.3 Turgor 86
5.4 Soil Drying 88
5.5 Phenomics 89
5.6 Genes 89
5.7 Abiotic Stress 90
5.8 Plant Responses 92
5.9 C_3 and C_4 Plants 95
5.10 Osmotic Stress 97
5.11 Water Transfer 100

5.12 Woody Plants 102

Chapter 6 Genome Assisted Breeding 103

6.1 Introduction 104
6.2 Physiological Features 105
6.3 Quantitative Trait Locus 105
6.4 Stomatal Function 107
6.5 Food Demand 109
6.6 Metabolome 112
6.7 Plant Adaptability 113
6.8 Genetics and Genomics 115
6.9 Molecular Markers 117
6.10 Rice 120
6.11 Cowpea 121
6.12 Hormone Signaling 124
6.13 Sugarcane 125

Chapter 7 VAM in Drought Tolerance 127

7.1 Introduction 128
7.2 Arbuscular Mycorrhizal Fungi (AMF) 129
7.3 VAM 131
7.4 Symbiotic Relationship 132
7.5 Interaction Between Fungi and Plants 133
7.6 Mycorrhizal Fungi 134
7.7 Bacteria 134
7.8 Leaf Adaptations 136
7.9 Soil Bacteria 138
7.10 Weathering 139
7.11 Symbiosis 143
7.12 Heavy Metal Stress 144
7.13 Biocontrol 147

Chapter 8 Quality Improvement in Rice 149

8.1 Introduction 150
8.2 Abiotic Stressors 151
8.3 Salinity Stress 153

8.4 Evapotranspiration 154
8.5 Rice Quality 157
8.6 Breeding 158
8.7 Marker Assisted Selection 160
8.8 Gene Expression Analysis 162
8.9 Databases 162
8.10 Rice Genomics 163

Chapter 9 Abiotic Stress Tolerance 167
9.1 Introduction 168
9.2 Messengers 168
9.3 Abiotic Stress 169
9.4 Proteins 171
9.5 Microarray 173
9.6 Climate Change 174
9.7 Tropical Forests 175
9.8 Ultra-Violet Radiation 176
9.9 Carbon Dioxide 178

Chapter 10 Nanotechnology Applications in Drought Stress Research 183
10.1 Introduction 184
10.2 Nanoclays 185
10.3 Food Packaging 186
10.4 Nanoparticles 187
10.5 Fertilizers 188
10.6 Nanofabrication 189
10.7 Nano-Biosensors 191
10.8 Advances in Nanotechnology 193
10.9 Applications of Nanotechnology 195

Chapter 11 Alternative Farming 197
11.1 Introduction 198
11.2 Bacillus Thuringiensis 199
11.3 DNA Based Markers 199
11.4 RFLP 201
11.5 Blast Disease 202

11.6 RAPD 203
11.7 Phytoremediation........ 203
11.8 Agrochemical Contamination 204
11.9 DDT 205
11.10 Contaminated Soils........ 207
11.11 Pesticides........ 207
11.12 Phytoaccumulation........ 209
11.13 Phytoremediation........ 209
11.14 Wetlands 210
11.15 Soil Erosion........ 214
11.16 Environmental Issues 216
11.17 Ecological Fertilization........ 217

Chapter 12 Drought Modeling........ 219

12.1 Introduction........ 220
12.2 Indices........ 220
12.3 Drought Assessment........ 222
12.4 Adaptation........ 224
12.5 Early Warning System 225
12.6 Water Scarcity 227
12.7 Insurance........ 228
12.8 Climate Change and Groundwater........ 230

References........ 233

Index 245

LIST OF FIGURES

Figure 1.1 Drought

Figure 1.2 Meteorological drought

Figure 1.3 Desertification

Figure 2.1 Drought stress

Figure 2.2 Drought and salinity

Figure 2.3 Plant responses

Figure 2.4 Seed priming

Figure 2.5 Plant priming

Figure 2.6 Chlororespiration

Figure 2.7 Arrhenatherum elatius

Figure 3.1 Biochemistry of drought tolerance

Figure 3.2 H2O2 and drought

Figure 3.3 Tall fescue

Figure 3.4 Protein phosphorylation

Figure 4.1 Water-stressed conditions

Figure 4.2 Calcium in drought stress

Figure 4.3 Drought and salinity stress

Figure 4.4 ABA in drought response

Figure 4.5 Late flowering plants

Figure 4.6 Soil water deficiency

Figure 4.7 Drought resistance

Figure 4.8 Phytoremediation

Figure 4.9 Heavy metals

Figure 4.10 Endosphere microbiome

Figure 4.11 Functional plasticity

Figure 5.1 Latest approaches

Figure 5.2 QTL

Figure 5.3 Soil drying
Figure 5.4 Arabidopsis
Figure 5.5 C3 and C4 plants
Figure 5.6 Osmotic stress
Figure 5.7 Plant metabolic networks
Figure 6.1 Physiological features
Figure 6.2 Stomatal function
Figure 6.3 Demand for wheat
Figure 6.4 Transcriptomics
Figure 6.5 Metabolome
Figure 6.6 Multiparent advanced generation inter-cross (MAGIC) population
Figure 6.7 Chilling injury
Figure 6.8 Hormone signaling
Figure 7.1 VAM
Figure 7.2 Symbiotic relationship
Figure 7.3 Interaction between fungi and plants
Figure 7.4 Salinization of arable land
Figure 7.5 Soil bacteria
Figure 7.6 Weathering of rocks
Figure 7.7 Symbiosis and Pi
Figure 7.8 Heavy metal stress
Figure 8.1 Oryza glaberrima
Figure 8.2 Salinity stress
Figure 8.3 Evapotranspiration
Figure 8.4 Marker-assisted selection
Figure 9.1 cDNA microarray
Figure 9.2 Effect of UV-B
Figure 10.1 Nanoclays
Figure 10.2 Nano-biosensors
Figure 11.1 RFLP
Figure 11.2 Blast disease
Figure 11.3 Contaminated soils
Figure 11.4 Phytoaccumulation

Figure 11.5 Wetlands
Figure 11.6 Soil erosion
Figure 11.7 Ecological fertilization
Figure 12.1 Drought assessment

PREFACE

Drought is one of the disastrous events witnessed by humans for ages. In recent years, researchers are in the process of estimating the actual cause of drought and its mitigation. The drought conditions can exist for decades if not months and this will damage the economic fabric of the society by bringing down the balance across communities. Due to climate change, drought vulnerability is being increased on a regular basis. Drought can be categorized as environmental stress and it can occur due to several factors. The molecular, biochemical, morphological and physiological characteristics of the plants will be affected due to prolonged drought conditions. High temperature stress affects the transpiration rates in the plants. Molecular breeding techniques have provided a path to overcome drought stress in plants and it is quite successful. The drought stress can be studied along with salinity stress due to various reasons as both are integrated. Genotypic variations among the plants will help them in sustaining drought periods. The hydraulic conductivity, size and depth of the roots and phenology remain the important indicators to know plant stress status. The activity of mesophyll and stomata can be altered due to changes in the natural settings if disturbed due to drought. The selection of good varieties through conventional breeding programs was expensive and this problem can be solved by using genome-assisted breeding. Quantitative trait loci can be used in the identification of the natural variations that occur in epidermal and stomatal characteristics in rice, barley and wheat crops. Vesicular Arbuscular Mycorrhiza can be used by plants in drought and harsh conditions. VAM can induce certain features in the plants that can help the plants in surviving with high temperatures, metals, salinity and drought. The quality improvement in the crops has been a challenging task due to its genetic diversity and this can be attributed to the complicated polygenic structure.

This book mainly focuses on the role of drought in reducing the yields and measures to mitigate the losses. Advanced technologies like genome-assisted breeding and nanotechnology are discussed in this book. This book can serve as a handbook for undergraduates and postgraduates who are interested in working on development of drought-tolerant plants through conventional and modern breeding approaches.

INTRODUCTION

The identification and characterization of the systems that allow plants to resist abiotic stressors have made significant progress. The identification of many transporters, as well as the understanding of metabolic fluxes and the key restrictions responsible for the synthesis of suitable solutes, open the prospect of genetic engineering in crop plants with better stress tolerance. Environmental stress has a significant impact not only on current crop species, but also on the introduction of agricultural plants into areas that are not already used for agriculture. Stresses are anticipated to exacerbate the severity of plant problems in the near future. Drought, salinity, severe temperatures, pollution, UV radiation, and other abiotic stressors are all experienced by plants. Abiotic stress has a negative impact on crop production over the world, resulting in lower yields for most main crops. Drought is considered the most dangerous of the numerous abiotic stresses. Drought may cause a significant food crisis by 2050, when the world's population is predicted to surpass 10 billion people, due to an expanding global population. This condition could deteriorate as a result of global climate change, which could increase the frequency, duration, and severity of water shortages. We need to understand the complex mechanisms behind drought tolerance and to design exceptional crop varieties that are drought tolerant while maintaining other agronomic and quality characteristics. The discovery of new genes involved in drought tolerance in crop plants will help us better understand the molecular mechanisms behind drought tolerance. The identification of novel genes, the investigation of their expression patterns in response to drought, and the identification of their possible functions in drought adaptation will serve as the foundation for effective breeding techniques to improve agricultural drought tolerance.

CHAPTER 1

INTRODUCTION TO DROUGHT

CONTENTS

1.1 Introduction 4
1.2 Desertification 6
1.3 Drought 9
1.4 Breeding 10
1.5 Drought Tolerance 12
1.6 Plant Adaptations 14
1.7 Environmental Conditions 18

1.1 INTRODUCTION

Droughts, after hurricanes, are the most expensive weather events. Droughts, unlike other severe weather disasters like hurricanes, tornadoes, and thunderstorms, are notoriously difficult to predict when they begin and end. Because the early consequences of a drought might be difficult to detect, it could take weeks or months to confirm that a drought has begun. For the same reason, the end of a drought is difficult to predict. A dry spell might last weeks, months, or even years. Drought conditions can last for a decade or more in some areas. The more time a drought lasts, the more negative consequences it has on people (Ahanger, Siddique, & Ahmad, 2021).

Drought (Figure 1.1) is a natural hazard that is now one of the world's least understood and managed events. Drought vulnerability is increasing as the global climate changes, the human population grows, water resources are increasingly under pressure for alternative uses, and people use water in so many different ways and apply so many different demands in relation to human consumption including drinking water and sanitation, industrial uses, agricultural production systems, and environmental requirements that there is no universal definition of drought. Droughts are not solely a physical phenomenon defined solely by climate fluctuation. A drought is not simply defined as a period of low rainfall; if it were, much of the world's arid and semiarid regions, for example, would be in a state of near-perpetual drought. Drought is thus largely governed by the balance between water availability and demand. A drought's impact is denoted by the interaction of a natural event and the demand placed on the water supply, with human activity typically aggravating the effects (Coles & Eslamian, 2017).

A meteorological drought (Figure 1.2) isa period of excessively dry weather that lasts long enough to generate a major hydrologic imbalance in the afflicted area. Drought in agriculture is a climatic event characterized by a lack of precipitation that has a negative impact on crop or range output. A period of below-average water content in streams, reservoirs, groundwater aquifers, lakes, and soils is known as a hydrological drought. Drought in the socioeconomic and environmental sense refers to a period when the supply of water falls short of demand, causing human activities and ecosystem functions to fail. It can be linked to elements of meteorological, hydrological, and agricultural drought.

Figure 1.1. Drought.

Source: https://www.ucsusa.org/sites/default/files/styles/original/public/images/1200-corndrought.jpg?itok=EAJuxQvZ

Droughts are thus natural occurrences, with intensity and severity influenced by local topography, water demand, and climatological factors, and can end nearly as quickly as they begin if moisture or water deficiencies are minor. Droughts in eastern Australia, for example, may be influenced by global phenomena such as the Southern Oscillation Index, which is driven by oscillations in the surface temperature. This level of influence allows for some limited forecasting and preparations, but they are not yet considered absolute assessments of the length and severity of the drought that may emerge in each climate zone, which is influenced by highly distinct large-scale, dominant weather and climate patterns.

Figure 1.2. Meteorological drought.

Source: https://smartwatermagazine.com/sites/default/files/styles/thumbnail-830x455/public/drought_ground_pixabay_2.jpg?itok=tSTaAgAo

1.2 DESERTIFICATION

Water erosion, wind erosion, and sedimentation by these agents are examples of these processes, as are long-term reductions in the amount or diversity of natural vegetation, where applicable, and salinization and sodification. Desertification, (Figure 1.3) according to this UN definition, refers to the interactions between humans and their environments that have an impact on human welfare in arid and semiarid areas. This term, however, does not lend itself to easy quantification and requires further explanation. Initially, it was thought that desertification was caused by a decrease in biological output. Desertification is now more widely characterized as land deterioration, as previously stated. While definitional issues have previously resulted in global overestimations of the degree of desertification of up to 66%, serious hazards can occur if environmental problems are minimized(Squires, 2017).

Figure 1.3. Desertification.

Source: https://www.esa.int/var/esa/storage/images/esa_multimedia/images/2007/06/environmental_issues/9795684-9-eng-GB/Environmental_Issues_pillars.jpg

Drought is a diachronically repeating feature of nature's climate variability on a regional scale. Drought is a serious natural calamity affecting a variety of areas of the economy, society, and environment. Drought is a type of extreme climatic event that is produced by a lack of precipitation occurrences in a region over a period of time. Meteorological drought refers to the early phases of accumulated precipitation deficits, which is an area-specific natural phenomenon because the atmospheric conditions that cause precipitation deficits are very diverse from region to region. Many different drought indicators and indices are utilized all around the world.

Furthermore, monitoring and assessing meteorological droughts, as well as substantial drought impacts, are taken into account.

The World Meteorological Organization recommends that the Standardized Precipitation Index be used worldwide, but it may not be considered as the only and absolute index. Other metrics should be considered based on specific regional and meteorological variables. Drought in the land phase of the hydrological cycle is characterized as a considerable reduction in all sources of water availability. A hydrological drought is a shortage of precipitation, resulting in the drying up of lakes, reservoirs, and rivers, as well as the depletion of groundwater. There is a global need to continue developing newer, potentially better drought indices that can account for a changing climate with potential shifts in temperature and precipitation regimes. Drought impacts, on the other hand, are not yet routinely recorded globally, and there are data restrictions, as well as a lack of sufficient and appropriate data, for predicting drought hazards. Despite these limitations and challenges, the international community is interested in tackling drought risk and building global drought risk models. International initiatives to define and implement drought quantification and monitoring standards, as well as systematically account for drought losses and impacts, are currently a crucial first step toward assessing drought risk. Furthermore, catastrophe insurances are unquestionably important components of drought risk management and reduction.

Every year, new local foci of desertification processes emerge, and it is most efficient to deal with them and the causes of their growth as they emerge. Desertification areas that are left untreated because of a lack of dedication, concern, political will, technology, or financial resources spread, sometimes at an accelerated rate. Dealing with the many facets of desertification and drought is a difficult task. Some elements, especially in developing countries' drylands, occur as a by-product of some usually crucial life-sustaining human activity. Desertification and drought, on the other hand, erode the all-important resource basis on which an ever-increasing population must progressively rely in the long run. Few environmental challenges demand immediate global attention as much as the impact of drought and desertification, which is now adding to the life-threatening pressures on residents of drought-prone areas in particular (Dalezios, Dunkel, & Eslamian, 2017).

1.3 DROUGHT

Drought is the most expensive natural calamity. It disturbs more people than any other natural calamity. Droughts have historically put a strain on the economies of many developing and industrialized countries around the world. During low-flow seasons, surface and subterranean water resources are put under a lot of stress. Changes in the natural environment, such as climate change, land use change, and population growth, have considerably increased the demand for water supplies all over the world. If a country or institution is not prepared for catastrophic events like droughts, it will face significant hardships. As a result, an organization should be prepared with the required instruments for assessing and interpreting low flows prior to an extreme event. A robust knowledge base and good administrative abilities are necessary for streamflow drought hazard mitigation.

Climate extremes, primarily induced by anthropogenic climate change and rising climate variability, have a negative impact on agriculture. Furthermore, agricultural production hazards and agriculture's vulnerability may become a problem in numerous parts of the world, since droughts and heat waves are anticipated to exacerbate crop failure. Drought is one of the most serious natural disasters, wreaking havoc on the environment, society, agriculture, and economy(M. K. Goyal, Gupta, & Eslamian, 2017).

Drought is the leading source of agricultural loss around the world, posing a serious danger to food security. Plant biotechnology is now one of the most promising fields in terms of producing crops that can provide high yields in water-scarce environments. The key response pathways to drought stress have been discovered through studies on *Arabidopsis thaliana* plants, and numerous drought resistance genes have already been put into crops. So far, most plants with increased drought resistance have had lower agricultural yields, indicating that new techniques to uncoupling drought resistance from plant development are still needed. Brassinosteroid (BR) hormone receptors use tissue-specific pathways to mediate various developmental responses during root growth. We discovered that boosting BR receptors in vascular plant tissues offers drought resistance without compromising growth in Arabidopsis, providing a unique opportunity to examine the mechanisms that confer drought resistance in plants with cellular specificity. In addition, current genome editing tools could aid in the discovery and manipulation of novel genes that potentially confer drought resistance. We anticipate that collaborative efforts will identify sustainable strategies for increasing agricultural yield in water-stressed regions in the next few years.

1.4 BREEDING

In traditional breeding, drought tolerance is introduced into lines with adequate agronomic features by identifying genetic diversity among crop varieties or sexually compatible species and introducing this tolerance into lines with suitable agronomic traits. Although conventional breeding for drought resistance had some success and continues to do so, it is a laborious process that is hampered by a lack of relevant genes for the same. Drought and cold stress have an impact on plant morpho-biochemical systems. Various biotechnological approaches have been used to explore the plant's response to these challenges. Drought resistance in plants is increased by several genes involved in the plant's response to abiotic stressors. Several drought-tolerant genes have been identified in several plant species, although their function is yet unknown. As a result, reverse genetics approaches are suitable for identifying genes associated with abiotic challenges; nevertheless, cloning and recognition of gene-families implicated in stress resistance are required. Plant responses at the biochemical level have received little attention.

Multiple genes and signaling pathways that regulate critical plant responses have been discovered in response to abiotic stressors. Plants use a variety of signaling chemicals, such as abscisic acid (ABA), to help them cope with abiotic stressors at the molecular level. These molecules regulate a wide range of stress-related genes. Different proteins, such as molecular chaperones and enzymes, are produced by these unique genes. Traditional plant breeding approaches for developing drought-tolerant crops have played an important role in recent decades; however, several issues have arisen because this approach is time-consuming, labor-intensive, and expensive. In addition, classic plant breeding techniques are used to change unwanted genes. In addition, to get the desired result, numerous breeding and selection cycles may be required. Traditional breeding strategies have had limited success because of the fact that the drought-resistant trait is regulated by multiple separate genes, and there is a strong relationship between the genes for stress resistance and those involved in yield potential. As a result, more effective methods for genetically modifying plants to boost drought resistance are required. As a result, the transgenic technique only allows for the transfer of desirable genes and the enhancement of features in the shortest time possible. Based on the studies of transgenic Arabidopsis, genetic engineering applications are being extensively studied to boost plant resistance to abiotic stresses by modifying regulatory genes or employing recently found genes to improve stress tolerance(Ahmed et al., 2019).

Drought conditions harm plants, and this is one of the most serious issues they confront. Water usage efficiency (WUE) for plants is being improved through several methods. We can control the amount of water utilized by the crops by changing the irrigation time and genotyping of plants. Plants use a variety of techniques to keep water in their bodies, including the formation of deep roots, salt pumping into and out of cells, increased water-resistive diffusion, and the formation of small succulent leaves to prevent water loss. During drought conditions, osmolytes such as polyols, metabolites, organic acids, glycine betaine, and proline, as well as many other amino acids, play a crucial role in maintaining the normal functioning of plant cells. Many enzymes and chemicals, such as polyamines and citrulline, act as antioxidants to mitigate the effects of drought. Genetic engineering and marker-based selection methods are all techniques that researchers have developed to make plants drought resistant. Low precipitation and high evapotranspiration cause a reduction in plant cell division, root proliferation, stomatal closure, changes in the relationship between plant and water uptake efficiency, and high production of Abscisic acid, which further reduces evapotranspiration by controlling stomatal pores. Because plant growth is a reversible process with numerous metabolic pathways, all phases of development, growth, cell differentiation, division, elongation, volume increase, and weight intensification are irreversible and influenced(Butt, 2017).

When the weather isn't cooperating, the Alfalfa plant uses less water than maize. Carbon and oxygen make up the majority of the plant's dry matter. These are heavier than hydrogen atoms, which are formed during the photosynthetic process of water hydrolysis. Through controlled opening and closure of stomata in appropriate situations, the waxy cuticle and stomatal opening limit water loss. Environmental elements such as humidity, temperature, air quality, and CO_2 concentration have an impact on water utilization. As humidity levels drop, the rate of transpiration rises as a result. Plants with increased drought tolerance were formed, despite the fact that it was a tough method because plants require water as a basic need. The findings of the experiments suggest that osmotic adjustment might be beneficial without sacrificing yield under optimal conditions. The importance of deep rooting was also deduced from the fact that deep-rooted trees are more tolerant than grasses with small, shallow roots. While the cuticle plays an important function in boosting plant tolerance, the variance was explored for plants with stomata on both sides and was linked to the thickness of the cuticle layer. The wax content of rice was lower than that of paddy rice. Epicuticle wax was applied to grasses in the United States to make them more tolerant.

1.5 DROUGHT TOLERANCE

Drought tolerance involves processes that operate at several geographical and temporal dimensions, ranging from fast stomatal closure through crop production maintenance. We look at how short-term mechanisms are controlled for maintaining shoot water potential, as well as how long-term processes have been constrained by evolution or breeding to fit into drought acclimation techniques. These short- and long-term feedback processes play a role in the trade-offs between carbon accumulation and the danger of harmful soil water depletion. Depending on the drought condition, corresponding characteristics and alleles may have a beneficial or negative impact on crop yield. The comparative interests of combinations of alleles are then estimated using a combination of modeling and genomic prediction, depending on drought circumstances. As a result, drought tolerance is studied probabilistically by calculating the benefit and risk of each gene combination(Chave et al., 2014).

In dry and semi-arid tropical areas affected by climate change, water scarcity is the fundamental limitation on crop output; as a result, agricultural systems must be adjusted. To improve crop yield and quality, several concepts and tactics should be addressed, particularly in vulnerable locations where such environmental changes pose a danger of food insecurity. Natural genes have evolved over time and are preserved in crop wild relatives and landraces for drought tolerance breeding using conventional and molecular methods. With advances in genomics and phenomics, a number of new tools are now available to aid in the discovery of drought resistance genes in crop wild relatives and landraces, as well as their relatively simple transfer into advanced breeding lines, thereby speeding breeding progress and generating drought-resistant varieties. Marker-assisted selection, genomic selection, and targeted gene editing are some of the technologies available. The combination of these two major strategies, including advances in conventional and molecular breeding for drought tolerance of conventional staple crops, as well as the introduction of drought-tolerant neglected and underutilized species into existing production systems, has the potential to

Knowledge of plant abiotic stress responses has evolved rapidly in recent years, thanks to the availability of comprehensive information on a few plant genomes as well as numerous genomics and proteomics technologies. The documentation of genes affected by stress is particularly extensive in the model plant Arabidopsis, and should be completed shortly in rice. Despite this, abiotic stress signaling pathways have been shown to be quite complicated, with much more to learn.

Only a few genes that play key roles in plant stress adaptation have been identified with certainty so far. It's also frequented in plant cells for receptor activation to result in the production of second messengers, which are intracellular signals that are translated from the main external signal. Other signaling components interpret these intracellular messengers, resulting in the activation of downstream pathways with diverse outputs. Reversible protein phosphorylation is frequently involved in these mechanisms. Protein phosphorylation may cause the activation of transcription factors that induce the production of stress-responsive genes. Additional components that recruit and build signaling complexes, target signaling molecules, and regulate their lifespan are frequently required for signal transmission. In many circumstances, the signaling pathways that may have been triggered by the same stress signals govern these components as well. Salinity, a key environmental stress that restricts agricultural production, has a negative impact on plant growth and development. The early phase of growth reduction is caused by an osmotic impact and is similar to the initial response to water stress. There are few genotypic variations in this phase. The incapability of a salt sensitive species or genotype to prevent salt accumulation in leaves to lethal levels distinguishes it from a more salt-resistant one in the second phase.

During germination, most crop plants are salt-tolerant, but they become salt sensitive during emergence and vegetative development. Salinity inhibits root and shoot growth; however, supplementary Ca partially alleviates the growth inhibition. The Ca effect appears to be linked to the maintenance of plasma membrane K over Na selectivity. Although salt stress can hasten reproductive growth, inhibit spike development, and reduce yield potential in wheat, low yield is primarily associated with a reduction in tillers, and in some cultivars, sterile spikelets, whereas in the more salt-sensitive rice, low yield is primarily associated with reduction in tillers, and in some cultivars, sterile spikelets. Plants with higher salt tolerance must flourish in saline fields with a variety of extra challenges. Salinity interacts with a variety of stressors, including boron toxicity, but the mechanisms of salinity-boron interactions remain unknown.

CHAPTER 2

DROUGHT STRESS TOLERANCE

CONTENTS

2.1 Introduction.......22
2.2 Drought Stress.......23
2.3 Plant Health.......23
2.4 Drought and Salinity.......25
2.5 Evolution of Plants.......27
2.6 Breeding.......28
2.7 Drought Sensing.......30
2.8 Climate Change.......31
2.9 Photorespiration.......33
2.10 ABA.......37
2.11 Gibberellins.......37
2.12 Ethylene.......38
2.13 Jasmonic Acid.......38

2.1 INTRODUCTION

Drought is an environmental stress, and it can occur for a variety of reasons, including inadequate rainfall, salinity, extreme heat and cold, and high light intensity, among others. Drought stress affects plants in a variety of ways, including physiological, morphological, biochemical, and molecular alterations. Many plants have improved their drought resistance mechanisms, although these processes vary and are dependent on the plant species. Typically, systems involved in drought tolerance in plants follow a similar plan: maintain cell homeostasis in water-short situations by boosting water intake to the cells. Another typical drought resistance strategy in annual plants is drought avoidance. Escape from stress circumstances is the key strategy for plant development in drought conditions with this mechanism (Salehi-Lisar & Bakhshayeshan-Agdam, 2016). Drought is one of the environmental variables that restricts plant growth and development. Plant drought tolerance is mediated by a variety of physiological and molecular processes. Drought stress memory and priming are two important methods by which plants might strengthen their tolerance to future droughts. Multiple physiological, proteomic, transcriptional, and epigenetic processes are involved in stress memory (Liu, 2014).

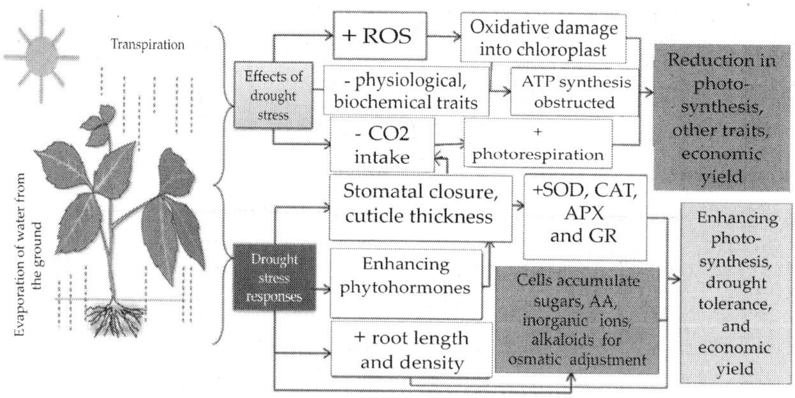

Figure 2.1. Drought stress.

Source: https://www.mdpi.com/plants/plants-10-00259/article_deploy/html/images/plants-10-00259-g002.png

2.2 DROUGHT STRESS

Drought stress (Figure 2.1) causes major physiological, cellular, biochemical, and molecular changes in plants, limiting their development and output. Plant hormone studies have made significant progress in explaining how they contribute to or influence whole-plant drought responses. During drought stress, developments in transcriptomics and mutant analyses have suggested that particular hormones govern processes like leaf senescence, antioxidant metabolism, carbon balance, and gas exchange. Hormones such as abscisic acid, auxins, cytokinins, gibberellins, ethylene, salicylates, and Jasmonates may regulate these plant responses independently by regulating transcription factors and inducing or repressing stress-responsive genes downstream. Hormone-to-hormone or hormone-to-sugar cross-talk, on the other hand, may aid drought responses. The current state of knowledge about hormone-to-hormone or hormone-to-sugar cross-talk in the regulation of physiological responses to drought stress is also examined. There are suggestions for further research, such as which molecules detect a certain hormone and then activate signal transduction cascades in response to drought stress. Cross talk signals produced from hormone-to-hormone, hormone-to-sugar, or secondary messengers like calcium or ROS are not well understood and should be explored further. Research into the unresolved concepts of hormone signaling perception and cross-talk between hormones and other metabolites will shed more light on the molecular processes that influence hormone regulation of drought tolerance in plants(Burgess & Huang, 2016).

2.3 PLANT HEALTH

Plants' health and productivity suffer significantly when they are subjected to drought stress. Agricultural businesses are very interested in methods to prepare plants for drought stress by priming them with exogenous therapies prior to water restriction, as this may help to minimize plant damage and crop loss. Exogenous priming treatments on plants can have a variety of physiological consequences. The type of priming used, the manner of application, the concentration and frequency of application, and the type of priming used all influence the observed results. The environment and plant type are other important considerations in deciding whether or not a priming approach is effective. Many researchers have looked at various compounds that could be employed as priming agents, which has not only enhanced our understanding of the priming compound's natural functions but also opened up new avenues for product and applied technology development (Merewitz, 2016).

Plants, on the other hand, use a variety of techniques, including osmotic adjustment, which can help sustain water uptake, cell turgor, and stomata functioning by accumulating huge amounts of osmolytes. The genetic diversity of osmotic adjustment in a number of crop species has paved the way for the use of osmotic adjustment–related genes and molecular markers in breeding to increase agricultural drought tolerance via marker-assisted selection (Zivcak, Brestic, & Sytar, 2016).

Plants are constantly exposed to a number of environmental stressors. Glutathione (GSH) has long been known to play a role in plant stress tolerance and management, in addition to its important role as a master antioxidant in living systems. However, it is yet unknown how GSH interacts with other stress-related phytohormones such as salicylic acid (SA), jasmonic acid, ethylene (ET), and abscisic acid. These lines have been shown to have significant abiotic stress tolerance capability i.e., drought and salt. Furthermore, transcriptome and proteomic profiling of these transgenic lines revealed the genes and proteins that were changed in the presence of increased GSH and are likely regulated by GSH. Several SA-related genes such as PR1, GLS, as well as proteins such as HSP 70, ADC, NBS-LRR, CA, and PR10 were discovered. The expression level of SA-related genes in the NtGB line was further validated by quantitative reverse-transcriptase polymerase chain reaction analysis. Interestingly, other key phytohormones such as ET, in addition to SA, have been found to interact with GSH. GSH interacts with SA and ET to counteract environmental stress conditions in plants, according to a comprehensive investigation at the transcript and protein levels (Chattopadhyay, 2016).

Plants are sessile organisms, meaning they are always exposed to changing environmental variables, unlike all other organisms. Drought, cold, salinity, and high temperatures are all connected in terms of their effects on plants. Heat-shock proteins are a highly conserved group of polypeptides synthesized by cells from all animals, including plants, to effectively respond to stressors (HSPs). Induced protein expression is a widespread occurrence in all living organisms. According to their molecular weight, these proteins are divided into Hsp100, Hsp90, Hsp70, Hsp60, and tiny heat-shock proteins (sHsps). There's a lot of evidence that these stress response proteins are important in plants in both normal and stressful situations. In reality, Hsps are now known to influence signaling mechanisms, host defense mechanisms, translation, glucose metabolism, and amino acid metabolism through the genome (Augustine, 2016).

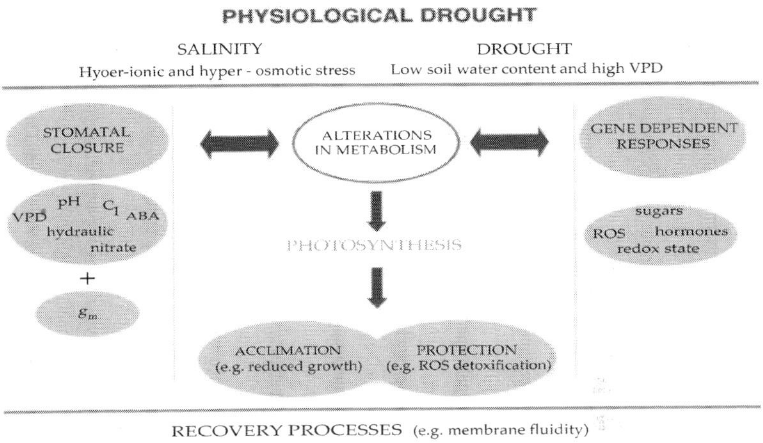

Figure 2.2: Drought and salinity.

Source: https://www.researchgate.net/profile/Carla-Pinheiro-2/publication/23133543/figure/fig1/AS:394524850442241@1471073462294/Direct-effects-of-drought-and-salinity-on-stomata-and-mesophyll-g-m-conductance-as_W640.jpg

2.4 DROUGHT AND SALINITY

Drought and salinity (Figure 2.2) impair redox equilibrium by interfering with essential metabolic activities including photosynthesis, respiration, and photorespiration, among others, resulting in nonspecific oxidative damage. Low molecular weight antioxidants paired with numerous other antioxidative enzymes act as redox buffers, combating the negative alterations. Antioxidants have been revealed to provide critical information on the redox status and to alter gene expression in response to dehydration and salinity stress, allowing for protection and stress adaptation(Chakrabarty, Aditya, Dey, Banik, & Bhattacharjee, 2016). In the context of global climate change, the major threat to global food security is crop loss caused by abiotic pressures such as drought and salinity. Significant progress in understanding how abiotic factors affect plant development and yield, as well as how plants respond to and adapt to these challenges. In times of drought, stomata gradually shut, resulting in a decrease in water-use efficiency and net photosynthetic activity. Aside from other factors, changes in photosynthetic pigments are strongly linked to drought resistance. Self-protective reactions to stress at the leaf level must thus be triggered swiftly to prevent irreversible damage to the photosynthetic machinery. Drought tolerance relies on the scavenging of ROS by enzymatic and non-enzymatic systems, cell membrane stability, the production of stress-responsive genes, and proteins. Furthermore, metabolite changes play a key role in drought adaptation, with polyphenols, lipophilic compounds, and some TCA cycle metabolites all playing a role in defense and protection, while carbohydrates, amino acids, and polyols play a role in osmoregulation.

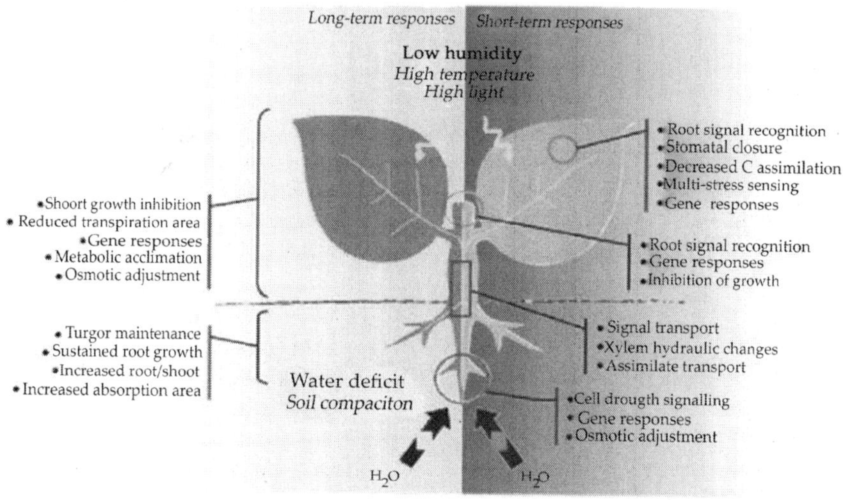

Figure 2.3. Plant responses.

Source: https://www.researchgate.net/profile/Sheel-Bansal/publication/269574336/figure/fig2/AS:614264469463041@1523463470147/Whole-plant-responses-to-drought-stress-Left-long-term-or-acclimation-responses-right_W640.jpg

Plant responses (Figure 2.3) to salinity follow a biphasic model, with an early phase resembling drought and a later phase causing ion toxicity. In the first phase, development slows dramatically, and stomata close in response to a drop in water potential. Ion accumulations, notably Na+, disrupt photosynthetic components like enzymes and pigments, and increase oxidative stress in the second phase. ROS serve a dual role in the salt stress response, acting as both a toxic by-product of stress and a signal molecule activating various pathways that lead to defensive responses similar to those seen in drought stress. Plant-associated microbes are important components of ecosystems and play an important role in plant adaptation and evolution to environmental stressors (such as drought and salinity). Microbes can protect plants against the harmful effects of drought and salt through a variety of processes, including nutrient solubilization, IST, and phytohormone synthesis, such as IAA, cytokinin, ABA, and GA, EPS, and ACC deaminase. Given a better understanding of the mechanisms involved in plant-PGPM interactions, it's projected that the use of PGPM in agricultural fields will skyrocket in order to help plants adapt to environmental changes. However, it is unclear if the mechanisms involved in PGPM-induced drought stress alleviation vary from those involved in salinity stress alleviation. More study is needed to show the underlying similarities and differences in microbe-induced drought and salinity resistance, based on advances in mimicking natural microbial interactions(Ma, Dias, & Freitas, 2020).

Under moderate and severe drought conditions, wheat cultivars showed diverse metabolic characteristics in terms of ROS generation, oxidative damage, antioxidant capability, and synthesis of osmotically active solutes. Two tolerance mechanisms were used in response to drought stress: the first involved upregulation of antioxidant enzyme activity and production of non-enzymatic antioxidants, and the second involved the accumulation of soluble sugars, free amino acids, and proline to aid osmotic adjustment. The wheat plant's metabolic system allows it to resist and survive water shortages. Because of its better ability to scavenge ROS and respond osmotically, the more drought-tolerant cultivar showed stronger photosynthetic stability during drought and faster recovery after drought. Furthermore, plants recovered from stress better during the tilling stage than other stages, underscoring the importance of drought timing in determining productivity. The ability of the plant to retain physiological processes during drought and recover after re-watering throughout vegetative stages is critical for determining final wheat productivity(Ullah et al., 2018).

2.5 EVOLUTION OF PLANTS

Plants have evolved in nature to withstand drought stress with a variety of morphological, physiological, and biochemical adaptations, as evidenced by the enormous diversity of plant species growing across climatic zones with extreme dry conditions. Drought resistance (DR) is a broad term that refers to plant species that have evolved adaptive traits that allow them to prevent, avoid, or withstand drought stress. Drought escape refers to a plant's capacity to complete its life cycle before drought strikes. Rapid phenological development refers to plants that grow quickly and produce a small number of seeds before the soil water runs out, with no specific morphological, physiological, or biochemical adaptations. Plants with developmental plasticity mechanisms grow slowly in the dry season, producing few flowers and seeds, but they grow indefinitely in the wet season, producing a great volume of seed. Drought avoidance refers to plants' ability to sustain higher tissue water content despite lower soil water content. This is accomplished by a range of adaptive features that include water loss minimization and water absorption optimization. Under drought stress, water spenders acquire higher tissue water status by maintaining water uptake via enhanced roots, hydraulic conductance, and other mechanisms. Drought tolerance (DT) refers to a plant's ability to withstand low tissue water content due to adaptive characteristics. Maintaining cell turgor by osmotic adjustment and cellular flexibility, as well as enhancing protoplasmic resistance, are among these adaptive features(Basu, Ramegowda, Kumar, & Pereira, 2016).

2.6 BREEDING

Plant breeders and crop physiologists face a difficult task in improving DR in agricultural plants since it is a complex genetic characteristic involving several mechanisms. The pyramiding and interaction of various mechanisms, characteristics, and genes that are relevant to individual crops and their growing environments are thus required for the effective development of drought-resistant agricultural plants. In drought-prone locations, success in this direction not only increases the growing area of crop plants but also ensures a constant output. The initial step toward developing drought-resistant crop plants is to identify genetic variation for DR. Wild species and adapted genotypes that have evolved under natural selection generally show such variety, and these are the finest sources of DR characteristics. The key to enhancing DR is to evaluate these resources in the field using integrated phenotyping and genotyping technique, as well as to identify features that are directly connected with yield. Comparative omics investigations of various germplasm should aid in a better understanding of the variety of drought-stress crop adaptations. Furthermore, understanding the broad basis of complex traits like DR requires investigation of specific genes focused on enhancing DR while stabilizing output. Climate change forces us to produce high-yielding crops that maintain yield stability amid drought and other environmental challenges. Drought-resistant plants should have a superior root system, stomatal control, WUE, and hormonal balance while avoiding negative grain yield consequences under normal and drought stress circumstances. As a result, a comprehensive crop development plan should include the use of a combination of morpho-physiological, biochemical, and anatomical adaptive responses to drought stress, as well as the deployment of high crop yield potential.

Drought causes physiological, biochemical, and molecular reactions at the cellular and molecular levels to combat the detrimental effects of the drought and sustain plant function. Drought tolerance is a multi-step process controlled by genes. Drought upregulates genes involved in osmolyte metabolism, secondary metabolite synthesis, and hormone synthesis, making it easier to induce response mechanisms and crosstalk quickly. Drought tolerance techniques include strengthening diffusive resistance, establishing a prolific and deep root system for increased water intake, and developing tiny and succulent leaves for reduced transpiration water loss. Mineral ions such as potassium help with osmotic adjustment and hence contribute to tissue water content management, whereas silicon improves water retention capacity by silicifying endodermal cells. Low molecular weight osmolytes are necessary for sustaining drought-induced damage and metabolic protection. Auxins, gibberellic acid, ABA, salicylic acid, jasmonic acid, and ethylene are among the growth hormones that help to modulate the drought response. Polyamines and secondary metabolites have antioxidant properties that help to mitigate the negative consequences of drought. Under drought conditions, phytohormones such as Abscisic acid (ABA) operate as crucial regulators, controlling either ABA-dependent or ABA-independent pathways.

Drought's negative impacts are mostly due to the failure of ROS scavenging mechanisms as a result of severe damage to macromolecules including proteins, lipids, and nucleic acids, which impairs the function of important organelles like mitochondria and chloroplasts. Ion leakage is caused by ROS-induced peroxidation of polyunsaturated fatty acids, leaving cells lacking in essential mineral ions for metabolism. The loss of cellular ion homeostasis has a direct impact on metabolic pathways, causing structural and functional changes in cells and the entire plant. Drought resistance is aided by the antioxidant system, suitable osmolytes, and secondary metabolites. Modulating the above-mentioned tolerance mechanism increases plant growth and function under drought and other conditions, according to molecular genetic studies. Non-enzymatic antioxidants like ascorbate, glutathione, and tocopherol play a critical role in maintaining redox equilibrium and ensuring the regular functioning of activities like photosynthetic electron transport. Modern research methodologies can help us comprehend these tolerance mechanisms even better.

2.7 DROUGHT SENSING

Before extending to other crops, the gene sets implicated in local and systematic drought sensing must be discovered in model crops. Our understanding of molecular genomics relevant to drought responses and tolerance mechanisms in crop plants has improved. Genome-wide association studies, which are used to investigate natural genetic variations, can identify genes and alleles involved in drought tolerance and metabolic regulation, paving the way for the mapping and cloning of gene loci regulating drought tolerance via controlled hormonal regulation, seed development, and water relations maintenance. Exogenous administration of biostimulants and phytohormones has been investigated for their efficacy in enhancing water usage efficiency and mitigating drought-induced oxidative effects, in addition to conventional management strategies such as mineral fertilizer application and compost use. Several researchers have addressed how molecular genetic techniques and physio-biochemical analyses have shown several critical routes for enhancing drought tolerance in plants, particularly food crops(Ahanger et al., 2021).

Water stress changes the anatomy of a leaf as well as its ultrastructure. Reduced leaf size, lower aperture, and fewer stomata, cell wall thickening, cutinization of the leaf surface and developed conductive system, stomata submersion in succulent and xerophytic plants, and the formation of tube leaves in cereals are some of the changes that occur in drought-stricken plants. In addition, when there is a lack of water, premature leaf senescence increases. Plants require proper leaf area development and stomatal opening to achieve optimal photosynthesis. As a result of the limited leaf area, increased resistance to gas exchange in stomata, and increased leaf senescence, net photosynthesis in water-stressed plants is reduced. One of the most important causes of the decline of plant size and biomass output is a poor photosynthesis rate. A reduced chlorophyll content is a common indication of drought stress, and it can affect plant shape. Plants stretch their roots and form a ramified root structure to maximize water uptake when they are dehydrated. In drought settings, increased biomass allocation to roots, as well as an expansion of the plant's root system in general, result in a larger capacity for water uptake. As a result, despite the fact that a slight water deficit reduces shoot growth, it has no effect on root growth. As a result, when plants are dehydrated, their root-to-shoot ratios normally increase, but their total biomasses decrease significantly.

2.8 CLIMATE CHANGE

Climate change will alter not only the air temperature but also the amount and distribution of precipitation, perhaps leading to more frequent droughts in the future. Drought is one of the most dangerous threats to plants, because a shortage of water affects plant–water relationships at all levels, from the molecular to the cellular to the organ to the whole plant. Drought inhibits plant growth and development, resulting in smaller organs and poor flower and grain output. Following a drought, stomata gradually close, reducing net photosynthesis and water consumption efficiency. In addition to soil water conditions, a complex combination of intrinsic and extrinsic variables influence stomatal conductance. One of the primary mechanisms driving lower plant growth and production during drought stress is the formation of reactive oxygen species in organelles such as chloroplasts, mitochondria, and peroxisomes. When reactive oxygen species (ROS) are formed in excess, they cause the peroxidation of cellular membrane lipids, as well as the destruction of enzyme proteins and nucleic acids. At the cellular, tissue, organ, and whole-plant levels, drought tolerance is mediated by a variety of physiological and biochemical pathways.

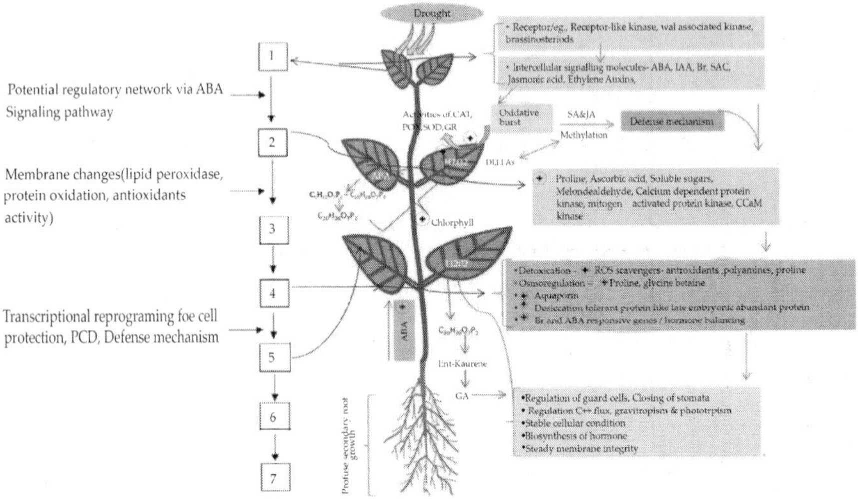

Figure 2.4: Seed priming.

Source: https://www.mdpi.com/ijms/ijms-21-08258/article_deploy/html/images/ijms-21-08258-g001.png

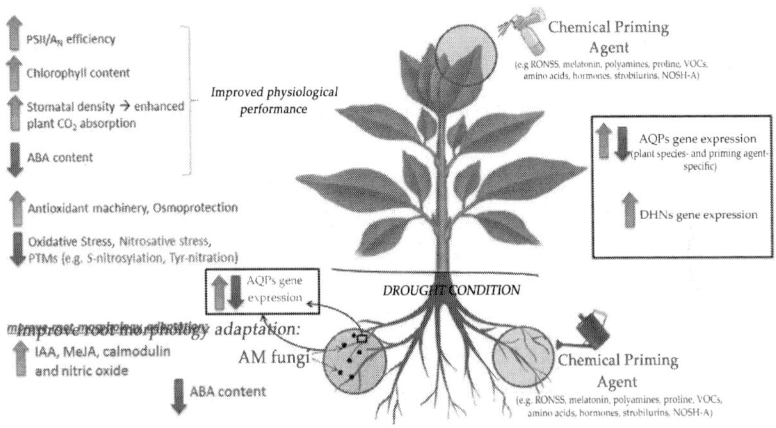

Figure 2.5: Plant priming.

Source: https://static.cambridge.org/binary/version/id/urn:cambridge.org:id:binary:20181019093905133-0654:S0021859618000126:S0021859618000126_fig1g.jpeg

Although both seed priming (Figure 2.4) and plant priming (Figure 2.5) can increase stress tolerance, they are not the same. Priming is a process in which plants are made more tolerant of subsequent stress occurrences after being exposed to mild stress for the first time. Seed priming is the process of partially hydrating seeds prior to sowing without permitting radicle emergence in order to improve germination rate, stress tolerance, and seedling establishment. Plant priming, as opposed to seed priming, is a process in which plants are exposed to biotic stress early in their lives, increasing their tolerance to later abiotic or biotic forces. Acclimatation or hardening is a comparable process that uses abiotic stress as the first exposure rather than biotic stress. Plant priming is a term used in some studies to describe the early exposure to abiotic stress that favors tolerance to later stress. Plant priming causes stress memory, which is mediated by proteins, transcription factors, and epigenetic alterations, and is invariably accompanied by a reduction in plant growth. Modifications at morphological, metabolic, subcellular, proteomic, and transcriptional levels, as well as changes in microRNA expression, can help plants adapt to drought. A vast range of physiological and molecular programs evolved to adapt to dry circumstances, resulting in morphological alterations. Smaller leaf areas to reduce transpiration and a bigger root system to increase water uptake capacity are common morphological acclimation strategies, and both are linked to water-use efficiency. Some studies have found that when water consumption drops while the photosynthetic rate remains high, WUE rises, implying that moderate dryness often results in increased WUE. It has also been shown that water supplies well below the ideal level resulted in a fall in WUE due to stress effects, whilst water supplies above the optimum level resulted in the opposite trend. Except in the case of early maturing wheat cultivars, WUE reduced in several cultivars even in response to water deficit stress during early growth stages, and it cut down significantly when water was withheld during grain filling.

2.9 PHOTORESPIRATION

The augmented rate of photorespiration observed in plants after the commencement of drought stress can be interpreted as an acclimation process to prevent PSII over-excitation under more severe drought circumstances. Due to the higher efficiency of the open PSII reaction centers, photosynthetic electron transport is increased in cotton during the onset of drought stress. While photosynthesis remains constant or slightly declines, the extra energy is utilized to raise the rate of photorespiration. Chlorophyll fluorescence measurement has been shown to be a reliable and efficient approach for determining a plant's drought susceptibility index. It can be used to choose a drought-tolerant cultivar and compare photosynthetic electron transport amongst cultivars with different drought tolerance levels. The experiment with different wheat cultivars produced in different years revealed that modern and intermediate cultivars had more sensitive stomata to water deprivation, but that the PSII reaction center's lower activity allowed these cultivars to avoid photoinhibition damage. Chlororespiration (Figure 2.6) is a respiratory electron transport chain in the thylakoid membrane that interacts with photosynthetic electron transport, involving non-photochemical reduction and plastoquinone oxidation, as well as oxygen consumption. The chlororespiration and cyclic electron pathways were discovered to play essential roles in drought tolerance, and various response mechanisms to drought stress were identified in sun and shade plants. Furthermore, the mesophyll conductance and photosynthesis of rice in response to drought stress are dramatically affected by nitrate nutrition-induced chloroplast shrinkage(Peltier & Cournac, 2002). Recent proteomic and enzymatic investigations have revealed that the main regulatory mechanisms for apple drought tolerance involve sustaining Calvin cycle activity by boosting essential enzymes and stabilizing photosynthetic electron transport, resulting in an increase in net photosynthesis rate. Plants can also cope with drought stress via signal regulatory proteins and abiotic stress-responsive proteins.

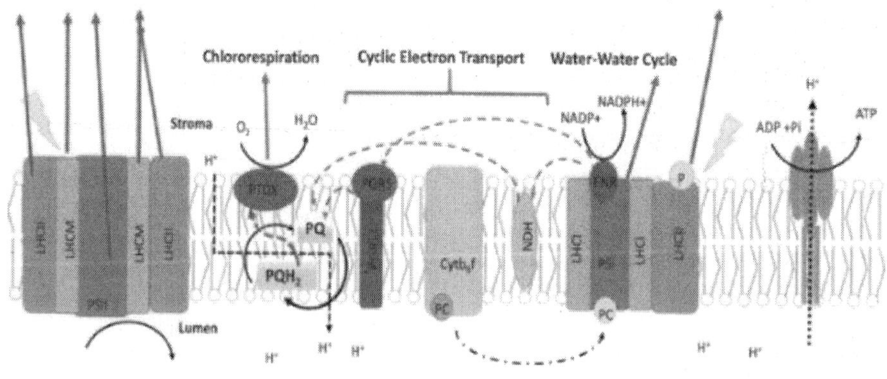

Figure 2.6: Chlororespiration.

Source: https://www.researchgate.net/profile/Chukwuma-Ogbaga/publication/334056673/figure/fig1/AS:775989793673217@1562021792484/Origin-and-location-of-different-photoprotective-mechanisms-associated-components-of_W640.jpg

Drought and other abiotic stresses are environmental challenges that plants face repeatedly throughout their lives. Plants can respond to later stress events in ways that are different from how they responded to the stress the first time they were exposed to it in order to survive the repeated shocks. Stress memory is defined as a distinct response to a similar stress. A coordinated reaction at the organismal, cellular, and genomic levels is thought to contribute to enhanced plant tolerance during this process. Seed tubers produced by plants grown under well-water and drought circumstances were seeded and exposed to identical watering treatments to assess the long-term stress memory on tuber yield and drought tolerance associated features in potatoes under drought. Primed plants yielded more tubers under both unrestricted and restricted water regimes. Water limitation resulted in a reduced fall in tuber output and carbon isotope discrimination in primed plants. Furthermore, it was discovered that long-term stress memory in potatoes appeared to be largely genotype-dependent.

Figure 2.7: Arrhenatherum elatius.

Sourcehttps://en.hortipedia.com/resources/imgs/6/67/Arrhenatherum_elatius_photo_file_4,72MB.jpg

The reactions of *Arrhenatherum elatius* plants (Figure 2.7) exposed to a second, later drought were compared to those of plants exposed to a single drought in grasslands(Grime & Curtis, 1976). Even after harvest, the first drought, the percentage of live biomass for plants that were subjected to drought earlier in the growing season was higher than for single-stressed plants after a late drought. When plants are exposed to subsequent abiotic stress, such as low temperature, drought, or heat, drought priming has been shown to maintain ROS homeostasis, increase photosynthetic rate, and increase grain production. Plants treated to a daily dehydration cycle show physiological and transcriptional stress memory: previously stressed plants showed partially closed stomata during a hydrated recovery period, aiding water saving during subsequent dehydration stress. Drought stress, which is due to lack of rainfall or deteriorating freshwater supplies for irrigation, limits the growth and production of many plant species in a variety of climate zones. Predictive models of global climate change have revealed that the frequency of precipitation events and net volumes have altered dramatically over the last century, implying that drought periods may occur more frequently and for longer durations in the future in certain places. Climate change's consequences, along with a growing global population, are anticipated to increase food insecurity, underscoring the need to investigate precise mechanisms underpinning drought tolerance in important and novel plant species.

Drought stress can also put non-crop species including trees, shrubs, and perennial grasses on the defensive. Senescence of the leaves is a critical developmental process that occurs naturally during plant maturation and is also a common effect of prolonged abiotic stress. Plant survival during short-term stress periods is improved by the coordinated breakdown and movement of leaf cellular constituents, and leaf senescence can be reversed if stress conditions are eased within a specified time period. Protein and RNA breakdown preempts chlorophyll degradation, directing amino acids and resources to other actively growing tissues or storage organs, increasing the likelihood of drought survival. There are also detailed reviews that detail the interaction between oxidative stress agents and antioxidative mechanisms that protect chlorophyll from cellular damage. During drought periods, oxidative stress may increase tocopherol, carotenoid, and glutathione, but ascorbate pools tend to diminish, in contrast to other abiotic stresses such as salt or UV-B radiation. Plants under drought stress may also reduce reactive oxygen species formation and delay leaf senescence by reducing cytochrome respiration and employing alternative respiratory pathways, while the impact of plant hormones on different respiratory pathways is unknown.

2.10 ABA

ABA has been identified as the major chemical signal beginning stomatal responses to drought stress among all classes of drought-responsive endogenous hormones now known to exist within the plant system. It is common knowledge that ABA levels rise in response to drought, and a series of chemical events involving carotenoids for ABA production and catabolism have already been identified through a series of experiments. Across practically all plant species, the auxin indole-3-acetic acid is produced in the rapidly dividing tissues of root and shoot apical meristems and young leaves. The biological link between auxin and stress-induced leaf senescence has been inconsistent in the past, depending on plant species or tissue maturity, and individual leaf responses may be reliant not only on auxin concentration but also on cellular responsiveness or sensitivity. In beans, leaf senescence develops as IAA concentrations fall to equal levels between stems and leaves, whereas exogenous IAA applied to the distal or proximal end of the abscission zone delays or promotes abscission, respectively. Exogenous or endogenous treatments lower the rate of natural or stress-induced leaf senescence, demonstrating that cytokinins regulate numerous biological functions throughout the plant system. They are also one of the most well-known positive regulators of senescence. Genes for CK synthase and adenosine phosphate isopentenyl-transferase (IPT) are downregulated as CK levels fall during stress-induced leaf senescence, while genes for CK oxidase are increased. Until recently, little was known about which genes directly control leaf senescence via CK signaling. CK signaling uses a histidine-aspartate phosphorelay system, in which histidine kinases (HKs) operate as cytokinin receptors, transferring a phosphoryl group to nucleus type-B response regulators (RRs), which activate the type-A RR primary response genes. While Arabidopsis has six distinct HKs, AHK2-4 are found in the endoplasmic reticulum, serve as CK receptors, and play different roles in numerous areas of plant growth and development.

2.11 GIBBERELLINS

Gibberellins are diterpenoid plant hormones with over several chemical structures currently known, though only a few are biologically active and influence a variety of plant growth and developmental processes such as seed germination, stem and root elongation, leaf expansion, transition from juvenile to adult phases, sex determination, and floral initiation. In comparison to other plant hormones, there is far less knowledge on the role of GAs in drought tolerance and how GA content changes as drought stress increases. Despite the fact that GA is typically classified as an antagonist to ABA, the few studies investigating GA's role in stomatal function suggested that GA may not be involved in stomatal movement, as exogenous GA had little or no effect on stomatal closure in Arabidopsis and GA-deficient mutants had similar transpiration rates during drought stress as wild-type plants. The direct effect of GA on stomatal mobility via exogenous applications or endogenous modification is an interesting research area. GA is the primary hormone that induces the production of stomata on Arabidopsis hypocotyls, and the GA-induced developmental signal is further controlled by interactions with auxin and ethylene. Though this has to be empirically confirmed, it would appear that leaf stomatal density, which is closely regulated by GA, is a primary regulator of transpiration rates and leaf water status during drought stress.

2.12 ETHYLENE

Ethylene, a gaseous plant hormone, causes a three-fold reaction in plant development. It is also linked to epicotyl growth and downstream consequences on a variety of organisms. According to the current ethylene signaling model, ethylene, a class of receptors that operate as negative regulators of the immune system, detect chemicals. The ethylene-responsive route is a pathway that responds to the presence of ethylene. It binds to receptors and inactivates the receptor-CTR1 complex, allowing it to function normally. To accumulate in the nucleus and express EIN3 and EIN3-like transcription factors ERF1 - 4 are transcription factor genes that control the activation or repression of hundreds of genes. During drought stress, networks may have a role in ethylene responses. Ethylene-mediated signaling pathways may possibly be involved. Stress-response genes are genes that regulate how the body reacts to stress.

2.13 JASMONIC ACID

Jasmonic acid and methyl jasmonate are lipid derivatives generated by fatty acid oxidation that play a role in the regulation of leaf senescence, ROS and NO signaling, antioxidant metabolism, and stomatal movement in plants. Jasmonates promote tolerance to abiotic stresses such as salinity, ozone, or

UV-B irradiance through downstream effects on antioxidant metabolism, similar to the current state of SA research, but less research has been done on the role of Jasmonates in antioxidant-facilitated drought tolerance. Nonetheless, some studies suggest that Jasmonates alleviate drought-induced oxidative bursts in the same way that they do other oxidative stresses, by increasing antioxidant enzyme activity. Recent research has discovered that hormone regulation of plant development and responses to drought stress not only involves interactions among hormones, but also interactions with other metabolites, such as sugars. Sugars may alter the production or response pathways of other plant hormones, such as those related to auxin or CK signaling, according to continuing and comprehensive research into crosstalk between several hormone classes. Sugars have long been recognized as important signaling molecules that regulate a variety of growth processes throughout the plant life cycle. Genes respond to specific sugars or sugar phosphorylations, it was discovered that Arabidopsis mutants with altered sugar responses exhibited phenotypes that were similar to plant-hormone biosynthesis or signaling mutants, implying that sugar and hormone signaling pathways are linked.

Initial similar mutant screens, as well as subsequent genetic and functional investigations, revealed that sugar, ABA, and ethylene signaling overlap extensively, preempting key downstream plant processes such as root formation. Mutants lacking genes encoding for ABA production or sensitivity are equally insensitive to high glucose concentrations, and ABI4, which encodes an AP2 transcription factor necessary for normal sugar response, could be the link permitting sugars and ABA-perception cross talk. Under drought stress, a sucrose synthase gene co-expressed with ABI3, as did ABI1 with one neutral invertase, two sucrose synthases, and one amylase, all of which could help to increase signaling capacity and phenotypic responses. Alternatively, mutants lacking genes encoding for ethylene perception are hypersensitive to glucose, whereas mutants with negative regulation of ethylene signaling are insensitive to glucose, suggesting that the antagonistic relationship between ethylene and glucose is mediated through repression of ABA biosynthetic genes. Arabidopsis mutants unable to catalyze glucose phosphorylation were resistant to exogenous auxin, insensitive to high glucose concentrations, and the hxk-based signaling negatively interacted with CKs, according to studies on ABA and ethylene mutants.

Although this has been a research field that has been investigated for many decades, there has been increasing data confirming the crucial functions of numerous plant hormones involved in regulating plant development and physiological responses to drought stress in the last decade. Leaf senescence, antioxidant metabolism, carbon metabolism, and stomatal motions are just a few of the drought responses that are directly or indirectly influenced by one or more hormones. Through transcriptome study and confirmation through the genetic transformation of overexpressing and silencing or altering certain genes, several transcription factors and downstream genes governing hormone synthesis, degradation, and responses or sensitivities have been found. Recent research is beginning to untangle signaling pathways of a single or many hormones, as well as interactions among hormones and between hormones and other metabolites such as carbohydrates, which co-ordinately impact drought responses, using transcription factors. However, the processes or molecules that initiate certain signal transduction pathways in response to the detection of a specific hormone are not totally known.

Crop loss owing to abiotic stresses, notably drought stress, has limited technical defenses in major agricultural businesses. Drought stress is a type of complicated stress because it frequently occurs in conjunction with strong light and high temperature stress to plant canopies. Plants are similarly prone to nutrient deficiency when they are deprived of water. During drought circumstances, major physiological activities including photosynthesis, transpiration, and respiration may be hampered, resulting in secondary stressors. Drought not only affects whole-plant responses, but it also severely limits cellular functions due to a lack of appropriate turgor pressure and other cellular damage. The key processes that affect plant growth and productivity are cell division, elongation, and differentiation. Drought sensitivity is high in several cellular activities. Plant breeding to improve drought tolerance is difficult due to the complicated impacts of drought stress on many plant systems and the quantitative nature of genes implicated in plant tolerance.

Breeding processes, whether traditional or biotechnological, are recognised for taking a long time to produce new germplasm for the market. Furthermore, commercially accessible germplasm is not the only solution to the problem of abiotic stressors in agriculture. Many locations around the world don't always have access to novel kinds, may lack the funds to do so, may have prohibitions on genetically modified materials, and certain new germplasm may not perform as expected in all conditions. Droughts are expected to last longer than usual as a result of global warming and climate change. Water stress is one of the key environmental conditions that restrict plant growth and productivity around the world. Plants are unable to escape poor environmental conditions due to their sessile nature. Plants interact with a variety of environmental elements, including biotic and abiotic stress factors, which all have an impact on plant growth and development, and so constitute a limiting factor for food supply and sustainability. The survival rate, biomass production, and crop yield are all affected by abiotic and biotic stressors. Drought, cold, salinity, and heat stress are among abiotic stresses that cause damage to plants by altering cellular structures and compromising important physiological functions. Stress tolerance is present in almost all plants; however, the level of this ability varies from species to species. Plants use a variety of molecular, physiological, and cellular responses to cope with abiotic stress. Drought is one of the major abiotic stresses, and it is widely regarded as the most disastrous environmental stress, resulting in lower plant productivity than any other. Drought stress is becoming more prevalent, and it is one of the most significant constraints to agricultural output. Water is becoming a more valuable and scarcer commodity.

Drought stress is a key environmental constraint that has a significant impact on plant growth and development. Drought stress reduces crop productivity, with almost 28% of the world's soil surface being too dry for typical crop yields. Plants adapted to a lack of water have acquired a wide range of morphological, physiological, and metabolic adaptations. A deeper knowledge of these systems could aid in the development of drought-tolerant crop plants. The significant role of sulfur-containing compounds in protection against biotic stress has been studied in depth in recent years. Sulfur and sulfur-containing substances have recently been proposed as having a role in abiotic stress defenses. The significance of sulfur-containing chemicals in stress tolerance, on the other hand, is not well recognised. During the drought stress response, a complicated balancing act is required to balance primary and secondary sulfur metabolism.

CO_2 levels in the atmosphere increased from around 280 ppm at the start of the industrial revolution to around 380 ppm now, and are anticipated to double by the end of the century. Climate change may be exacerbated by rising CO_2 levels in the atmosphere, which may result in changes in precipitation patterns and evapotranspiration. Drought risks may grow as a result of climate change in various locations. Crop yields are affected by seasonal variations in rainfall, which is one of the most important elements. Drought stress is a factor for traditional rain-fed agriculture in arid and semiarid areas, which cover around 40% of the world's land surface. Water supply is already a limiting concern for agricultural output in various parts of the world. Climate change and variability will have a considerable impact on agricultural output due to changes in precipitation patterns, temperature rises, and carbon dioxide levels. Climate change would have an impact on the hydrological cycle and the availability of water resources, implying that it has an impact on crop productivity. Climate change can hasten the hydrological cycle by changing rainfall, evapotranspiration, etc. The potential and actual evapotranspiration may grow as the temperature rises in the future climate.

CHAPTER 3

BIOCHEMICAL ASPECTS OF DROUGHT TOLERANCE

CONTENTS

3.1 Introduction 44
3.2 Temperature 45
3.3 Agricultural Productivity 46
3.4 Anthropogenic Actions 49
3.5 Agriculture and Climate Change 50
3.6 Stages 52
3.7 Drought Resistance 54
3.8 Water Uptake 54
3.9 Abiotic Stressors 56
3.10 Plant Functions 60

3.1 INTRODUCTION

Drought is the most damaging condition that limits crop output around the planet. Wheat is grown in poor countries in rain-fed conditions with no additional irrigation. Drought affects 50% of the wheat-growing area each year, according to estimates. Drought is most damaging to wheat growth during the seedling establishment, booting, and grain filling stages. Early growth phases of autumn seeded wheat are hampered by a lack of water. Low soil moisture during germination lowers germination percentage and delays germination, reducing plant maturity and yield. Early water constraint results in a weak plant and poor sink growth, both of which have a negative impact on production at maturity. Seedling stage drought may be more harmful to yield than stress during later growth stages, depending on the severity of the stress. As a result, it's critical to maintain appropriate seedling establishment and a strong plant stand as soon as possible following germination. Drought tolerance phenotyping at the seedling stage is a frequent practice in wheat, barley, triticale, maize, and rice. It is a quick, low-cost, and dependable way to assess a plant's performance. The necessity of screening for drought tolerance at the seedling stage has been shown by reports on the link between drought tolerance at the seedling stage and reproductive stage in wheat and cowpea (Figure 3.1). -

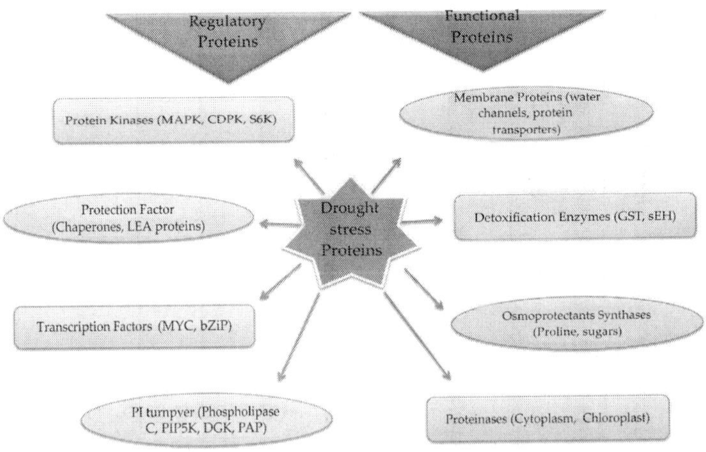

Figure 3.1. Biochemistry of drought tolerance.

Sourcehttps://link.springer.com/chapter/10.1007/978-981-10-9029-5_1

Seedling survival, dry weight, root shoot ratio and root length, relative water content, and seed reserve mobilization are the features that have been employed for drought tolerance screening of germplasm. Despite the fact that a considerable number of studies on wheat drought tolerance have been undertaken, screening wheat germplasm using the appropriate collection and evaluation technique has received little attention(Abid et al., 2018).

3.2 TEMPERATURE

Plants are exposed to high ambient temperatures, germinating seeds are exposed to soil warmed by absorbed infrared radiation from the sun, high plant transpiration followed by less water absorption, reduced transpiration capacity in certain plant organs, forest fires, natural gas blowouts, and other factors cause high temperature stress. Though extensive research has been done at the ultrastructural, molecular, and gene expression levels under various temperature extremes, the temperature perception and the molecules involved in it are not well understood. When an organism is exposed to extreme temperatures, all of its cells respond. When exposed to rapid increases in external temperatures, typically 5 to 10 degrees Celsius above normal growth temperatures for a few minutes to a few hours, all organisms produce an elite set of proteins identified as heat shock proteins (HSPs), which are absent or present in small quantities in unstressed organisms. These HSPs are a result of repair. In tropical, semi-arid, and dry parts of the world, understanding the mechanisms and development of thermotolerant plants is critical(Lamaoui, Jemo, Datla, & Bekkaoui, 2018).

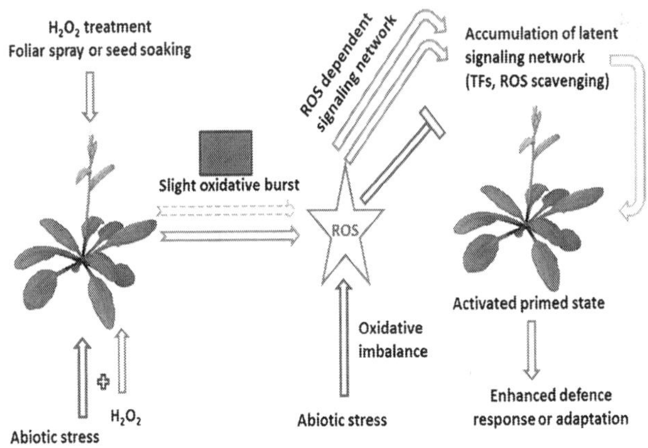

Figure 3.2. H_2O_2 and drought.

Source:https://www.frontiersin.org/files/Articles/141857/fpls-06-00420-r2/image_m/fpls-06-00420-g002.jpg

Drought stress has an adverse impact on the development of vegetable-type soybeans, which are still in their infancy in Africa despite their enormous nutritional benefits. The effects of severe drought stress on ascorbate peroxidase, guaiacol peroxidase, and glutathione reductase activities, as well as proline, total soluble sugars, and hydrogen peroxide contents, were investigated in five vegetable-type soybean cultivars at the flowering and drought caused large increases in proline, TSS, and malondialdehyde. Under drought stress, UVE8 and AGS354 had the highest H_2O_2 levels at blooming, while AGS429 had the lowest (Figure 3.2). However, under drought stress, AGS429 was the only cultivar with a significant rise in H_2O_2. Drought stress also resulted in large increases in AGS429's APX, GPX, and GR activities during flowering. For all antioxidative enzymes, AGS354 showed the greatest drop, whereas UVE17 only showed a decrease for GPX. Except for H_2O_2, all biochemical indicators were considerably greater at pod filling than at blooming. H_2O_2 had a significant positive association with total seed mass or total seed per plant at both stages, whereas TSS and proline had a significant positive link with total pods per plant. According to the findings, during drought, the tolerance responses of vegetable-type soybean, APX, GPX, and GR work together to reduce H_2O_2 production and lipid peroxidation, allowing H_2O_2 to play a role in the signaling events that lead to drought tolerance induction. The drought tolerance response of this crop is influenced by the induction of TSS at flowering and proline at pod filling(Moloi & van der Merwe, 2021).

3.3 AGRICULTURAL PRODUCTIVITY

Drought is a factor impacting global agricultural productivity. Horsegram was chosen as a drought-tolerant legume to compare and decipher the biochemical pathways of drought stress tolerance. Drought stress and control conditions were applied to 25-day old plants of a drought resistant and a sensitive species of horse-gram for 48 hours. The Horsegram plants' leaf and root tissues were collected for biochemical and antioxidant enzyme testing. Under drought stress, the tolerant variety's Relative Water Content, proline, and phenol content were found to be considerably greater. Under drought stress, the sensitive variety's protein and malondialdehyde concentration were found to be considerably greater. Peroxidase, Superoxide Dismutase, Catalase, and Glutathione-s-transferase were among the antioxidant enzymes that increased significantly in the tolerant variety compared to the sensitive one under drought stress. However, the activity of Glutathione Reductase was shown to be lower. Abiotic stress, such as drought, is a prevalent abiotic stress in wheat. PEG-induced drought stress resulted in a significant decrease in morpho-physiological features in Bijoy but not in BG-25, implying that drought tolerance mechanisms exist in BG-25. In drought-stressed roots of BG-25 and Bijoy, semi-quantitative RT-PCR demonstrated an increase of TaCRT1 (calreticulin Ca2+-binding protein) and DREB1A (dehydration responsive transcription factor) transcripts, albeit to a lesser level. These findings suggest that enhanced TaCRT1 expression may be linked to wheat plant survival during drought(Kabir et al., 2021).

Water is released from the cytosol and vacuoles to the apoplast during drought stress, resulting in cellular dehydration. Variations in stomatal conductance, osmolyte accumulation, and gene expression are all examples of plant responses to water stress. Due to its quick accumulation in extreme conditions and participation in physiological and biochemical processes that allow plants to survive this challenge, abscisic acid (ABA) is defined as the principal stress hormone in these processes. Maintaining root development in this situation would improve drought tolerance by increasing water intake capacity. Plant functions at low water potential and their recovery after rehydration, on the other hand, both contribute to good yield during cyclic drought times. Plant recovery following rehydration is a critical feature for survival, reflecting a balance between the repair of damaged structures and the restoration of appropriate metabolism. As a result, detailed research into plant responses to rehydration is critical. It is generally understood that the formation of reactive oxygen species is linked to typical metabolic processes like aerobic metabolism and photosynthesis. However, under abiotic stress conditions, its synthesis is boosted by mechanisms such as the suppression of NADP+ regeneration, the Mehler reaction, and photorespiration. Under these circumstances, ABA build-up causes superoxide anion and hydrogen peroxide generation in chloroplasts by inhibiting CO_2 absorption and altering the transport electron chain. In addition, H_2O_2 has been identified in the ABA response as a modulator of stomatal closure or inhibition. Under stressful situations, the antioxidant system, which contains both enzymatic and nonenzymatic substances, is capable of detoxifying ROS. When ROS generation surpasses antioxidant activity, the cell experiences oxidative stress, which manifests as lipid peroxidation and oxidative damage to proteins and DNA.

Most legumes, including peanuts, have unique characteristics in response to drought, such as lower nodulation rates and biological nitrogen fixation. Drought stress raises soluble sugar levels and lowers solute potential in cells of nodules, keeping the turgor at a low water potential. Three components have been studied in relation to drought stress and biological nitrogen fixation: carbon metabolism, nodule permeability to oxygen, and nitrogen feedback. The genotype of nodulated peanut demonstrated various tolerance responses to drought stress circumstances; however, the effects of water shortage and subsequent rehydration on the symbiotic relationship peanut-Bradyrhizobium sp have yet to be investigated(Furlan, Llanes, Luna, & Castro, 2012). Tall fescue (Figure 3.3) is a popular cool-season turf grass that is susceptible to water stress. In many plants, melatonin has been shown to boost abiotic stress tolerance. While drought stress lowered shoot height and fresh weight of tall fescue seedlings, melatonin pre-treatment increased them when compared to control plants. When compared to untreated seedlings, tall fescue seedlings pre-treated with melatonin had lower levels of reactive oxygen species, electrolyte leakage, and malondialdehyde, but higher levels of antioxidant enzyme activities and total chlorophyll content. Melatonin treatment also reduced leaf water loss and raised leaf water content, resulting in better plant development under drought stress. Melatonin also increased root development in tall fescue seedlings under osmotic stress. Exogenous melatonin improved drought tolerance in cool-season tall fescue. As a result, melatonin could potentially be employed to protect plants from the negative impacts of drought or water scarcity(Moloi & van der Merwe, 2021).

Figure 3.3. Tall fescue.

Source:https://www.noble.org/globalassets/images/news/noble-news-and-views/2018/08/hero/drought-tolerance-an-important-characteristic-for-tall-fescue-persistence.jpg

3.4 ANTHROPOGENIC ACTIONS

Anthropogenic actions in the past and present have caused global warming and, as a result, drought has affected both plants and animals. Plants that are sessile are more susceptible to drought stress, which reduces plant development and output. It is critical to determine the plant response mechanisms to drought stress in order to lessen the impacts of drought stress on plants. Drought response mechanisms encompass morphological, physiological, biochemical, cellular, and molecular processes that occur in drought-stressed plants. Improvements in the root system, leaf structure, osmotic adjustment, relative water content, and stomata management are among these processes. Drought stress is also combated by calcium and phytohormone signaling pathways, as well as the scavenging of reactive oxygen species. Furthermore, microbes such as bacteria and fungi have an essential role in improving drought tolerance. Quantitative trait loci, transgenic approaches, and the use of exogenous chemicals are all important in further elucidating and improving drought tolerance in plants. As a result, the current research combines diverse mechanisms of drought tolerance in plants(Ilyas et al., 2021).

The development of drought-resistant crops is critical for food security. Plants have evolved sophisticated regulatory networks to cope with drought, including drought escape, drought avoidance, drought tolerance, and drought recovery after stress, as part of the evolutionary process. The main processes for improving plant drought resistance are drought avoidance and drought tolerance. Drought avoidance refers to a plant's ability to respond to drought by controlling morphological and physiological properties such as root growth, leaf shape, stomatal movement, photosynthesis, and other factors in order to maximize water uptake and decrease water loss. Furthermore, the use of chemical fertilizers in agriculture has resulted in enormous energy waste and environmental damage. As a result, new crop drought resistance strategies that are both successful and environmentally beneficial are urgently needed.

Drought reaction necessitates the use of abscisic acid. Exogenous ABA therapy has been found in a number of studies to be an effective strategy to improve plant drought resistance. Exogenous ABA treatment, for example, can regulate energy, amino acid, and lipid metabolism; promote the accumulation of flavonoids, betaine, and other substances; improve enzymatic and non-enzymatic antioxidant regulation systems; and improve photosynthetic performance and RWC of plants to promote plant growth and improve drought resistance in a variety of crops, including maize, wheat, sweet potato, pearl millet, and other ploughed grains. According to certain studies, treating plants with ABA plus jasmonate or benzyl aminopurine at the same time improves drought resilience. Although ABA is an agrochemical that can be used directly, its use has been limited due to its quick conversion to physiologically inactive metabolites. As a result, different types of ABA-mimicking agents have been created to circumvent the limits of ABA use in diverse agricultural applications. Pyrabactin was the first synthetic ABA agonist identified, and it has various ABA-like properties, including seed germination suppression, stomatal closure induction, and root hydraulic conductivity increase. The structural optimization of quinabactin resulted in the discovery of AM1 fluorine derivatives with stronger agonist properties than the parent molecule quinabactin.

3.5 AGRICULTURE AND CLIMATE CHANGE

0Agriculture and climate change are intertwined, and the world's climate is changing faster than it has ever been. Water scarcity is a major constraint in worldwide crop growth, development, and output, and climate change is the primary cause, exacerbating the problem and having negative consequences for agriculture in most parts of the world. The degree and duration of drought stress have an impact on plant growth and development. Plants evolve a variety of tactics, defenses, and adaptation processes to cope with and overcome drought stress, including morphological, physiological, biochemical, and molecular changes. These alterations have been discovered to play a critical function in plant survival when exposed to drought. Through the application and manipulation of various approaches, the effect of drought stress on plants can be mitigated. Breeding procedures, exogenous application of phytohormones and osmoprotectants to plants or seeds, and drought-hardening are some of the ways and adaptation strategies that give drought tolerance in plants. Because early seedlings are flexible, their drought tolerance can be increased by exposing them to low water stress earlier on. As a result, drought-hardening is a practicable and practical strategy that entails pre-conditioning seedlings to drought stress by reducing or partially irrigating them. Drought-hardening not only enhances drought resistance, but it also gives tolerance to other abiotic stimuli in plants, such as cold stress in tomato seedlings, chilling stress in cucumber seedlings, and cross adaptation to heavy metals in spring barley. Drought-hardening in several plant species has been found to help mitigate the detrimental effects of drought stress and give drought resistance to plants in several studies.

Drought-hardening enhanced plant drought tolerance through morphological manifestations such as increased root to shoot ratio, leaf extension, and root surface area, physiological adaptations such as transpiration rate, photosynthesis rate, relative water content, osmotic potential, chlorophyll content, biochemical adaptations such as improved antioxidant defense system, proline content, soluble sugars, and polyamines, and molecular manifestations such as increased expression of genes. The seedlings' survival could be increased by exposing them to various durations and intensities of water stress prior to implantation. Drought-hardening time intervals were shown to be variable in different investigations. In different plant species, the length of water stress for drought-hardening varies from a few days to a week.

4Drought stress resulted in a larger concentration of reactive oxygen species, which is harmful to plant growth. Plants induce stronger antioxidant enzyme activities and higher expression of their related genes to minimize the effects of ROS, which aids in drought stress tolerance and adaptability. Apart from the regulation of antioxidant enzyme activities and their related genes expression in response to drought stress, other genes such as osmolytes biosynthesis-related genes, signaling genes like SnRK2, and transcription factors like WRKY, AREB, and DREB were also regulated and highly expressed, resulting in drought stress tolerance and adaptation in plants. These genes could be used to emphasize the importance of drought-hardening in tobacco drought tolerance(Khan et al., 2020). Drought stress reduces plant growth and, as a result, poor yield. Potassium is a necessary macronutrient for increased growth and yield in drought-stricken areas. K is vital for osmotic adjustment, stomata opening and closing, and enzyme activation. K is a component of plant structure that increases plant growth and yield by optimizing several physiological and biochemical processes. The influence of drought on plant growth, physiological and biochemical properties, the role of K in plant growth, and physiology under drought stress are important to study. By modifying the physiological and biochemical processes, K is also important for improving the growth and yield properties of crops during drought stress circumstances (Ahmad et al., 2018).

Drought stress is having a significant impact on maize production, as it is with other crops. Maize is a drought-sensitive crop; therefore, it is affected by a lack of moisture at every stage of growth and development. Drought occurs frequently during the seedling period, resulting in poor crop stand and, in extreme cases, complete seedling failure. In response to prolonged mild drought stress, plant metabolism shuts down, followed by plant mortality owing to stomatal closure and restricted gaseous exchange. In the case of maize, the reproductive growth stage is comparatively more sensitive to drought stress, and infertile ear production may result in severe drought. The importance of maize around the world, as well as the negative effects of drought on maize, prompted breeders to create drought-tolerant maize germplasm. Drought-responsive characteristics and adaptive mechanisms must be understood before drought-tolerant maize stock can be developed. In maize genotypes, genetic variation is evident based on adaptation mechanisms such as drought escape, drought avoidance, and drought tolerance.

Under mild drought stress, the requirement for photosynthates and energy is reduced in leaves due to reduced leaf area caused by leaf rolling or curling. Photosynthetic assimilates from leaves are sent toward roots, causing them to lengthen in order to maximize water intake. Roots serve as the principal sensor of water scarcity in the soil, sending messages to the aerial portions to control growth and development. Chemical and hydraulic vectors carry signals from the roots to the aerial parts. Reduced water and nutrient intake raise the pH of the xylem, which sends ABA-mediated signals to the leaves to reduce stomatal closure and water loss. Drought stress causes a reduction in root growth, as well as a reduction in cell division and cell elongation. Because microtubules are involved in cellular morphogenesis, embryo development, organogenesis, stomatal conductance, and organ twisting, they are essential for cell division and cell elongation. Under dehydrated conditions, reduced root turgor promotes ABA build-up and plasmolysis. The microtubule backbone and cellular geometry are severely harmed by plasmolysis.

3.6 STAGES

The blister stage, soft dough stage, milking stage, etc., are all part of the kernel development process, which is very essential in terms of productivity. During the blister stage, high moisture content is required for grain filling, and drought stress at this point results in low-quality kernels. The moisture demand during the soft dough, milking, and hard dough stages is significant enough that drought stress can affect kernel quality and yield at these periods. Drought stress during the hard dough stage leads the cobs to hang prematurely. Although the water requirements of the dent stage of kernel development are lower than those of the pre-dent stage, drought stress at this point can nevertheless result in yield and quality losses. Kernel development in maize is divided into three stages: lag phase, during which sink capacity is developed, water content increases, and biomass accumulation decreases; effective grain filling stage, or linear phase, during which maximum biomass accumulation occurs and kernel size is determined; and physiological maturity, during which maximum dry weight is gained and the grain enters a quiescent phase. For grain filling, sink capacity and source strength interact with one another. Grain weight differences are caused by differences in the source sink ratio. Photosynthesis and carbohydrate absorption determine source strength, whereas sink capacity is determined by sink activity.

Drought stress lowers photosynthesis and photosynthetic assimilate transfer, resulting in diminished grain filling. Drought stress in maize reduces source strength and sink capacity. Reduced remobilization of photosynthetic assimilates causes grain size reduction. Under drought stress, grain filling is also reduced due to the lower activity of sucrose and starch producing enzymes. Pre-anthesis phases determine the number of kernels, whereas post-anthesis stages affect the weight of the kernels. Kernel weight loss is caused by drought stress during the post-anthesis stages. The final kernel volume is determined by the interaction of water and biomass during kernel formation. The water content of the kernel increases during the early stages of growth, then decreases as the kernel matures, followed by an increase in biomass accumulation. Drought stress reduces the final kernel volume because biomass build-up is dependent on source strength and sink capacity, both of which are severely reduced by drought stress.

Drought stress has a lot of unpredictability and uncertainty because it reduces production, yield potential, and performance over time. Selection of genotypes with higher yield under drought circumstances, on the other hand, is an efficient method for reducing drought stress. The irregular and unpredictable behavior of stress is being exacerbated by heterogeneity in the nature of drought stress, varying impacts in location and time, and the degree and severity of stress. Plants have been given the ability to adapt to adverse environments in order to survive and produce. Plants have a variety of morphological, physiological, and biochemical characteristics that enable them to adapt to or resist stress. Resistance is defined as the lowest yield reduction under drought stress compared to normal water availability. Escape, avoidance, and tolerance are all forms of resistance. Drought resistance is defined as a variety or species' ability to live and reproduce in the face of reduced water availability in evolutionary terms.

3.7 DROUGHT RESISTANCE

Drought resistance refers to plant systems that contribute to the lowest yield losses under drought conditions compared to higher yields with normal water availability. It was discovered that there is genetic heterogeneity for drought resistance among various agricultural plants and variations of the same species. Drought resistance was determined using yield stability in wheat, maize, rice, barley, and sorghum. Differences in yield, yield stability, and drought resistance demonstrated that this genetic diversity may be used to advance production, yield stability, and drought resistance. Drought tolerance in crop plants can be improved by a variety of methods, including cultivated varieties, land races, and the production of transgenes. Landraces and wild relatives may be a viable approach for obtaining drought resistance genes and incorporating them into contemporary cultivars to improve drought resistance. The wild relative of maize, *Zea mexicana* or *Tripsacum floridanum*, is a rich source of new genes for improving drought tolerance and other challenges. If drought resistance genes are present in non-crossable parents, transgenes can be formed(Khanna-Chopra & Singh, 2015).

Several physiological and metabolic processes are not affected by water stress and continue to function normally in the absence of drought. Drought avoidance is defined as an estimate of tissue water status under drought stress expressed as turgor water potential. Reduced transpiration or increased water intake are used to maintain plant water status. Drought resistance in maize is aided by leaf rolling, leaf firing, canopy temperature, stomata closure, leaf characteristics, and root features. Stomata play a role in both transpiration and gas exchange. Plants conserve water by closing their stomata and preventing water loss. Leaf rolling and leaf firing are essential features in maize that are used to determine drought resistance. Leaf drying is caused by insufficient leaf transpiration, which results in leaf firing. In maize, there was a negative association between leaf senescence and yield. Under drought stress, canopy temperature is also inversely linked with maize yield.

3.8 WATER UPTAKE

Water uptake can be maintained with the assistance of a large root system. For successful drought avoidance in maize, structural or phenotypic, as well as functional or hydraulic root characteristics, are critical. In maize, architectural or phenotypic traits such as apical zone length, basal zone length, branch inter-space length, maximal number of branches, initial root elongation, insertion angle, root radius, standard deviation of the root tip heading, and tropism type are used to estimate water uptake pattern using mathematical models. For modeling of water absorption and identifying responses to water stress, hydraulic root features such as axial conductance between crowns of maize nodal roots, non-homogeneous soil water potential, transpiration during the day, water potentials in the root, and hydraulic conductance are being investigated. Increased water availability in the root axis caused higher transpiration, according to these structural and functional models(Comas, Becker, Cruz, Byrne, & Dierig, 2013).

Osmotic adjustment is defined as the formation of a water gradient to increase water input while reducing the osmotic potential to maintain turgor. Osmotic adjustment aids in the preservation of tissue water status. Drought damage is mitigated by the build-up of solutes in the cytoplasm and vacuole of cells. The turgor potential and physiological processes are maintained through osmotic adjustment, which provides protection. Water potential, osmotic potential, turgor potential, and relative water contents are used to determine plant water status. Drought tolerance is estimated using relative water contents as an integrative metric. Under drought stress, stomata shut, resulting in a decrease in CO_2 build-up as a result of lower relative water content. Sustainable management of photosynthetic rate and turgor potential also maintains photosynthetic assimilate translocation to growing kernels. The first group of osmoprotectants includes nitrogenous compounds such as proline, polyols, polyamines, and glycine betaine, whereas the second group includes hydroxy compounds such as polyhydric alcohols, sucrose, and oligosaccharides. Osmolytes, also known as osmoprotectants, are chemical substances that are non-toxic to plants. These osmolytes help to preserve cellular proteins and membranes from the drying effects of drought stress.

Glycine betaine is an osmoprotectant that is organic, water-soluble, and non-toxic to plants. It plays a significant function in protecting plants from drought, salinity, cold, and heat stress. Glycine betaine serves as an osmoprotectant and protects the photosynthetic machinery. It also helps to stabilize cellular proteins and reduces the load of reactive oxygen species. Certain proteins, collectively known as stress proteins, play a key function in conferring drought tolerance. Stress proteins are mostly water-soluble, and hydration of cellular contents ensures tolerance. Certain proteins, such as late embryogenesis abundant (LEA)-dehydrin, heat shock proteins (HSP), cold shock proteins, aquaporins, and cyclophilins (CYP), play a key role in improving drought tolerance. Drought resistance is conferred by aquaporins, which are water channel proteins that increase water intake by providing an open water channel.

3.9 ABIOTIC STRESSORS

When abiotic stressors are prevalent, LEA proteins naturally accumulate in pollen grains, seeds, and vegetative tissues. Desiccation resistance in crop plants is linked to the accumulation of LEA proteins. Based on amino acid sequences, particular domains, different motifs, and peptide composition, LEA proteins are classified into a variety of traditional groupings. Among these several categories, 1, 2, and 3 are the most important, as they contain a huge number of LEA proteins. Proteins in Groups 1 undergo a variety of post-translational modifications, such as acetylation, phosphorylation, deamination, and methylation, which reveal their seed-specific function. Group-2 and group-3 LEA proteins are known for their anti-aggregative capabilities. Membranes are protected by Group 2 LEA proteins, which inhibit denaturation. Dehydration tolerance is connected to Group 3 LEA proteins, which limit cell shrinkage caused by water loss. The representative LEA proteins of groups 1, 2, and 3 for assessing functional features in maize are EMB564, MLG3, and RAB17, respectively(K. Goyal, Walton, & Tunnacliffe, 2005).

Heat shock and cold shock proteins are chaperones that evade cellular contents from denaturing and folding during stress. The larger HSP groups, which indicate tolerance to drought and heat stress, are made up of small heat shock proteins. Cold shock proteins or CSPs build in response to stress and serve as chaperones to protect RNA from malfunction. Bacterial CSPA and CSPB genes were transfected into maize, increasing tolerance to drought, heat, and cold stress by protecting photosynthetic, vegetative, and reproductive growth phases. Multi-year, multi-location, and multi-hybrid yield trials revealed that yield rose in CSPA and CSPB transgenic maize. In maize lines, there were qualitative changes in heat shock protein production. Plant adaptive responses are divided into three categories. They are osmotic adjustment or homeostasis; stress damage control, detoxification, and repair; and growth regulation. Drought stress signaling is divided into three functional groups based on plant responses. They are osmotic stress signaling to restore cellular homeostasis; detoxification stress signaling to prevent and repair cell damage; and signaling to maintain cell division and cell expansion to maintain growth. To maintain growth, homeostasis and detoxification signals confer drought tolerance and regulate stress responses.

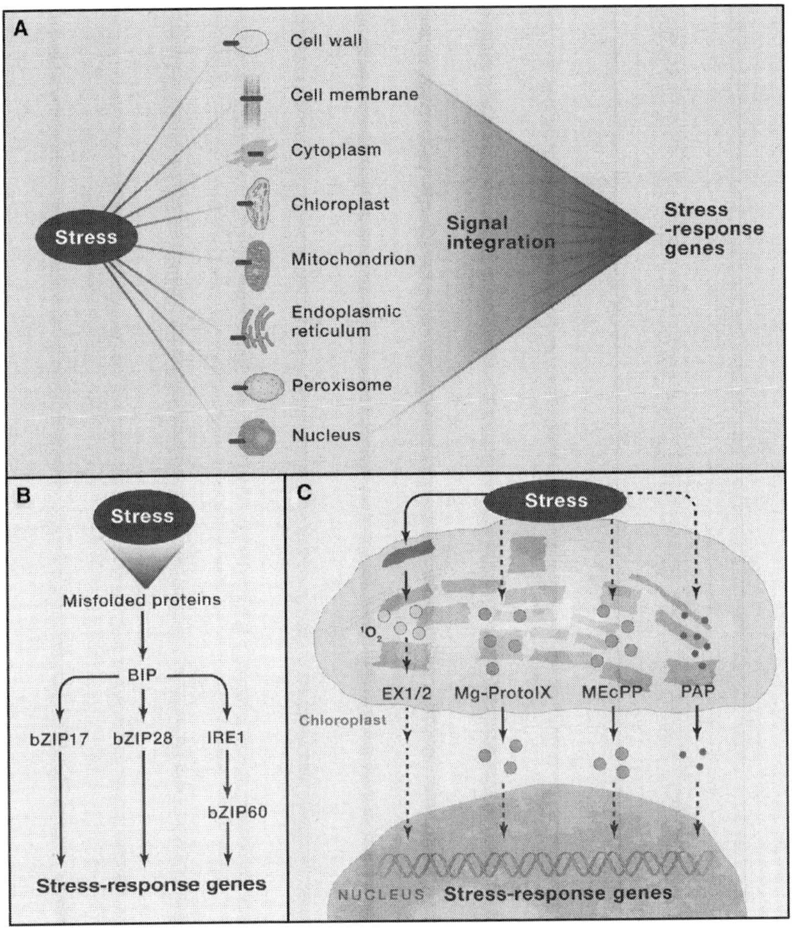

Figure 3.4. Protein phosphorylation.

Source:https://els-jbs-prod-cdn.jbs.elsevierhealth.com/cms/attachment/b9778048-f6db-4b1d-ad66-d1a07e7bfec2/gr1.jpg

Protein phosphorylation is a side effect of osmotic stress signaling (Figure 3.4). In reaction to osmotic stress, protein kinases are activated. Calcium signaling increases calcium-dependent protein kinases (CDPK), which affects downstream reactions in response to osmotic stress. The expression of specific genes that are susceptible to ABA, cold, and osmotic stressors is regulated by constitutive overexpression of CDPK in maize protoplast. In response to osmotic stress, the transcriptome for protein kinases such as MAPK, MAPKK, MAPKKK, and histidine kinase increases. Understanding signaling pathways necessitates a comprehensive understanding of the pathway's inputs and outputs. Osmotic stress, changes in turgor, and subsequent injury are inputs to the osmotic signaling system, whereas osmotic adjustment by osmolytes, damage prevention, and damage healing are outputs. In reaction to osmotic stress, membrane phospholipids produce several signaling molecules and offer structural support to cells. Phospholipids operate as messengers and regulate downstream responsive genes at low levels of osmotic stress, but excessive levels of phospholipids indicate cellular damage at high levels of osmotic stress. The category classification of phospholipid signaling systems is based on phospholipases and other messengers. The secondary messenger inositol 1, 4, 5-trisphosphate (IP3) causes stomatal closure by raising Ca^{+2} ions in the cytoplasm. Drought-tolerant maize hybrids for temperate zones were developed. Experiments revealed that selection has caused significant changes in the behavior and reaction of germplasm over time(Sah, Reddy, & Li, 2016).

Drought stress that occurs during the flowering stage is thought to have the greatest impact on maize kernel output. Different hybrids from various historical times were studied and contrasted. Drought stress management is used in a variety of ways around the world to help crop plants reduce output losses. Saving irrigation water through various water management measures, utilizing agronomic practices to increase crop performance under drought stress conditions, and developing drought-tolerant germplasm are the key instruments that agronomists and breeders use. Water conservation and cultural practices are considered to be inconvenient, costly, and time-consuming. Drought-resistant germplasm development has been shown to be an effective, efficient, and feasible method for increasing production in drought-prone areas. Numerous biological treatments increase various strategic features, allowing plants to escape, avoid, and withstand drought stress. Screening germplasm for tolerance variations, generation of tolerant genotypes through traditional breeding, mutation breeding, molecular breeding, and transgenic techniques are all viable options that can be used to improve things further.

The formation of eukaryotes from prokaryotes, followed by divergence toward the production of plants, animals, fungi, bacteria, viruses, etc., have diverged the biological diversity at various levels of organization. The vast diversity that exists is classified into taxonomic levels such as species, genus, order, class, phylum, and kingdom. A wide range of varieties, strains, cultivars, and lineages make up a species. For a variety of criteria, different groupings of populations, even within species, have genetic variances.

Drought resistance is one of the traits for which there are many genetic differences among species. Drought stress resistance could be used to examine these genetic differences. The plants were able to escape drought stress by developing a large root system and preventing water loss. Drought tolerance in plants is achieved by maintaining normal physiological mechanisms while producing a suitable yield. Drought resistance, avoidance, and tolerance have genetic heterogeneity among maize genotypes, which can be obtained by appropriate germplasm screening. In germplasm screening, different distinctive factors that are strongly linked to these three mechanisms are emphasized. Maize germplasm screening could take place in a growth room under controlled settings or in the field under natural conditions. Because the ultimate goal of screening is to increase grain yield, the parameters utilized for germplasm screening must be linked to grain yield. In screening, traits that are highly affected by drought stress are targeted.

It is necessary to evaluate the real long-term water consumption of CO_2-enriched crops in field conditions. Plants that are larger as a result of CO_2 enrichment have a larger leaf area, which tends to enhance transpiration. They'll presumably have a bigger, more vigorous root system and be able to extract more water, resulting in reduced water stress and higher yields, but maybe a higher total water use. Any reductions in transpiration will free up more energy for soil evaporation. Nonetheless, we believe that as the crop canopy closes, the higher stomatal resistances associated with higher CO_2 concentrations will result in higher overall canopy resistances, resulting in lower total water use. Increases in water use efficiency, when combined with increased production, are likely to be tremendous, with major benefits for both irrigated and dry-land agriculture. Plants adapt to a wide range of conditions. Because plants lack the mobility required to avoid severe situations, flexibility is critical. Because brief exposure to harsh climatic conditions can significantly reduce the production of cultivated plants, altering the plants' surroundings to limit damage appears to be desirable. Sprinkling the lawn, covering the tomatoes on a cold night, misting houseplants, and carefully placing particular house plants in sunny or shaded positions are all examples of small-scale environmental modification. The use of large-scale plant environment modification to alleviate the effects of severe high temperatures or drought has had mixed results.

3.10 PLANT FUNCTIONS

Many critical plant functions are influenced by the temperatures of various plant segments, including phenological development, photosynthesis, respiration, translocation, water uptake, stomatal conductance, and disease resistance. As a plant grows, not all of its parts are at the same temperature, nor are they all near the same temperature as the air. Leaves with many stomata encounter temperatures ranging from 5°C or more below air temperature to 5°C or more above air temperature on a regular basis(Deva, Urban, Challinor, Falloon, & Svitákova, 2020). Flowers that lack stomata may experience temperatures that are 10°C or higher than the ambient air temperature. Because meristematic tissue remains below ground for much of the vegetative development period in many grasses, such as corn or sorghum, soil temperatures are likely to be more relevant than air temperatures in early plant development. At extremely high temperatures, sterility in reproductive organs is prevalent.

When wind speeds are low, non-transpiring plant components exposed to high quantities of solar radiation are particularly sensitive to high temperatures. Sprinkling these organs may minimize their vulnerability to high temperatures and may be beneficial even when the crop has plenty of water for transpiration. Stomatal closure can result in higher plant temperatures if plant water stress develops in hot conditions. Independent of temperature and stomatal effects, water stress can alter a variety of plant functions, including cellular expansion, nitrate reductase activity, respiration, and possibly photosynthesis. The temperature of plant components can also have a significant impact on the rate at which pests that can harm a crop develop. In some conditions, altering the plant habitat may be enough to delay the establishment of a pest long enough to limit damage to acceptable levels. Evaporation within the substomatal cavity causes evaporation from a leaf with a dry surface. The quantity of latent heat leaving the leaf surface is less than that of a leaf with a moist surface because the stomatal opening imposes a diffusion resistance to water passage between the leaf and atmosphere. A leaf with a dry surface will have a greater temperature than a wet leaf due to this resistance. A reduction in transpiration requirement is another possible benefit of keeping the leaf surface wet. When the transpiration demand is high enough to produce water stress in well-watered plants on extremely hot, dry, windy days, soaking the leaf surface reduces the transpiration requirement. By chilling the soil surface, intermittent sprinkling can have a potentially beneficial effect on the plant.

CHAPTER 4

MOLECULAR BREEDING

CONTENTS

4.1 Introduction 62
4.2 Water Stress 63
4.3 Plant Adaptations 64
4.4 Flowering 69
4.5 Abiotic Stimuli 70
4.6 Soil-Water Deficiency 72
4.7 Phytoremediation 74
4.8 Heavy Metals 75
4.9 Microbiome 77

4.1 INTRODUCTION

Salinity, a key environmental stress that restricts agricultural production, has a negative impact on plant growth and development. The early phase of growth reduction is caused by an osmotic impact and is similar to the initial response to water stress. There are few genotypic variations in this phase. The incapacity of a salt-sensitive species or genotype to prevent salt accumulation in leaves to lethal levels distinguishes it from a more salt-resistant one in the second phase. During germination, most crop plants are salt-tolerant, but they become salt sensitive during emergence and vegetative development. Salinity inhibits root and shoot growth; however, supplementary Ca partially alleviates the growth inhibition. The Ca effect appears to be linked to the maintenance of plasma membrane K over Na selectivity. Plants with higher salt tolerance must flourish in saline fields with a variety of extra challenges. Salinity interacts with a variety of stressors, including boron toxicity, but the mechanisms of salinity-boron interactions remain unknown. Future studies should focus on crop tolerance to a mixture of pressures to better understand crop tolerance in saline field environments.

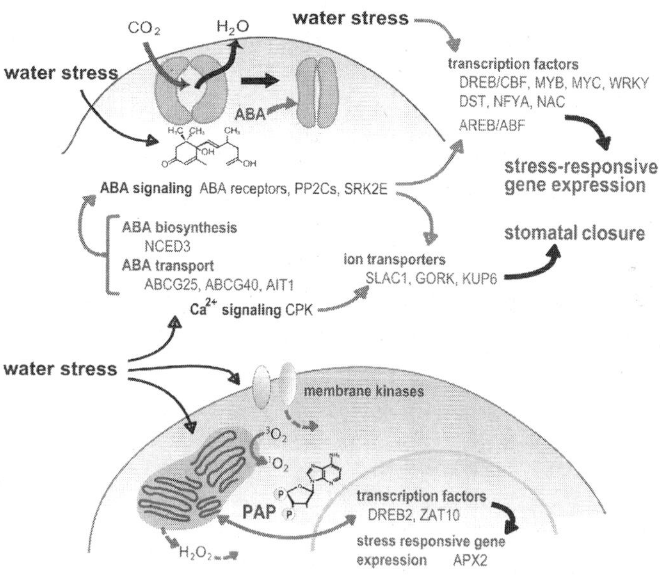

Figure 4.1. Water-stressed conditions.

Source: https://www.frontiersin.org/files/Articles/76566/fpls-05-00086-r2/image_m/fpls-05-00086-g002.jpg

4.2 WATER STRESS

Under water-stressed conditions, (Figure 4.1) root growth and function are critical for agricultural output, but direct root improvement through plant breeding has been slow. One challenge is seeing and quantifying root systems in conditions that are representative of field environments. Identification and selection of specific loci that could optimize the acquisition of water from the soil profile is another problem. However, research into the regulation and evolution of root growth is progressing. The use of a biophysical model of cell expansion helps to focus the debate on themes related to growth and development regulation. Growth is a complicated process, and new knowledge adds to our understanding of how roots sense and respond to changes in the environment. It's critical to deciphering these adaptive processes so that it's evident how changing one process affects the overall function of the plant, and so that the impact on the final output and water consumption can be predicted. Because of this complexity, simple linear models are insufficient as explanatory tools, and a systems approach is required to account for the web of interconnected signaling and response pathways. The actual difficulty is figuring out how to boost root growth, providing breeders with realistic methods for identifying and introducing superior alleles in crop species, and ensuring that findings lead to increased output in the field(Tracy et al., 2020).

Plant life in a changing environment depends on the management of gas exchange at the leaf level. Controlling stomatal opening and closure is an evolutionary approach for balancing water loss with CO_2 uptake and yield in this context. Increased levels of abscisic acid, a well-known stress hormone, are often associated with a decrease in leaf-root water potential caused by soil drought. Drought-stress sensing requires ABA, which leads to efficient stomatal regulation, which prevents yield losses during stressful conditions. Different yield-optimization tactics must be used depending on the crop species or the growing environment. In an interdependent network, ABA effects are controlled by the levels of and sensitivity to other hormones. Understanding the intricate regulatory processes of stomatal regulation involving hormones, second messengers, ion channels, and other classes of involved proteins may provide fresh insights into how to design plants to make the most of available natural resources. Either initiate an earlier stress response without reducing yield, or dampen the plant stress response to promote absorption. Overexpressing or downregulating the expression of certain genes involved in the complex and possibly redundant signaling network of stomatal responses might result in the desired characteristics.

4.3 PLANT ADAPTATIONS

Plants have a waxy coating that protects their aerial organs from excessive water loss. The cuticle has long been thought to be a non-responsive adaptation that simply acts as a barrier to water loss, however, cuticle metabolism is extremely responsive to environmental stimuli. Changes in cuticle chemistry and gene expression in drought and salt-exposed plants have established the cuticle's sensitivity. Changing cuticle features through breeding and technological approaches could help improve drought and salt tolerance in crops. The plant cuticle is a hydrophobic covering formed of a cutin polyester membrane interlinked and coated with free waxes that serves as the outermost barrier over virtually all aerial plant organs, and whose primary purpose is assumed to be for plant water conservation across the plant kingdom. A long-held belief has been that a thick cuticle provides a more efficient barrier to water loss, and hence can be classified as a unique xeromorphic characteristic. While cuticle does have a considerable impact on plant transpiration, current research calls into question the simplistic view of the cuticle as a simple lipid coating whose barrier characteristics are determined solely by its thickness. Many changes occur in the biology of the plant cell as a result of soil-water deficiency stress, beginning with the sensing of the stress and progressing to alterations that facilitate adaptation to the stress. Plant researchers are still baffled as to how plant cells translate the physical stress of water loss into biochemical alterations in the cell. However, understanding the purpose and significance of many of these reactions remains unsolved. Researchers are optimistic that the lessons acquired from Arabidopsis plants' molecular responses to stress can be applied to increase crop development in less-than-ideal field settings and reduce the requirement for irrigation in locations where water availability for agriculture is diminishing.

Many diverse species have evolved to flourish in dry habitats, demonstrating features that allow them to thrive in the absence of water. Adaptation is thus a function of the plant's genetic make-up, which may be seen in morphological traits and exhibited in all areas of the plant, growth, and development. Heritable differences govern adaptation to the environment, allowing some species to function better under soil water deficits due to their constitutive characteristics. When faced with a water shortage in the soil, many species will acclimate to the absence of water. When the root medium has a low water potential, acclimation allows plants to continue to develop or survive by adjusting their physiology through inducible reactions. Arabidopsis seedlings, for example, adapt to the media's lower water content. When Arabidopsis seedlings are exposed to low water potential in a PEG-conditioned agar media, they acclimate in numerous phases and patterns over the course of 96 hours. Plants adapt to changes in the environment by activating signal transduction cascades that control and coordinate physiological and biochemical responses. Abiotic stress response in plants has been linked to a number of signaling channels that act together as a network. Calcium signaling was discovered to be involved in a variety of signaling pathways during abiotic stress responses, including those to heat, cold, drought, and salt. Calcium signals, such as calmodulins and CaM-binding proteins, calcineurin B-like proteins, and other Ca^{2+}-binding proteins, are generated and transduced through an array of Ca^{2+} activated proteins and downstream components, including calmodulins. Important enzymes or proteins involved in diverse cellular metabolism and physiological adjustment could be targeted as calcium signaling effectors.

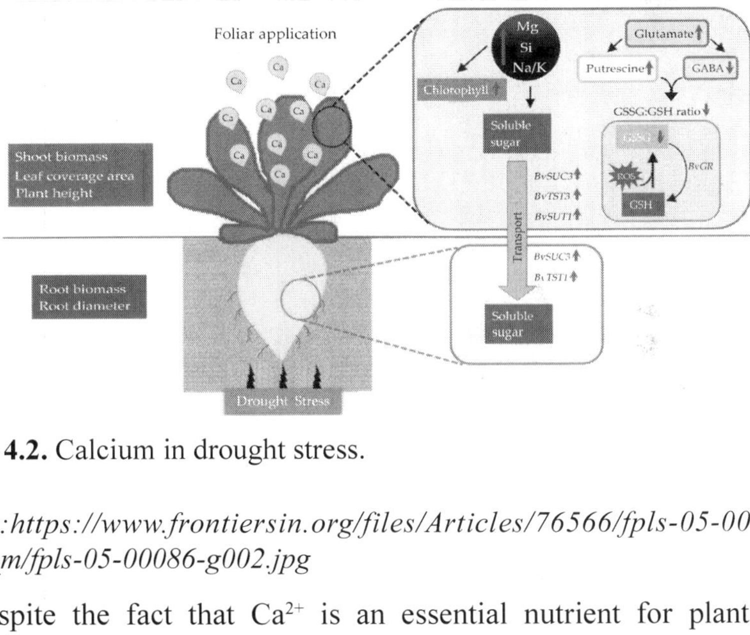

Figure 4.2. Calcium in drought stress.

Source:https://www.frontiersin.org/files/Articles/76566/fpls-05-00086-r2/image_m/fpls-05-00086-g002.jpg

Despite the fact that Ca^{2+} is an essential nutrient for plants, large Ca^{2+} concentrations in the cytosol are harmful (Figure 4.2). The calcium concentration in soil is usually greater than 1.27 mM, however, Ca^{2+}-ATPases and Ca^{2+}/H^{+} antiporters help plant cells maintain a low cytosolic calcium level. These enzymes transfer cytosolic Ca^{2+} into apoplasts or the lumens of intracellular organelles including the vacuole and endoplasmic reticulum (ER). Rapid and substantial increases in Ca^{2+} during cell response to biotic and abiotic stress are facilitated by low cytosolic calcium concentrations. Abiotic stressors always result in an increase in the concentration of free calcium in the cytoplasm. Plants exposed to NaCl treatments repeatedly have long-term changes in cytosolic and apoplastic Ca^{2+} concentrations. When plants are exposed to hypo-osmotic stress, they produce more Ca^{2+}. Using aequorin as a reporter, researchers discovered that cold and wind can cause different Ca^{2+} signals in tobacco seedlings. Reactive oxygen intermediates are frequently produced as a result of environmental stressors (ROIs) (Tuteja & Mahajan, 2007). Many conditions cause minor phospholipids such phosphatidic acid (PA) and phosphoinositides (PIs) to grow transiently in plants. These alterations are the first signs of a plant's stress response. Lipid mediators influence cellular processes by interacting directly with proteins and altering the structure of cell membranes. Protein phosphatases and kinases involved in membrane trafficking and the cytoskeleton have all been identified as lipid targets in plants. The involvement of lipids in signaling, intracellular trafficking, and cytoskeletal architecture in plant drought and salinity tolerance is crucial. Cell membranes play a crucial role in plant responses to environmental stresses as the physical barrier that divides the interior of a cell from its surroundings. The first site of cellular perception of stress stimuli is the membrane. Membranes are frequently involved with subsequent steps in signaling cascades, such as activation of effector proteins, formation of second messengers, and changes in cellular metabolism. While much research on membrane-associated signaling events has focused on proteins, recent advances have revealed that membrane lipids and their derivatives are critical contributors in the signaling network of plant responses to stress, such as drought and salinity.

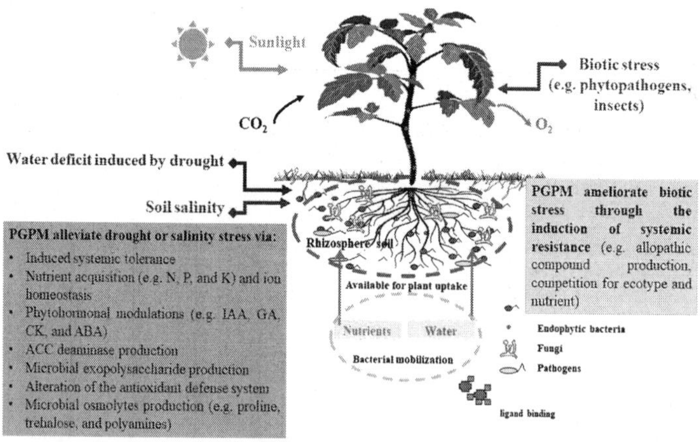

Figure 4.3. Drought and salinity stress.

Source:https://www.frontiersin.org/files/Articles/591911/fpls-11-591911-HTML/image_m/fpls-11-591911-g002.jpg

Drought and salinity (Figure 4.3) are two important environmental variables that affect plant growth, productivity, and distribution. Plant responses to these stressors rely heavily on lipid-mediated signaling. Information on the development of putative lipid mediators under various growth circumstances is continuously expanding. However, our understanding of the precise involvement of particular lipid signaling processes in plant adaptation to drought and salt stress is currently restricted. The function of various lipid signaling reactions can be elucidated by genetic modification of enzymes that generate lipid mediators. The distinct phenotypes that occur from the deletion of different PLDs suggest that some PLDs have specific roles in plant response to various osmotic stressors. These differences could be linked to the location and time of PA production controlled by different PLDs. Because of their limited mobility within the cell, spatial and temporal regulation is necessary for all signaling processes, but it is especially vital for intracellular lipid messengers. Other phospholipid-signaling enzyme groups, such as PLCs, DGKs, PI kinases, and phosphatases, are also likely to have distinct functions. When discussing the involvement of distinct types of lipid signaling enzymes in a certain physiological response, it is critical to specify the exact genes and enzymes involved. This data will also be required for biotechnological modulation of lipid signaling networks to improve plant stress tolerance.

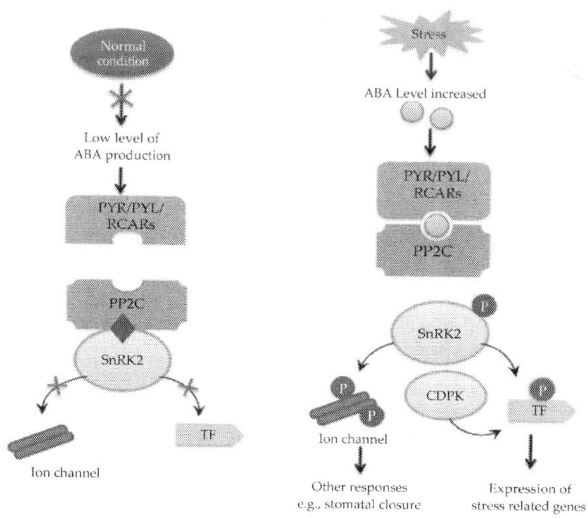

Figure 4.4. ABA in drought response.

Source:https://www.researchgate.net/profile/Abid-Ullah-17/publication/312107590/figure/fig2/AS:447783648927744@1483771349021/ABA-mediated-signaling-pathway-during-normal-and-stress-conditions-Under-normal_W640.jpg

Abscisic acid (ABA), (Figure 4.4) a plant stress hormone, performs a number of important roles in plant stress tolerance and response. The role of ABA in the activation of stress-responsive genes and the regulation of guard cell mobility has been extensively researched. ABA has also been shown to regulate root adaptation to drought stress more recently. To date, there has been limited success in controlling plant ABA activity in order to increase plant drought resistance. Exogenous ABA experiments as well as plant mutants deficient in ABA production or responses were used to deduce the role of ABA in several physiological systems. Other methods, such as looking at ABA's spatial and temporal localization, have been used to deduce cellular activities that may be governed by ABA. These methods may not be able to disclose all of ABA's functions. Furthermore, when there is a lack of ABA, the cellular machinery becomes more sensitive to it.

In these biosynthetic mutants, the deleterious effects of ABA shortage may have been greatly reduced. As a result, it's conceivable that undiscovered cellular activities require ABA, or that ABA is required for plant growth and development. A complete loss of ABA or ABA signaling in the latter situation would be fatal. It's possible that this is one of the reasons why the ABA receptor was missed in genetic tests. Under normal unstressed conditions, all ABA deficient mutants have dramatically reduced vigor, implying that ABA is essential for specific cellular activities that aren't connected to stress adaptation. The role of ABA in seed maturation and dormancy is well recognised. Seed maturation necessitates the production of storage proteins and the development of desiccation tolerance so that the embryo may survive high dehydration, as seen in dried seeds. Seed maturation may require even less ABA than dormancy inception and regulation, while ABA activates genes involved in both processes. The importance of ABA in seed development is demonstrated by the fact that when ABA levels are inhibited immunologically, embryos are unable to establish desiccation resistance and are hence not viable when desiccated. The premature germination of maize ABA deficient mutants and the suppression of exogenous ABA on seed germination, on the other hand, clearly demonstrate the significance of ABA in seed dormancy.

Figure 4.5. Late flowering plants.

Source:http://clemsonhgic.wpengine.com/wp-content/uploads/2018/03/drought-stress-in-tomato-plants-can-cause-flowers.jpeg

4.4 FLOWERING

Late flowering plants, (Figure 4.5) as well as specific ecotypes within a plant species, such as Arabidopsis, grow more slowly than rapidly flowering plants. Drought stress is frequently induced by long-term water scarcity in the soil, which is insufficient to meet plant transpiration demands. Different plant species may use different drought-resistance techniques. Plants, for example, may respond to drought difficulties in one of three ways: by reducing water consumption, increasing water intake, or mitigating the detrimental effects of water scarcity. These tasks can be completed in a variety of ways. First, when a guard cell experiences drought, its stomatal pores seal, reducing transpiration water loss. Second, a batch of stress-response genes is turned on. These genes' products play a role in drought tolerance, either directly or indirectly. Third, there are some developmental changes that may make plants more adaptable to drought stress in the long run.

Understanding gene control pathways is critical for crop abiotic stress resistance genetic improvement. Plants use a variety of gene control mechanisms in response to developmental and environmental signals. MiRNAs are made from MIR gene transcripts, whereas siRNAs are made from dsRNA made from heterochromatin DNA repeat transcripts, mRNAs made from natural cis-antisense gene pairs, and ssRNA/mRNA cleavage driven by miRNA. Small RNAs direct mRNA cleavage, translational repression, chromatin remodeling, and DNA methylation to control the expression of complementary genes(Guleria, Mahajan, Bhardwaj, & Yadav, 2011). Plants have been found to have several stresses responsive short RNAs, and their role in oxidative stress tolerance, osmolyte accumulation or osmoprotection, and food deprivation response has been established. Stress-upregulated miRNAs may downregulate their target genes under abiotic pressures, which are probable negative regulators of stress tolerance, whereas stress-downregulated miRNAs may result in accumulation of their target gene mRNAs, which may positively regulate stress tolerance. Overexpression of miRNA-resistant target genes may assist overcome post-transcriptional gene silencing, resulting in improved transgenic plant expression of engineered traits. Understanding the functions of short RNAs in transcriptome homeostasis, cellular tolerance, phenological and developmental plasticity, and abiotic stress recovery in crop plants can aid genetic engineering of abiotic stress resistance.

4.5 ABIOTIC STIMULI

To resist all abiotic stimuli, plants have evolved some stress-specific tolerance mechanisms as well as some common tolerance mechanisms Plants control the rate of transpiration by altering stomatal number, distribution, sunken stomata, stomatal hairs, and the opening and closing of the stomatal aperture. Under short-term drought, plants minimize water loss by closing their stomata, whereas under long-term drought, plants use non-stomatal mechanisms such as deposition of epicuticular wax in the epidermal cells of stems and leaves, radiation reflectance, reduction in transpiring surface i.e., reduction in leaf area, leaf rolling, folding, and senescence. Plants can withstand osmotic strains through osmotic adjustment or active solute accumulation under osmotic conditions. Chaperone proteins safeguard cellular machinery, osmoprotectants, changes in plasma membrane and organellar membrane composition, and other processes are among them. Antioxidants detoxify reactive oxygen species, which is an important abiotic stress tolerance mechanism. Plants use processes to maintain cellular ion and osmotic homeostasis, as well as damage control and repair mechanisms when they are stressed by salt. Abiotic stress tolerance mechanisms include oxidative stress management and osmoprotection.

Sugars provide the energy and carbon that plants require for numerous growth and development processes. Sugars produced during photosynthesis are delivered to sink tissues and transformed to lipids, starch, protein, sucrose, and fructans, or directed to respiration. As a result, it's likely that the metabolic processes involved are influenced by sugar concentrations. Low sugar levels in source tissue promote photosynthesis, nutrient mobilization, and export, whereas high sugar levels in sink tissues promote growth and storage. Sugar build-up in source tissues suppresses photosynthesis and maintains sugar balance. Due to a decrease in photosynthesis, a change in metabolism, and a necessity for high maintenance respiration, abiotic stressors affect the sugar balance of cells. Sugar status is used by plants to modify source-sink activity and, as a result, growth and development in response to numerous environmental stimuli.

Figure 4.6. Soil water deficiency.

Source:https://www.2pressrelease.net/wp-content/uploads/2019/12/Soil-Water-Deficit-1024x682.jpg

4.6 SOIL-WATER DEFICIENCY

Soil water deficiency, (Figure 4.6) salinity, and temperature fluctuations all contribute to osmotic stress. Osmotic stress is caused by a decrease in soil water availability due to drought or a drop in the water potential of soil solution due to salt, which results in decreased water uptake and turgor loss. Low temperature stress is frequently coupled with osmotic stress, as low temperatures can decrease root water intake, while freezing-induced ice formation in the apoplast causes water to migrate from the cytoplasm to the apoplast, resulting in severe cellular dehydration. Osmotic adjustment refers to the reduction of osmotic potential as a result of the net accumulation of compatible solutes in response to osmotic stressors, and it is conserved in both halophytic and glycophytic plants. At low water potentials, osmotic adjustment aids in maintaining cell turgor. Furthermore, solute build-up may aid in cell freezing tolerance by lowering the freezing point of the cells. Compatible solutes like proline, betaine, polyols, sugar alcohols, and soluble sugars protect plants against stress by detoxifying reactive oxygen species and stabilizing proteins and membranes, in addition to their osmotic role. Abiotic stress tolerance was improved in some transgenic plants modified to overproduce osmolytes(Giri, 2011).

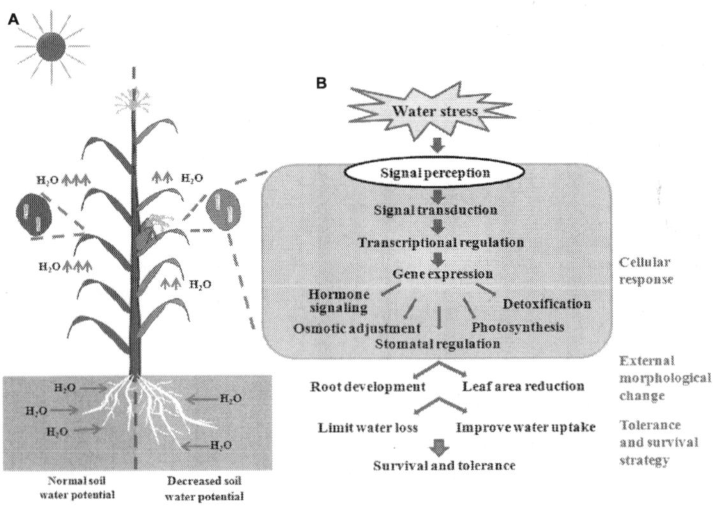

Figure 4.7. Drought resistance.

Source:https://www.frontiersin.org/files/Articles/273925/fpls-08-01110-HTML/image_m/fpls-08-01110-g001.jpg

Drought resistance (Figure 4.7) is defined as a plant's ability to produce its highest economic product in a water-stressed environment. Plant species, age, growth stage, genotype, and drought intensity all play important roles in the plant's drought response system. Drought resistance included methods for drought escape, drought avoidance, and drought tolerance. Drought tolerance is influenced by heritable factors at various stages and includes cellular adaptation, phenotypic adjustment, and physiological acclimatization. Reduced osmotic potential, turgor pressure, and chlorophyll content are all symptoms of cellular adaptation. Under drought stress, the key physiological events are increased stomatal conductance, reduced rate of transpiration to limit excessive water loss, and early maturation for improved production. Under a water-stressed environment, morphological adaptations include increased root length and thickness, a thick cuticular layer on the leaf surface, reduced leaf size and area, fewer epithelial cells, and late leaf senescence. Drought avoidance is described as a plant's ability to keep its tissue water potential relatively high despite a lack of soil moisture. Rice types that can withstand drought have deep and coarse roots, as well as a higher root to shoot ratio, flexibility in leaf rolling, early stomatal closure, and increased cuticular resistance. Drought escape is described as a plant's capacity to complete its whole life cycle before experiencing significant soil and plant water deficiencies. Quick phenological growth, developmental flexibility, and remobilization of pre-anthesis assimilates to the grain are all part of this system.

Drought and salinity are difficult environmental restrictions to crop yield around the world. Both of these stresses are fairly sensitive to the cultivated tomato, Lycopersicon esculentum Mill., throughout its ontogeny, including seed germination, seedling emergence, vegetative growth, and reproduction. For abiotic stress tolerance, there is limited diversity in cultivated tomatoes; nevertheless, related wild tomato species provide a rich source of genetic variation that can be leveraged for crop improvement. This variation has been used to characterize the physiological and genetic bases of tolerance to various abiotic stresses, such as salinity and drought, throughout the last several decades. Abiotic stress tolerance is a complex phenomenon influenced by uncontrollable environmental conditions and controlled by several genes. Analysis of response and associated response to selection, as well as the identification of quantitative trait loci (QTLs) conferring tolerance at various stages, have all confirmed this. Transgenic techniques have also been used to better understand the genetic and physiological roots of salt and, to a lesser extent, drought tolerance in tomatoes, as well as to generate transgenic plants that are more stress tolerant. Despite extensive classical genetics and physiological research, as well as recent molecular marker and transgenic studies in tomatoes, no commercial cultivar with salt or drought tolerance has been reported.

4.7 PHYTOREMEDIATION

Our natural resources are becoming significantly degraded with numerous inorganic and organic substances as a result of fast-expanding population and related anthropogenic activities.

Phytoremediation (Figure 4.8) is the most efficient and environmentally friendly of the various remediation processes available for the removal of contaminants. Phytoremediation is a technique in which toxic chemicals are removed from polluted sites using plants and associated bacteria. The rhizosphere microbiome, which includes bacteria, endophytes, and AMF, is important for plant growth and development. Plants benefit from the rhizosphere microbiota because it improves nutrient availability, emits growth-promoting phytohormones, and protects them from abiotic and biotic stressors. The interaction of soil-borne bacteria with plants aids in the uptake, sequestration, and detoxification of pollutants from polluted areas(Tiodar, Văcar, & Podar, 2021).

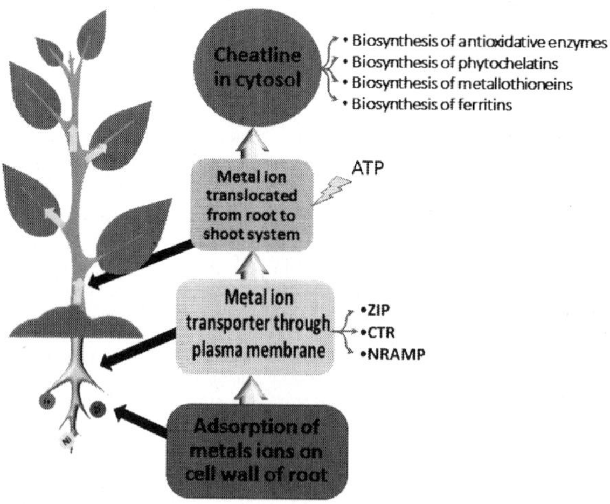

Figure 4.8. Phytoremediation.

Source:https://www.researchgate.net/profile/Mohammed-Nagah/publication/333824257/figure/fig1/AS:770773220327427@1560778064627/Mechanism-of-phytoremediation_W640.jpg

Rapid global population growth has resulted in a massive increase in global energy consumption, which has resulted in resource overexploitation, urbanization, industrialization, and land use changes, among other things. All of these human activities have resulted in a significant increase in harmful pollutants released into the environment. Polycyclic aromatic hydrocarbons (PAHs), heavy metals, pesticides, petroleum hydrocarbons, furans, dioxins, halogenated hydrocarbons, and other contaminants have been found in the environment. The main causes of soil, air, and water pollution are pollutants emitted by a variety of human activities such as crude oil production, unintentional oil spills, incorrect waste management methods, excessive pesticide use, and so on. Heavy metals are very harmful as mutagens and carcinogens, causing damage to the central nervous system, mental functioning, and different important organs such as the lungs, kidneys, and liver. Long-term heavy metal exposure may cause Alzheimer's disease, Parkinson's disease, muscular dystrophy, sclerosis, and other diseases.

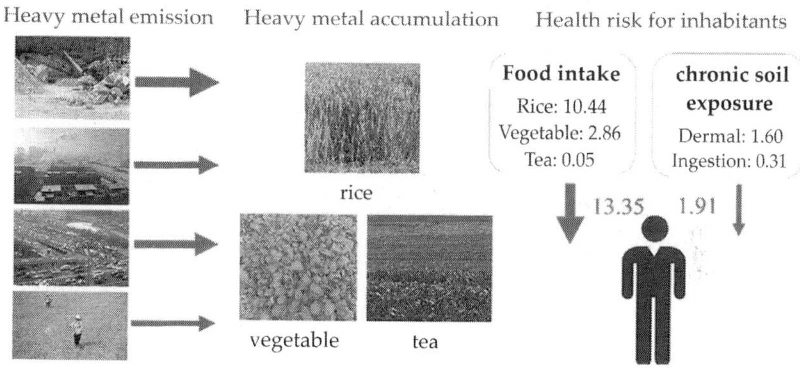

Figure 4.9. Heavy metals.

Source: https://ars.els-cdn.com/content/image/1-s2.0-S0301479717310691-fx1_lrg.jpg

4.8 HEAVY METALS

Heavy metals (Figure 4.9) like cadmium, chromium, copper, lead, mercury, nickel, arsenic, and zinc are released into the environment as a result of mining, landfilling, industrial solid waste disposal, industrial effluent disposal, airborne industrial wastes, and the use of chemical-based pesticides and herbicides. In addition to the use of leaded gasoline for land transportation, the combustion of non-renewable energy resources, and nuclear wastes have all contributed significantly to heavy metal-based air pollution. Apart from the negative impacts on human health, the presence of these harmful chemicals in the environment has resulted in serious ecological degradation, such as habitat destruction, biodiversity loss, air quality deterioration, poor vegetation development, and ecosystem imbalance. Soil is thought to be a repository for a wide range of microscopic organisms, including bacteria, fungi, protozoa, and algae. All of these microbes contribute to the ecological balance in different ways. Bacteria are discovered to be the most common among these many living species. The majority of bacteria present in soil are located in the rhizosphere, which is the area around the plant's roots. The presence of nutrients such as amino acids, sugars, organic acids, and other nutrients in the exudates released by the roots causes a dense population of bacteria in the rhizosphere. As a result, the rhizosphere has enormous potential for preserving plant health and nourishment("The Rhizosphere - Roots, Soil and Everything In Between | Learn Science at Scitable," n.d.).

In this ecology, the plant kingdom does not exist in isolation; it is surrounded on all sides by microbes that are constantly interacting with it. These bacteria and viruses are part of the Holobiont, which is made up of plant biochemistry. Apart from being part of plant nutrition and nutrient intake, the bacteria in this Holobiont offer plant health and stability by providing resistance to abiotic and biotic stress, as well as antibiotics and medicines. However, it still has a level of intricacy that has to be investigated. The plant's microbiome evolves as the environment changes, as do the microorganisms in the microbiome, which adapt to the environment by causing structural changes. The presence of VAM fungi demonstrates a mutualistic interaction between the host plant and the microbe, which provided nutritional strength and physical stability to the host plant, which was important in the establishment of terrestrial habitat. Their presence also represented genetic diversity or changes in their genetic arrangements as a result of environmental changes, which resulted in plant evolution and diversification.

The study of the genome of an endophytic bacterium at the genomic and molecular level, as well as the study of signaling behavior, resulted in the development of molecular markers that aided in the transcriptome analysis of any system. Thus, molecular applications based on genomic sequencing and protein analysis have greatly improved our understanding of the diverse interactions of the plant microbiome; however, this understanding is still insufficient to explain the functionality of the plant microbiome in light of vast ecological diversity. The microbiome of that plane is determined by a combination of root-associated bacteria and soil microorganisms, much as it is in the rhizosphere. There can be systemic resistance as a result of the microbiome's temporary association with the host plant, or there can be a lateral transfer of resistivity from soil to seed and then to plant, where the physical environment, such as soil properties, pH, humidity, and temperature, act as growth chambers for the microbiome. It also highlights the fact that, rather than the plant genome, the microbial community determines the plane's microbiome in major part, with the host genome having a role in microbiome selection due to its evolutionary significance.

4.9 MICROBIOME

Another element that impacts the microbiome is the separation of bacteria in distinct spheres and planes, i.e., it's difficult to separate rhizosphere microbes from soil microbes occupying the rhizoplane. When a bacterium enters the tissues from the rhizosphere or from the air, it reaches the endosphere. Arbuscular mycorrhizal fungi, which penetrate deeply, are first in the endosphere microbiome, followed by bacteria, and finally archaea. Unlike the rhizosphere microbiome, the endosphere microbiome is extremely specialized and primarily selected to interact with the host. In comparison to the rhizosphere, the endosphere has less diversity. When taxonomic units of the endosphere and rhizosphere microbiomes were compared, it was discovered that microbial diversity dropped from the rhizosphere to the endosphere. The rhizosphere microbiome is less structurally variable and has a smaller ecological niche than the well-structured endosphere microbiome, according to studies on field-grown popular trees on structural variability and niche differentiation in the rhizosphere and endosphere bacterial microbiome. As a result, the endosphere microbiome contributes to the host plant's greater adaptability and strength(Dastogeer, Tumpa, Sultana, Akter, & Chakraborty, 2020).

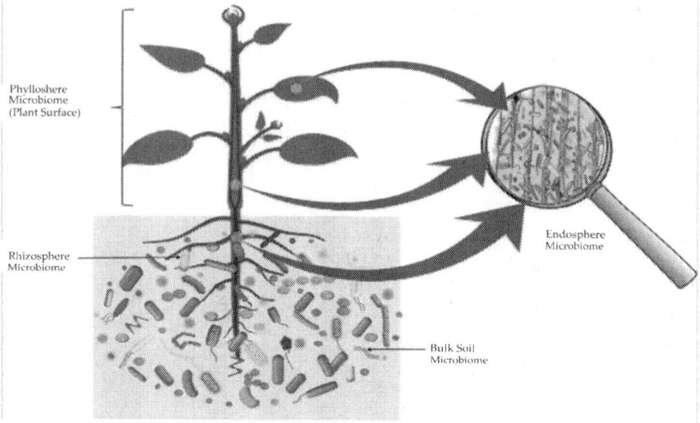

Figure 4.10. Endosphere microbiome .

Source:https://upload.wikimedia.org/wikipedia/commons/1/1b/The_plant_microbiome.jpg

The endosphere microbiome (Figure 4.10) can also infiltrate leafy tissues and even reproductive organs, most likely for vertical or lateral transfer to the next generation. These microbial inhabitants live either on the surface of the aerial plant tissues or within them. Few reports have been found of microbiomes in reproductive organs such as anthers and carpels, indicating vertical transmission from seed to young plant. The existence of a considerable number of microbiomes on the surface of the host leaf and in the internal tissues of the leaf indicates that they have grown in distinct environments, resulting in a very different microbial population in the endosphere and ectosphere of plant tissues above and below ground. It has also been reported that the microbiome on the leaf surface, or phyllosphere, can include archaea such as methylotrophs because they are exposed to much harsher conditions than the rhizosphere due to desiccation, UV radiation, and oligotrophy; as a result, the microbiome is more structurally defined and has a better ecological niche. They have the ability to produce unique proteins such as microbial rhodopsins, which absorb green light and operate as proton pumps to aid photosynthesis. In comparison to other microbiomes, these methylotrophs and photo harvesters promote the competitive exclusion of plant diseases, allowing them to survive in space or have a larger niche. As a result, the phyllosphere can have an impact on the Holobiont.

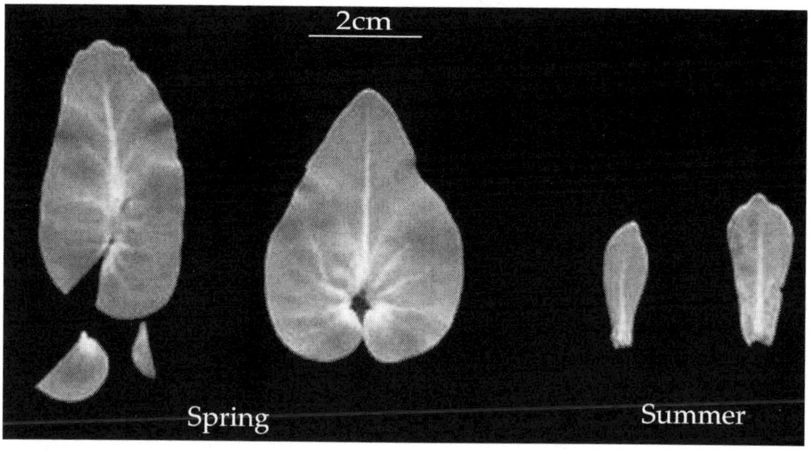

Figure 4.11. Functional plasticity.

Source:https://media.springernature.com/lw685/springer-static/image/art%3A10.1038%2Fs41467-020-17875-1/MediaObjects/41467_2020_17875_Fig1_HTML.png?as=webp

Plants have functional plasticity (Figure 4.11). The Holobiont microbiome was formed following screening and competitive exclusion of bacteria that were weak. As a result of these interactions with the environment, weaker populations are continually eliminated and a healthier microbial flora on the phyllosphere develops, resulting in diversification. Because of this diversity, even the same plant species can have diverse ecological niches, resulting in different plant strains. The Holobiont of multiple strains of the same plant species allows different activities such as abiotic stress survival, heterogeneity, mutualism, resource accessibility, and the formation of novel habitats. It is vital to highlight the phrase core microbiome, which is created from a plant community. Again, diverse genomic and transcriptome analyses will aid in understanding the core microbiome's functions. A core microbiome will be made up of bacteria from distinct Holobiont families that have similar genomic and transcriptome analyses. Microbial diversity is a vital part of any ecosystem. The important steps in developing innovative technologies for the successful exploitation of microorganisms to solve sustainable environmental and economic concerns and increase the quality of life without compromising yields are intellectual strategy and careful selection of bacteria. Appreciation of these elements could help to improve public perceptions of microorganisms by demonstrating that the long-term utilization of microbial diversity offers ecological and economic benefits.

Recent technical improvements in microbial diversity exploration have revealed that a huge percentage of microorganisms remain unexplored, and their ecological roles are mostly unknown. Several microbial applications are well-known in resolving important agricultural problems such as crop production, plant health protection, and soil health maintenance, as well as environmental challenges such as organic and inorganic pollution and bioremediation of soil and water. They also offer a number of favorable characteristics that can be applied to a variety of environmental domains. They require fewer nutrients for optimal growth throughout the year and can withstand extremes in temperature, pressure, and other factors. Bacteria have garnered attention in environmental microbiology because they are simple to produce and research and have an important role in the improvement of our environment. Because most early bacteriological studies focused solely on medical bacteriology to determine the etiology of prevalent human diseases, the focus has now switched to improving our deteriorated environment. Biopesticides, biofertilizers, biocontrol, and PGPRs are just a few of the biological agents that are becoming increasingly important for environmental management. Environmental microbiologists have shifted their focus from pathogenic bacteria to helpful bacteria, such as those used in the treatment of industrial wastewater, in order to protect and preserve the polluted environment as progress has been made in controlling disease-producing bacteria. Apart from that, attempts are being made to boost agriculture productivity by using microbes in agricultural fields. Some bacteria fix nitrogen in symbiotic relationships with plants, while others live freely in soil and aquatic environments.

According to their roles and how they affect soil quality, plant growth and health, and yield, soil microbiologists and microbial ecologists have classified soil microorganisms as constructive or destructive. Microorganisms that fix atmospheric nitrogen, detoxify pesticides, decompose organic wastes and residues, suppress plant and soilborne pathogens and diseases, improve nutrient cycling, and produce bioactive chemicals like hormones, vitamins, and enzymes that stimulate plant growth are all examples of constructive microorganisms. Despite the fact that bacteria and fungi are linked, the problem with fungi is their complexity when compared to bacteria. Excessive use of chemical and inorganic fertilizers in agriculture has led to a set of environmental issues, including eutrophication, poor soil health, a drop in natural microflora, and ozone depletion, as well as negative health consequences. Biofertilizers have been found to be useful and efficient in overcoming these issues. Microorganisms are increasingly frequently used in nature farming and organic agriculture since they have been found to be effective in removing problems connected with the usage of artificial fertilizers and pesticides. Certain microbe species are commonly employed because they have unique qualities that allow them to produce natural products and act as a decent substitute for artificial fertilizers. Biofertilizers are believed to be a safe alternative to chemical fertilizers, with the goal of minimizing or eliminating environmental disturbance. They are both cost-effective and environmentally beneficial, and may be produced in large quantities on the farm. They aid in the decomposition of organic debris and the release of growth-promoting compounds.

They are a low-cost source of nutrition and a good supply of microelements and micronutrients. They can assist to boost crop output by 10–40%, fix nitrogen by 40–50%, and improve soil texture, pH, and other soil qualities(Nadeem & Farooq, 2019). In cultivable fields, they also prevent soil-borne infections and maintain soil structure. Bacteria, fungus, and cyanobacteria are the most common biofertilizers. Biofertilizers allow plants to use all of the food available in the soil and air, reducing the amount of artificial fertilizer used by farmers. Phosphate and nitrogen molecules are found in the environment and are necessary for plant growth. Phosphate affects crop stress tolerance and maturity, as well as nitrogen fixation, either directly or indirectly. Decomposition, nitrogen fixation and denitrification, nutrient mobilization and mineralization, and nutrient and water storage are all aided by rhizospheric bacteria.

CHAPTER 5

PLANT RESPONSES TO DROUGHT STRESS

CONTENTS

5.1 Introduction 84
5.2 Drought Adaptation 85
5.3 Turgor 86
5.4 Soil Drying 88
5.5 Phenomics 89
5.6 Genes 89
5.7 Abiotic Stress 90
5.8 Plant Responses 92
C_3 and C_4 Plants 95
5.10 Osmotic Stress 97
5.11 Water Transfer 100
5.12 Woody Plants 102

5.1 INTRODUCTION

The most prominent abiotic factors that limit the production of food crops are drought and salinity. Knowledge of physiological features, as well as new molecular methods for identifying critical genes or providing molecular markers, has the potential to boost output beyond current levels. The application of molecular knowledge necessitates careful experimental design and phenotyping. For progress toward crop development on dry and saline terrain, it is critical to test the usefulness of physiological features and key candidate genes. The combination of novel phenomics approaches and quickly improving molecular tools has created a significant incentive for identifying critical stress tolerance features and genes, as well as new means for introducing these genes into important food crops (Figure 5.1).

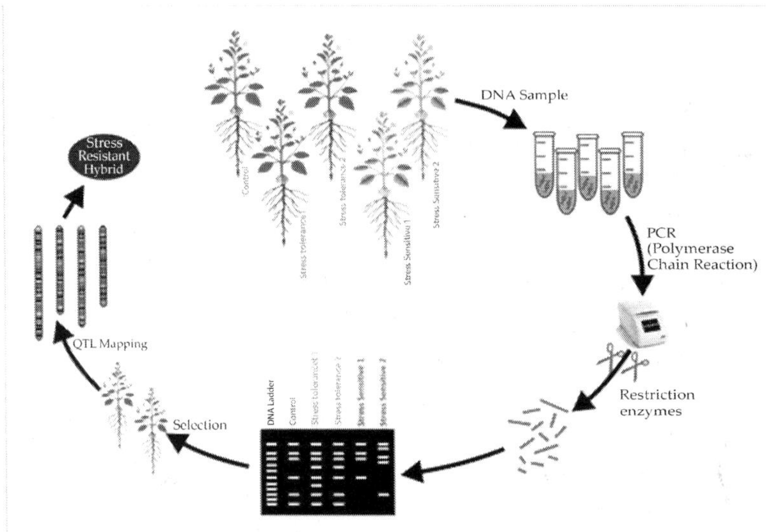

Figure 5.1: Latest approaches.

Source: https://www.mdpi.com/plants/plants-09-01374/article_deploy/html/images/plants-09-01374-g001.png

Soil salinity, like drought, lowers the soil water potential and plants' ability to absorb water, slowing the pace of cell development in growing tissues. The reduced flow of assimilates to the plant's meristematic and developing tissues, including leaves and roots, is due to the slower formation of photosynthetic leaf area. Leaves are often more affected than roots. Simultaneously, water and salt stress lower stomatal conductance in older leaves, limiting their photosynthetic activity. Because the carbohydrate status of the growing tissue is usually unaffected and sometimes higher, indicating unused assimilation, it's difficult to tell whether this causes an additional reduction in growth rate or not. Salt may have an additional effect on growth over time. If too much Na or Cl gets into the plant, it can build up to toxic levels in the older transpiring leaves. The passage of carbon compounds to meristems and growth zones in leaves will be further restricted due to injury, which will be compounded by a reduced leaf area. It's possible that the plant will die before the seed is formed. The rate of leaf death is critical to the plant's survival. If the pace at which new leaves are formed exceeds the rate at which old leaves die, the plant will have enough photosynthesizing leaves to produce flowers and seeds, albeit in smaller numbers.

5.2 DROUGHT ADAPTATION

Many of the features that explain drought adaptation in plants, such as phenology, root size and depth, hydraulic conductivity, and reserve storage, are connected with plant development and structure and are not stress related. However, a big element of a plant's drought tolerance is its ability to drain excess radiation, which is a stress in natural settings. Drought and salt stress have direct effects on photosynthesis, such as stomatal and mesophyll diffusion and changes in photosynthetic metabolism, or secondary effects, such as oxidative stress resulting from the superimposition of numerous stresses(Chaves, Flexas, & Pinheiro, 2009). The rate and degree of photosynthetic recovery, as well as the rate and degree of photosynthesis decline during water depletion, may influence a plant's carbon balance during a time of salt or water stress and recovery. Root cells modify their growth rate and differentiation in response to a soil water deficiency, and the root system architecture changes the degree of branching or rate of branch root elongation. They also send a message to the shoot. Although root elongation slows, it does so at a slower rate than shoot growth. In Arabidopsis, the proportion of roots as a fraction of total plant biomass can increase by at least 30% as the soil dries out. Roots in drying soil continue to spread into deeper, wetter layers in grapevines in the field, whereas irrigated plants' roots proliferate in the topsoil.

Drought stress reduced rosette expansion more than photosynthesis, whereas root growth was increased, according to an Arabidopsis study of the impact of water restriction on carbon absorption activity and 30 key enzymes in carbon metabolism(Muller et al., 2011). All C metabolites showed an increase in diurnal turnover when water was scarce, but no fundamental reprogramming, showing quantitative changes in C metabolism but neither stimulation or repression of novel genes. Grapevine genotypes respond to water stress in different ways, with isohydric or anisohydric behavior linked to stomatal and non-stomatal effects. Grapevine stomatal regulation is influenced by ABA and hydraulic factors, the latter of which is associated with stress development and recovery. Aquaporins are critical mediators of radial water transport across roots and into the xylem, as well as out of the xylem in leaves. They could explain why iso and anisohydric behavior differs. Grapevine stomatal sensitivity varies by genotype and can be influenced by irrigation techniques.

5.3 TURGOR

Turgor is required for cell survival, as well as for the growth of elongating cells and stomata. Osmotic adjustment is necessary for plants to adapt to low-water-potential soils, but it can have negative consequences for carbon allocation in the rapidly developing period of plants like wheat and barley. Compatible solute synthesis has an energy cost, and so may result in a growth drawback. For example, it takes about 7 moles of ATP in leaf cells to accumulate 1 mole of NaCl as an osmoticum, whereas it takes an order of magnitude more ATP to synthesize 1 mole of an organic solute(Zahran, 1999). These chemicals are synthesized at the expense of plant growth, but they may help the plant survive and recover from a period without water. Hormonal control of cell division and differentiation is evident in the emergence of smaller but thicker leaves, showing that cell size and shape have changed. Leaves from salt-treated plants have a larger weight: area ratio, indicating that they have a higher transpiration efficiency, which is a frequent feature in plants acclimated to dry and saline soil.

They can be complicated, but they require expensive equipment, such as a mass spectrometer to determine water use efficiency, which is measured by carbon isotope discrimination. Alternatively, they could be biologically straightforward but difficult to quantify, such as leaf rolling. A large number of genotypes can be screened if a trait can be simply assessed as a quantitative measurement. If genetic variation exists, a QTL study can be carried out by crossing the contrasting parents, screening the F2 generation for the trait, and mapping with molecular markers. A different population should be used to validate the QTL.

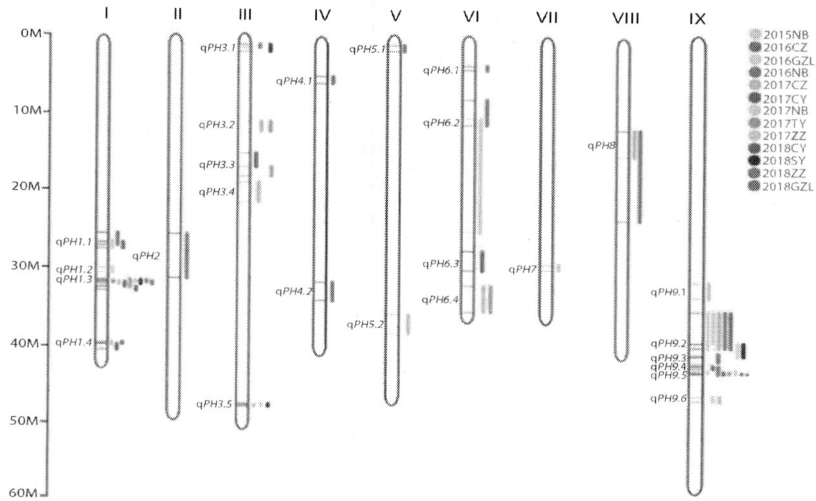

Figure 5.2. QTL.

Source:https://www.researchgate.net/publication/345040223/figure/fig2/AS:952564589817858@1604120504635/Quantitative-trait-loci-QTL-map-showing-location-of-QTLs-for-plant-height-at-different_W640.jpg

Fine mapping can be done with the help of a model species if the QTL (Figure 5.2) is robust and reproducible in diverse genetic backgrounds, and a candidate gene may be revealed. If this is not the case, the original connected marker can be utilized for breeding. Because of the rapid uptake of Na and Cl, osmotic adjustment is easy to achieve, and it is managed in species like wheat and barley so that it does not become toxic. Because the concentration of Na and Cl in the vacuole can be as low as 30 mM, compatible solutes are required to balance them. Sucrose, proline, and glycine betaine are the most often accumulated molecules, while other molecules can accumulate to high amounts in some species. Proline or glycine betaine is found in high enough concentrations in the leaves of many halophytes to contribute to the osmotic pressure in the cell as a whole. These solutes apparently have another function at low concentrations, such as stabilizing the tertiary structure of proteins and also act as osmoprotectants.

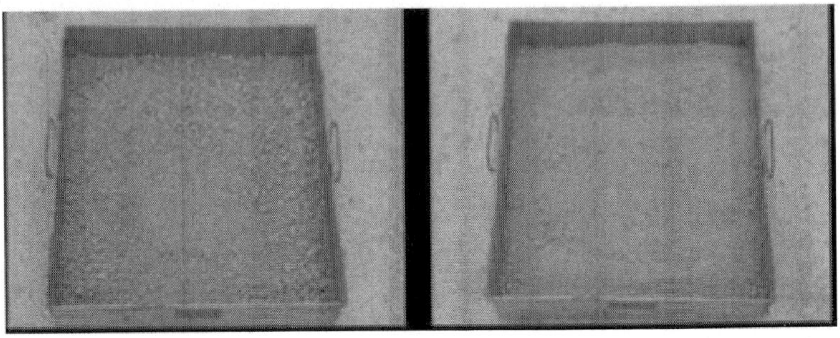

Figure 5.3. Soil drying.

Source:https://www.researchgate.net/profile/Rohaya-Alias/publication/334960892/figure/fig1/AS:788287170039810@1564953715536/Pre-drying-methods-of-soil-samples-a-Three-days-air-drying-at-37-C-b-24-hours_W640.jpg

5.4 SOIL DRYING

Soil drying (Figure 5.3) is difficult to manage, especially when comparing genotypes with varying levels of vigor or development rates. Even a drying soil, in addition to being difficult to keep at a uniform and consistent water potential throughout the entire soil profile, can have unique impacts; for example, at low soil water potentials, nutrient transfer is limited. Another issue with soil is that it has poor drainage and is readily saturated at the bottom. The use of inorganic soils, such as calcined clay, or the addition of vermiculite can help solve the problems of clay-rich soil that doesn't drain well or a primarily sandy soil that stores little water and releases it rapidly when the soil dries. Drainage issues are avoided with hydroponics. Mannitol and other non-ionic osmotica have been used to simulate a drop in soil water potential. However, some of these small molecules penetrate roots and travel via the xylem to the shoots, either through root fractures or membranes that aren't totally impermeable to neutral solutes of this size. Many early investigations that attempted to establish a regulated water deficit looked at high-molecular-weight PEG. Its main issue is that it has a high viscosity, which reduces O_2 transport to the roots, causing them to become oxygen-deficient. The latter can be solved by bubbling with O_2 instead of air; however, the tests must be limited in time because the PEG can penetrate the roots and reduce hydraulic conductivity.

Plants are less likely to take up any one ion to lethal amounts, hence concentrated mixed salts, such as the macronutrients employed in Hoagland's solution, are preferable to NaCl. When hydroponics is used to imitate a drought or a soil water deficit, the effect on plant growth might be completely different, and much less, than when the soil is dried up. Many drought and salt tolerance features have a distinct phenotype, and many of these traits have molecular markers, if not perfect markers, derived from candidate genes. However, others do not, or the process is too expensive and slow, such as the efficiency of transpiration. The measurement of plant growth rate is more cost-effective in this scenario. Also, because we may lack sufficient information of characteristics and their interactions, and plant breeders continue to select for yield, a measurement of growth rate, biomass, or yield, however empirical, may be the most effective.

5.5 PHENOMICS

As a result, modern phenomics methods can be used in selection approaches. In controlled situations and in the field, non-destructive, high-throughput crop evaluation is being developed. This allows for a high-resolution and fast correlation of gene function with plant performance and environmental response. Determining root function in soil and screening for root structure and growth optimization has long been a difficult task. For cereal species living on stored soil moisture, access to water at depth is critical for drought tolerance, and researchers are currently screening model species to identify genes important for root characteristics. Small, short-lifecycle crop models that are more appropriate for cereal species have recently been developed and are effective systems for phenomic screening. Deeper roots can obtain more water, sustain high stomatal conductance and thus photosynthesis, and be identified by cool canopies, a trait that is less time-consuming. Non-destructive phenotypes can be obtained via infrared thermography, canopy greenness, or carbon isotype differentiation. In the future, new non-destructive methods for measuring soil moisture, such as ground penetrating radar and electrical resistance thermography, should become available and become sensitive enough to reveal genotypic differences(Kim et al., 2020).

5.6 GENES

Gene discovery is another application of current phenomics. Even in Arabidopsis, a considerable fraction of the genome is classified as 'gene of unknown function' or annotated based on just loose sequence homology-based indications. Physiological understanding and realistic trials are essential for the future of gene-based approaches to boost growth and output under drought and salinity. Knowing the essential qualities is important since it may be used to find genes using a QTL. An unambiguous and quantifiable phenotype, as well as a realistic and repeatable experimental design, are required for this. Validation of QTLs in multiple populations is required. Traditional or insertion mutagenesis, as well as cDNA transformation, necessitate appropriate experimental design, and information is enhanced if a trait is hypothesized. Traits that have been established in a controlled environment chamber must be validated in the field. Many other elements are present in the field, including the atmosphere, particularly the vapor pressure deficit and temperature, and the soil, which will surely have other restrictions such as poor structure, either physical or chemical, such as acidity or sodicity. There are numerous forms of saline soil, each with its own mineral composition and salinization process, all of which impact plant responses to increased osmotic pressure or specific ion concentrations. Waterlogging and salinity can both be temporary, and the salinity–water logging combination can severely stifle plant growth.

5.7 ABIOTIC STRESS

Abiotic stress is a condition that occurs when plants are exposed to sub-optimal levels of environmental inputs such as water, light, temperature, or solutes. Abiotic stress, such as a lack of water or high salinity, causes photosynthesis, transpiration, and growth to decrease. Such strains are a key, global element restricting mankind's ability to generate enough amounts of plant food, fiber, and fuel for an ever-increasing global population. In addition to expanding knowledge, ongoing botanical research into plant responses to abiotic stresses may promote breakthroughs in our practical abilities to maintain and increase plant yield in difficult natural situations. The emergence of increasingly powerful analytical methods has marked the genomic era allowing for deep genomic, transcriptomic, proteomic, and metabolomic investigations. Heritable epigenetic modifications involving DNA methylation, histone modification, and microRNA have recently piqued researchers' interest. These advances in cell molecular biology are ushering in a post-genomic era, in which systems biology aims to combine knowledge of metabolic processes into holistic models of development, albeit mostly at the cellular level. For better knowledge and more successful adjustments of whole-plant responses to environmental change, parallel breakthroughs in research into transport and communication systems linked with development at the level of tissues, organs, and whole organisms are also required.

Excess salinity in the soil solution can come from geochemical sources such as seawater infiltration of coastal ground waters, sea water salts in the wind and rain, excessive fertilizer application to soils, or supplementary irrigation with salt-containing irrigation waters. An excessive salt build-up in the rhizosphere can cause growth inhibition, leaf necrosis, accelerated senescence, wilting, and death in plants. Various physiological systems may be at work. The accumulation of salts in the rhizosphere or the tiny volume of fluids in the apoplastic cell wall compartments is an osmotic process. This results in greater negative water potentials, which can reduce or even reverse the water absorption, turgor maintenance, and cell expansion gradients that are inwardly directed. Reduced water intake in cells can be gradually rectified by the cytoplasmic accumulation of extra solutes through the osmotic adjustment process, or by the establishment of larger negative xylem water potentials in the case of root water uptake. As the world's population grows, water resources become scarcer, and climate change worsens, food yields must be maximized while water and soil usages are optimized. On a global basis, salinity causes massive yield losses, particularly in dry areas. We can update our knowledge of how drought and salinity affect photosynthesis, a highly sensitive process to environmental conditions and a key predictor of plant growth and output. The effects of ABA signaling on stomatal control during water shortages are given special attention. The coordination of stomatal conductance with the CO_2 requirements of leaf mesophyll that may allow constant water usage efficiency in varied situations is being investigated. WUE has not been improved by enhancing Rubisco carboxylase capabilities to increase photosynthetic efficiency.

Abiotic stress conditions such as drought and salinity rarely occur in isolation in nature, and single-factor models rarely anticipate the combination of several stresses since synergistic, antagonistic, or overlapping effects might occur. The fact that critical temperatures for photosynthesis might increase in leaves of water-stressed plants relative to well-watered plants, as found in lupins and several Solanaceae family members, can be examples of antagonistic responses to stressors. Water deficiencies may protect against heat stress by boosting membrane stability. It is also true that one stress can lead to the emergence of a second. This is the situation of stomatal closure caused by dehydration and salinity, which stops leaf water loss. Stomatal closure, on the other hand, reduces leaf cooling by transpiration, resulting in superimposed heat stress, with leaf temperatures often climbing to 5 or 6 degrees Celsius above air temperature. In the case of excessive salinity, the negative effects on leaf physiology were shown to be dependent on exposure to sunlight, with more negative effects identified in sun-exposed locations than in shaded canopy sites.

5.8 PLANT RESPONSES

Plants respond differently to different pressures in terms of photosynthesis, respiration, and signaling. As a result, the importance of studying such interactions in natural settings cannot be overstated. Co-occurrence of distinct stresses might result in the activation of different response signaling pathways at the molecular level. For example, roughly 10% of the regulated genes in *Arabidopsis thaliana* plants exposed to both heat and drought coincided with those genes regulated by both types of stimuli when they were applied independently. In stress physiology investigations, several model plants have been utilized to allow molecular segmentation of stress-tolerance pathways in essential agricultural plants. *A. thaliana*, which is considered a model for dicotyledonous species, is one of the most well-known. Numerous Arabidopsis mutants have been isolated using forward and reverse genetic screens, which are useful methods for identifying genes involved in signaling networks and influencing plant response to environmental stress. Large-scale screens have identified mutants with abnormal stomatal characteristics or photosynthetic behavior, which have been used to study various aspects of stomatal regulation and photosynthetic mechanisms, as well as to identify genes implicated in relevant signaling networks, such as those of abscisic acid. T-DNA tagging, followed by phenotypic screening, has been found to allow the discovery of genes implicated in cold, osmotic, and salinity stress, as well as ABA-mediated gene expression. More recently, researchers discovered mutations in Arabidopsis seedlings that cause NaCl hypersensitivity and identified the major components of the SOS pathway, which regulates ion homeostasis and salt tolerance.

Figure 5.4. Arabidopsis.

Source: https://i0.wp.com/www.botany.one/wp-content/uploads/2019/08/mcy237.jpg?ssl=1

Arabidopsis (Figure 5.4) is sensitive to NaCl and has offered a wealth of knowledge on Na transport and tolerance. In Arabidopsis, many of the known transcriptional components of environmental stress-responsive genes that regulate and control stress reactions associated with drought or salinity have been identified. *Thellungiella halophyla* has recently emerged as a new model plant for investigations on plant responses to salinity. *T. halophyla* is related to Arabidopsis, but it can withstand high salinity and drought. The accumulation of suitable solutes in the cytoplasm to osmotically balance ions build-up in the vacuole during salt adaptation is an important feature of salt tolerance. *Thellungiella* is a proline-accumulator that can be utilized to investigate specific features of proline-related signaling pathways in determining fitness under environmental stress. Specific growing conditions and advancements in available equipment for leaf gas exchange or the application of imaging techniques, on the other hand, helped to overcome the constraints of the physiological characterization of mutants.

Another drawback of model plants is that the knowledge gained with them may not be readily transferable to distantly related crop species, particularly when it comes to long-term strategies for enhanced stress tolerance in the field. For example, in Arabidopsis and cereals, the relationship between Na tolerance and accumulation is distinct, with an inverse association commonly reported within cereal species but not in Arabidopsis ecotypes. As a result, extrapolating salinity tolerance data from Arabidopsis to cereal crops should be done with caution. Root architecture is another characteristic that influences tolerance to water and nutrient stress in the root system. *A. thaliana* is a Brassica with a tap root system, which poses a challenge for research of crop root systems that are normally deep. As a result, in order to better understand the mechanisms underpinning crop species' reaction to environmental stress, non-model species should be cultivated in field circumstances and examined over lengthy periods to simulate the crop's life span.

ABA was long thought to be the signal that communicated water stress from the root to the shoot. However, it was recently discovered in Arabidopsis that, whereas ABA production and signaling in the shoot are required for the water deficit response, the root's ability to synthesize ABA has no effect. Instead, a hydraulic signal appears to initiate the root-to-shoot transmission, which is followed by ABA production in the shoot and subsequent signaling, followed by stomatal closure. The coordination of several cellular processes, such as guard cell turgor, cytoskeleton organization, membrane trafficking, and gene expression, is required for ABA-induced stomatal closure, which is mediated by a complex guard cell signaling network of kinases/phosphatases, secondary messengers, and ion channel regulation. Forward and reverse genetic techniques have discovered a large number of signaling components that affect ABA-dependent stomatal closure. Guard cell signaling components are found in the majority of Arabidopsis mutants with defective ABA-mediated stomatal responses.

ABA production is stimulated largely in vascular tissues in response to a stress signal, and ABA is exported into other cells via specialized ATP-dependent transporters. This method permits ABA to be rapidly delivered to guard cells, where it causes stomatal closure by altering ion fluxes. They have an impact on ABA responses and are most likely implicated in the ABA response network. The transcriptional regulation of several genes is involved in the control of stomatal aperture by ABA signaling, although little is known about the transcription factors (TFs) involved. Despite the fact that the bZIP domain proteins are known to play a function in ABA signaling by interacting with gene promoters containing ABRE motifs, none of these transcription factors have been proven to alter stomatal motions under abiotic stress. Under low water availability or salinity, photosynthesis is primarily limited by the amount of CO_2 available to Rubisco's catalytic site as a result of stomatal closure, and hence by the rate of carboxylation. Under low CO_2 and high temperatures, Rubisco will utilize more O_2 as a substrate, resulting in increased photorespiration. This results in the loss of more than half of the carbon fixed by photosynthesis. This suggests that Rubisco activity will be crucial in determining carbon uptake by leaves under stress situations. Calvin cycle enzymes involved in RuBP regeneration, on the other hand, may alter the rate of carbon intake, especially when CO_2 is saturating, but even during severe drought.

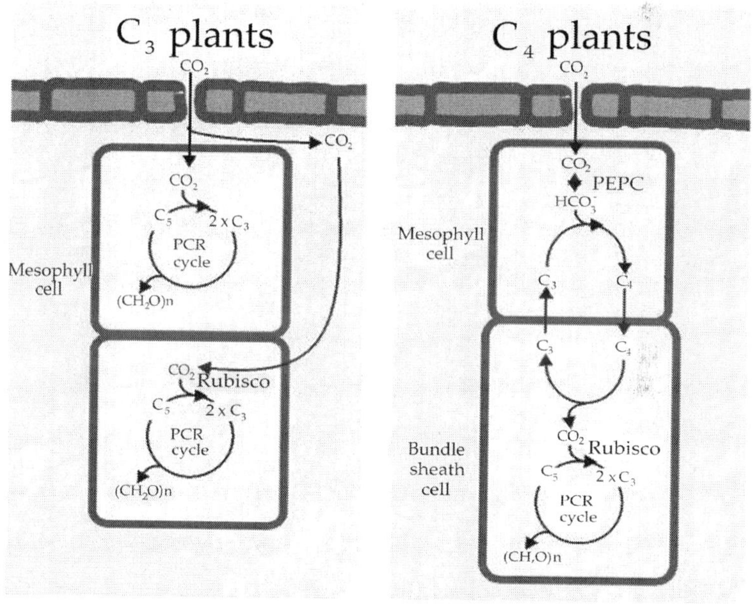

Figure 5.5. C_3 and C_4 plants.

Source:https://www.intechopen.com/media/chapter/18412/media/image2.jpg

5.9 C_3 AND C_4 PLANTS

These restrictions have been successfully solved through evolution with the advent of C_4 plants that can concentrate CO_2 at Rubisco and hence reduce photorespiration by being more efficient under lower stomatal conductance, strong light, and high temperatures (Figure 5.5). Indeed, unlike C_3, C_4 photosynthesis is CO_2 saturated in the environment, allowing it to deal with reduced intercellular CO_2 caused by stomatal closure. Limitations imposed by stomata and carboxylation due to water deficits can be partially overcome by growing plants in an enhanced CO_2 environment, as evidenced by a vast number of experiments conducted over the last two decades. Plants exposed to long-term CO_2 undergo a variety of growth and physiological changes, some of which mitigate the harmful effects of drought. Plant growth and development are influenced by the environment. Drought, heat, cold, and soil salinization are examples of abiotic stresses that limit plant growth and necessitate some degree of adaptation to such harsh circumstances. More research is needed on reprogramming physiological and metabolic activities during stress, regulation of growth and development under suboptimal settings, and plant adaptation to harsh environments as a result of climate change. Around 40% of arable land is arid, semi-arid, or influenced by soil salinity, which affects crop yield and has a greater ecological, agronomic, and economic impact. Because of the link between the increased frequency of extreme weather events and global warming, more efficient, ecologically friendly agricultural techniques are needed, including the adoption of new crop cultivars that are more resistant to environmental pressures.

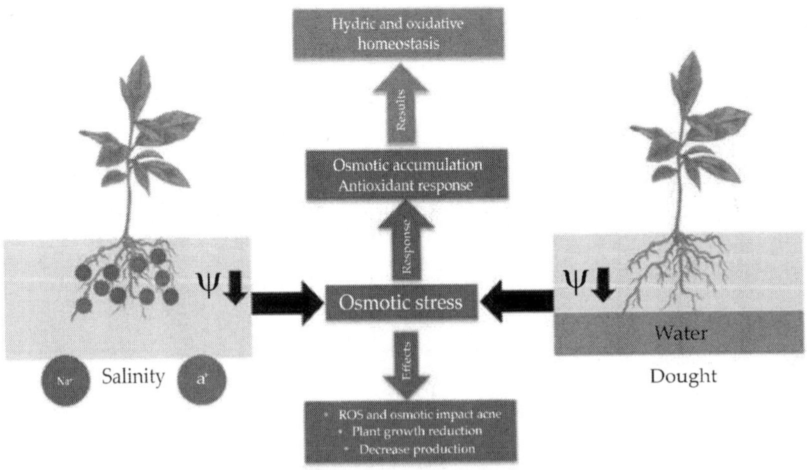

Figure 5.6. Osmotic stress.

Source:https://www.researchgate.net/publication/349299136/figure/fig1/AS:991097924907008@1613307567544/Schematic-model-of-osmotic-stress_W640.jpg

5.10 OSMOTIC STRESS

Water scarcity, harsh temperatures, or high salinity cause cellular water to be depleted, increasing cellular osmotic potential and causing osmotic stress in plants (Figure 5.6). Reduced shoot development and photosynthetic activity, build-up of reactive oxygen species (ROS), variations in ion transport and compartmentalization, and changes in metabolite profiles are all physiological responses to drought, cold, and salt stress. Metabolism is critical for plant growth and development because it provides building blocks and energy for plant structures and reserves, as well as generating signaling molecules for coordination and defense, and producing protective substances to deal with adverse environmental conditions. At a systemic level, metabolism is the most integrated action, reflecting both genotypic and environment-dependent regulators and resulting in phenotype. Energy intakes and outputs are primarily controlled by metabolic processes in order to sustain optimal growth and reproduction, which is a fundamental aim for all organisms. Plants face a difficult task in metabolic adjustment since they must adapt to changing environmental conditions that are commonly considered as stressors. Plant metabolic networks (Figure 5.7) are indeed constrained, and metabolic reconfigurations are found in response to abiotic stress.

The build-up of primary metabolites such as amino acids, non-structural carbohydrates, and organic acids is a common trend in many plant species exposed to high salt, poor water availability, and high or low temperatures. Osmotic potential regulation, thermo- and osmoprotection, chaperon-like roles of macromolecules, adjustments in the carbon/ nitrogen balance, scavenging of ROS or reactive nitrogen species, redox buffering, pH adjustment, carbon and nitrogen reserves deposition, and signal transduction have all been proposed for these stress-induced metabolic adjustments. Depending on the intensity and duration of the physico-chemical pressure, the impact of stress on the metabolic network is frequently coupled with the concurrent decline of growth and development. Protein stabilization under conditions that could lead to protein denaturation is one of the most essential functions of osmoprotective substances. The composition of the solvent influences the equilibrium between native and unfolded versions of proteins. Protein denaturation, protein aggregation, and rapid protein breakdown are all possible biological outcomes of such stress. The preservation of native folding of proteins, on the other hand, is critical for their function and can be aided by osmoprotective chemicals. By boosting the free energy of unfolded proteins, protective osmolytes have been shown to stabilize proteins and push the folding equilibrium towards native, actively folded proteins. The protein backbone, rather than the side chains, has been revealed to be the main target for osmolytes' stabilizing activity, explaining their ubiquitous protective impact.

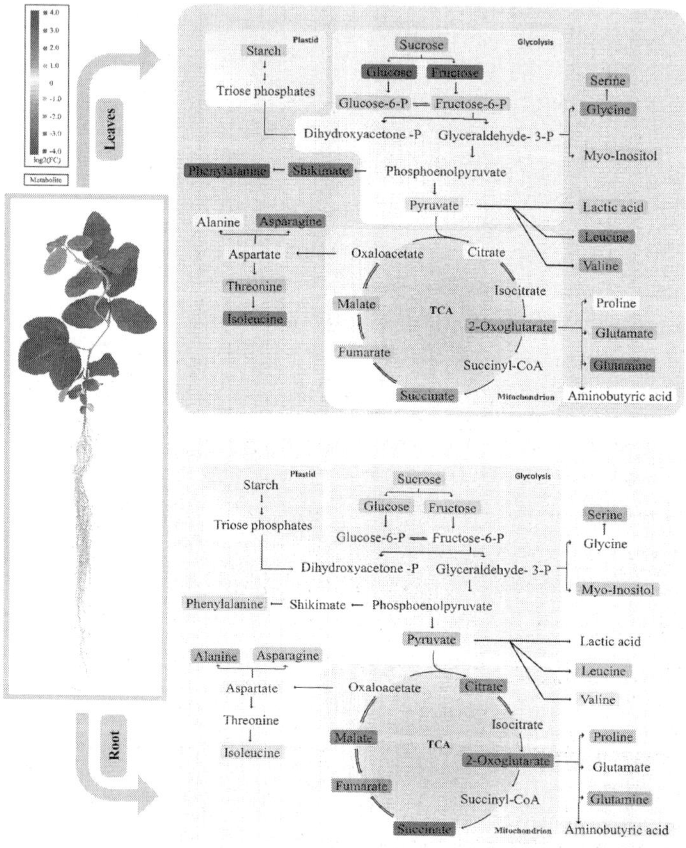

Figure 5.7. Plant metabolic networks.

Source: https://www.researchgate.net/profile/Xiaomin-Yu-5/publication/321755345/figure/fig7/AS:570870109097985@1513117448686/Primary-metabolic-pathways-responses-to-Mg-deficiency-in-leaves-and-roots-Upper-panel_W640.jpg

Protecting osmolytes advances protein thermodynamic stability via hydrogen bonding, which, unlike hydrophobic interactions, has no effect on other cellular activities during stress. Pinitol and myo-inositol, two cyclic polyalcohols, have been found to protect plant and bacterial enzymes from heat inactivation. Proteins are protected from aggregation and degradation by keeping their original shape, which is one of the main activities of osmoprotective substances. Different osmolytes are most likely acting independently of one another in their interactions with proteins, with no synergistic or competing effects. Plants do not suffer from a water shortage in this situation because they are able to adjust their vegetative and reproductive phenology according to the most favorable seasons. Dehydration avoidance methods are based on the capacity to keep tissue water potential as high as feasible through a number of adaptive features that include water loss minimization and water uptake optimization. Plants that can withstand low tissue water potential have tolerance methods that include osmotic adjustment and the production of more compact and rigid tissues.

The life cycle of most annual plants is accompanied by the completion of their reproductive cycle before the dry season, which is a classic example of avoidance. There is evidence that increased aridity can affect the transition between phenophases, depending on species-specific phenotypic plasticity, which is also regulated by the environment. In response to induced drought, desert plants shorten their development cycles with early seed maturation and senescence, but Mediterranean annuals do not change phenophases as quickly. To avoid dryness, flowering timing changes rapidly. Adaptation to dryness in other species, such as perennials and woody plants, is based on the evolution of specific organs, such as deep taproots or storage organs, or the emergence of specific habits. Shrubs play a major role in semiarid environments because they can thrive in places of extreme environmental stress where trees cannot. The leaf shedding models are based on plants' ability to photosynthesise at optimal times and allocate resources for growth, reproduction, or storage.

Drought is one of the most stressful situations for plants all around the world, decreasing their output. The ability of roots to absorb water is determined by their morphological, anatomical, and molecular characteristics. There is a stronger link between drought tolerance and root length density. The ability of a root to absorb water, on the other hand, varies along its length. Also, during droughts, apoplastic barriers form, reducing root water transport to some extent. Finally, the precise involvement of aquaporin in the modulation of root hydraulic characteristics during drought is still unknown. To understand the role of aquaporins in the control of root water uptake under drought conditions, subcellular localization and knockout experiments for each kind of aquaporin are required.

The formation of these apoplastic barriers can also be viewed as a method for reducing water loss from roots to soil and so preventing cellular death. However, it was discovered that soil coverings around the root developed during drought conditions were more important than exodermis formation in limiting water loss from roots to the soil in *Chloris gayana* plants. It has also been observed that when chickpea roots were exposed to dehydration, they lost their epidermal and cortical cells, but the surviving endodermal cells and stele were able to restart growth after soil rehydration, indicating that suberized endodermal cells prevented stellar cell failure. In addition to apoplastic barriers, dryness causes some plants, such as maize and rice, to grow aerenchyma tissue in their roots. This aerenchyma development reduces root water uptake capacity, which may improve or decrease plant drought tolerance depending on its intensity. It's also important to see how different Arabidopsis suberin mutants react to drought. Simultaneously, because water can flow to the xylem vessels via two separate paths in its radial movement, the role of aquaporins in the cell-to-cell path must also be considered(Steudle, 2000).

5.11 WATER TRANSFER

The transfer of water from wet to dry zones of the soil is one of the methods that plants use to cope with drought stress. This hydraulic lift not only improves the water status of drought-tolerant plants, but it also enhances transpiration, which can improve CO_2 assimilation and plant growth. Similarly, a rise in hydraulic lift has been seen in various crop species during drought, and this increase has been linked to the photosynthetic rate being constant during drought. Hydraulic lift aids in water absorption not just by the plant that is moving the water, but also by plants nearby. As a result, using plants with deeper root systems that reach groundwater and planting them away from the affected crops may help to delay the negative impacts of drought on those crops. This agronomical method can also be improved by shadowing the hydraulic lift plant. After rainfall episodes, water redistribution from moist to dryer soil layers can also occur from top to bottom soil layers. Despite the fact that hydraulic lift is critical for drought stress tolerance, the molecular processes underlying this action are unknown. The hydraulic lift is determined by soil and root conductivities, water potential gradients, and transpiration intensity. To date, the interconnections between these elements are not totally known.

Plants respond to drought in one of two ways: by reducing water loss or sustaining water intake. Osmotic adjustment (OA), a biological mechanism that helps plants acclimate to dry and saline conditions, aids this process within plant cells. The quantity of osmotically active chemicals in the cell increases as a result of OA. This increase in solutes causes a greater negative osmotic potential, which can help maintain turgor in leaf tissue and other metabolically active cells by improving cell hydration. In other words, if OA occurs, plants can survive longer in drying soil and retain metabolic functions. Under drought stress, OA has a favorable effect on crop cultivars' growth and output. Inorganic cations and anions, organic acids, polysaccharides, and amino acids are only a few of the chemicals that might cause OA. OA is frequently linked to the build-up of certain solutes that have protective properties. Sugars, cyclitols, proline, and glycine betaine are examples of suitable solutes with hydroxyl groups that can concentrate in the cytoplasm and protect cellular proteins, enzymes, and membranes against dehydration. Nonetheless, it is critical to remember that individual solutes play a little role in OA in many species, and that OA is mostly produced through the accumulation of a large number of solutes. Because OA necessitates the metabolism or uptake of solutes, it is a gradual and delicate process that is affected by the timing and severity of stress. Furthermore, there is evidence that studies have underestimated leaf relative water content (RWC) when quantifying OA, which adds to the inherent variability in OA expression.

Trees must overcome the constraints imposed by harsh dry land medium throughout the course of their long lives, especially given their considerable developmental plasticity. This is most likely one of the most important characteristics that makes them good models for understanding the intricacy of adaptation processes in plants throughout their life cycle. Forest tree species offer a wide spectrum of adaptive solutions to the problems of terrestrial media within the plant world. True trees with a homoisohydric lifestyle first appeared in the mid Devonian period. Since then, one of the key obstacles for these plants in order to live in areas where water availability is severely limited has been maintaining a proper water economy. For any tree species, maintaining tissue hydration within levels that are compatible with their functionality is critical. This system must strike a balance between conductance efficiency and security against the loss of functionality in severe situations.

Though non-stomatal processes such as leaf cuticle resistance have a role in regulating leaf water losses, stomatal regulation is likely the most critical point of control in water usage. Leaf stomatal closure is a frequent plant technique for conserving water during droughts, although it reduces net CO_2 absorption. Furthermore, stomata respond to a wide range of environmental stimuli, and even within the same forest ecosystem, a whole range of water economy strategies can be found under drought conditions, ranging from isohydric to anisohydric. However, in recent years, the notion that forest tree species' leaf stomatal regulation is controlled by the necessity to preserve hydraulic performance against an optimal carbon balance has gained traction. Maintenance of the hydraulic system within safety margins from water-stress-induced embolism is critical in secondary-growth species, particularly forest trees.

5.12 WOODY PLANTS

Woody plants exhibited that fine stomatal control of xylem pressures can prevent partial cavitation of their xylem. This regulation would be especially important in dry situations, when optimum performance in both tasks is threatened and constrained by biophysical constraints. As a result, either hydraulic failure or carbon deprivation has been proposed as the mechanistic foundation for the phenomenon of mortality following severe droughts. The specific approach of forest tree species involving stomatal management of water losses would be important in both processes. Overheating of leaves is a result of stomatal closure under water stress to prevent a catastrophic loss of hydraulic conductivity and to reduce dehydration of leaf tissues. A surplus of excitation energy cannot also be delivered to the photosynthetic electron transport chain. Non-consumed electrons in CO_2 fixation may react with O_2, forming reactive oxygen species and raising the risk of oxidative damage. The pH and de-epoxidation state of the xanthophyll cycle pigments both rise under these conditions, safeguarding the photosynthetic machinery by a mechanism that discharges excess light as heat. However, there is little indication of significant long-term photodamage in water-stressed plants, as seen by the lack of impacts of drought on maximum potential PSII efficiency even in severely stressed leaves.

CHAPTER 6

GENOME ASSISTED BREEDING

CONTENTS

6.1 Introduction........104

6.2 Physiological Features........105

6.3 Quantitative Trait Locus........105

6.4 Stomatal Function........107

6.5 Food Demand........109

6.6 Metabolome........112

6.7 Plant Adaptability........113

6.8 Genetics and Genomics........115

6.9 Molecular Markers........117

6.10 Rice........120

6.11 Cowpea........121

6.12 Hormone Signaling........124

6.13 Sugarcane........125

6.1 INTRODUCTION

Direct selection for grain yield in classical breeding programs for enhancing cereal abiotic stress tolerance is slow and expensive, requiring several years and field trial sites. Grain yield is largely determined by the characteristics and functions of the flag leaf, and is linked to the plant's ability to regulate water content and synthesize, store, and relocate carbohydrates from leaves to grains. Despite its importance in cereals, little is known about the genetic control of the flag leaf's cellular structure and development under stress. In plants, the stomata cells in the leaves regulate water loss through transpiration and photosynthetic CO_2 uptake. It is vital to investigate the processes of stomata management in order to maintain a high photosynthetic rate for higher yield under drought and saline conditions. Knowledge of the physiological and genetic factors that affect stomatal morphology and development in relation to grain yield is a fundamental challenge in breeding for abiotic stress tolerance. The genes that are subject to natural variation of stomatal traits in wheat, barley, and rice have been identified using quantitative trait loci (QTL) mapping. Furthermore, there is a lot of genetic variation in stomatal and epidermal cell features that could be employed for marker-assisted breeding and the development of novel beneficial qualities in cereals (Kole et al., 2015).

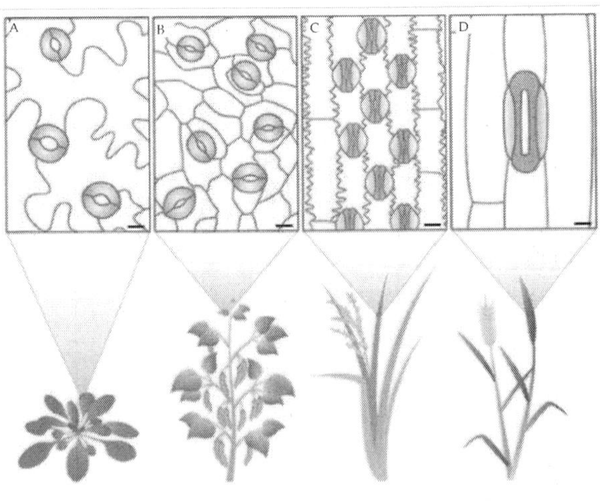

Figure 6.1. Physiological features.

Source: https://www.researchgate.net/profile/Robert-Caine-2/publication/331550421/figure/fig1/AS:733471119011841@1551884550704/Stomatal-

traits-vary-between-species-The-eudicots-A-Arabidopsis-thaliana-and-B_ W640.jpg

6.2 PHYSIOLOGICAL FEATURES

Physiological features of stomatal traits have been extensively studied in recent decades, primarily in the model plant Arabidopsis (Figure 6.1). At least 40 genes have been identified in Arabidopsis that affect stomatal regulation and development. The genetic control of stomatal size, density, and index in crops, which involves the assembly and modification of new leaves in response to changing environmental conditions, is less known. Plants' physiological responses to climate change and water availability are influenced by stomatal characteristics. Over the previous few decades, decreasing stomatal density has been linked to rising CO_2. Stomatal characteristics are most likely influenced by differences in photosynthetic demand, surface qualities, light penetration, and the interior architecture of leaves. Agronomically desirable alleles present at quantitative trait loci that alter plant stress responses can be identified using genetic and genomics-based techniques. Through its influence on photosynthetic rate and stomatal conductance, stomatal frequency is intimately linked to water-use efficiency. These are some of the traits that have been researched to see if they can be used for indirect yield selection or to see how they interact with other physiological characteristics. Wheat cultivars with significant genetic variation for stomatal conductance and photosynthetic rate demonstrated a positive connection with grain production. Despite the fact that such traits are critical for selecting tolerant plants, little is known about the genetic and genomic resources related to stomatal traits, genes, and genetic networks that affect biochemical and physiological pathways, signaling, synthesis, accumulation, transport, and efficient use of initial resources in cereals.

6.3 QUANTITATIVE TRAIT LOCUS

A quantitative trait locus is a place on the genome that is genetically linked to variation in the phenotypes of a quantitative trait. QTL mapping in a bi-parental mapping population can reveal the chromosomal location, closely related markers, estimated additive allelic effects, and percentage of phenotypic variance for stomatal traits. The introduction and development of molecular markers in quantitative genetics have made it much easier to investigate complicated quantitatively inherited features. The detection of poly genes for such qualities into individual Mendelian factors has been made possible by the construction of high-density genetic linkage maps for cereals. Fine mapping of the QTL influencing desired attributes can be exploited in marker-assisted selection using divided regions. Stomatal features have been identified genetically and phenotypically, however, the genetic mechanisms underlying these traits are unknown. The regulation in the genome does not end with genetic control of stomatal features, which is mediated through transporter and hormonal control of function. Recent data suggests that the epigenome, a second layer of control, is involved in stomatal growth and function. This is particularly significant when addressing the interplay between genotype and environment because there is evidence that the environment, particularly abiotic stress, can modify stomata via epigenetic regulation.

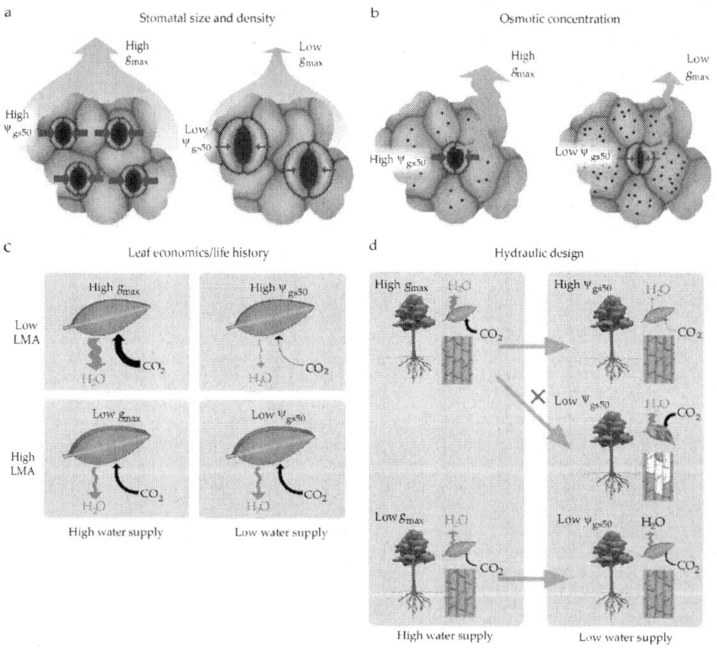

Figure 6.2. Stomatal function.

Source: https://www.researchgate.net/profile/Robert-Caine-2/publication/331550421/figure/fig1/AS:733471119011841@1551884550704/Stomatal-traits-vary-between-species-The-eudicots-A-Arabidopsis-thaliana-and-B_W640.jpg

6.4 STOMATAL FUNCTION

Stomatal function (Figure 6.2) in response to stress, as well as stomatal architecture can be influenced by modulating stomatal signaling and metabolism. Overexpression of the maize NAD-malic enzyme in tobacco, for example, resulted in plants with lower stomatal conductance but higher biomass per unit of water consumed, implying that modifying both stomatal and mesophyll processes could improve plant water use efficiency. Although stomatal qualities can be engineered, it's critical to understand how they interact with other variables in the chain. Plants should not become more vulnerable to environmental constraints as a result of reprogramming stomatal function. The type of stress, changes in stomatal behavior in various species, plant water status, and leaf age may all influence such approaches. Combining physiological and molecular genetic approaches with quantitative systems analysis can help us move closer to these goals. Additional evidence about quantitative kinetics and signal transduction mechanisms in plants will also be helpful. Improving water usage efficiency and crop resilience to climate change's adverse consequences is a hot topic in research and on the scientific agenda. Stomata serve an important function in decreasing water loss and enhancing photosynthetic rate, and they may be manipulated to improve stress tolerance. Quantitative genetic approaches combined with next-generation sequencing (NGS), RNA-sequencing data, and high-throughput non-invasive phenotyping platforms will be possible in the post-genomic era to identify genes responsible for stomatal behavior in response to fluctuating environmental conditions. The knowledge-base on stomatal features and behavior is expanding in tandem with other 'omics' methodologies such as transcriptomics, metabolomics, and proteomics, with the goal of boosting agricultural productivity and yield under ever-changing climate regimes.

Plants in nature are subjected to a variety of environmental pressures, both biotic and abiotic. Heavy metals, which comprise both necessary and non-essential metals, are toxic compounds that are discharged into the environment and cause a variety of deleterious effects on living creatures in the food chain through accumulation and biomagnification. By influencing physiological, biochemical, and molecular reactions, these metals have a negative impact on crop productivity and growth. Heavy metals are defined as metals or metalloids with a density greater than that of water and are dangerous even in small amounts. They gradually build up to the point where they endanger the physicochemical qualities of soils, resulting in a loss of soil fertility and agricultural productivity. Various anthropogenic and geological activities, such as the use of pesticides and fertilizers in agricultural fields, the burning of fossil fuels, metal mining and smelting, industrial production of batteries and other metal products, sludge, and the disposal of municipal and sewage waste, raise heavy metal concentrations to levels that are harmful to both animals and plants. Heavy metal accumulation and bioavailability in plants are determined by the biology of the plant species. Certain contaminants, such as arsenic (As), can last for a long time in the environment. Arsenic poisoning is most common in South America, Asia, and highly populated flood basins and deltas of South and Southeast Asia, where As polluted drinking water is used. Groundwater As pollution was first recorded in India in 1983 in West Bengal. As polluted water above the acceptable limit of 50 g/L has been found in a number of other states due to the use of hand tube wells(Shaji et al., 2021).

Lead (Pb) is found in conjunction with other heavy metals such as Cd and Zn, and it is a major source of concern due to its widespread distribution, which harms human health and pollutes the environment. Cadmium (Cd), a very hazardous heavy metal with a high mobility and water solubility, is quickly taken up and accumulated by plants. The copper (Cu) ion is a necessary component of numerous proteins and enzymes involved in development activities such as mitochondrial respiration, photosynthesis, cell wall metabolism, and the hormone signaling pathway. Zinc (Zn) is essential for plant metabolic processes including glucose metabolism, protein synthesis, and enzyme activation.

Many economically important quantitative qualities are under polygenic control and are directly selected for. As a result, there should be a method for indirect selection that is unaffected by the environment. Identification of linkage between genes for quantitative traits and marker loci can help tremendously in this area. Quantitative trait loci are DNA regions containing genes that affect a certain quantitative trait. The connection of trait variation with genetic variation is used to identify QTL, and a significant correlation between phenotype and genotype observed at a QTL shows that the QTL can be useful for determining trait expression. The genetic process driving phenotypic complexity is studied via QTL mapping based on a high-density molecular linkage map. QTL mapping is a strong genetic approach for determining the number, location, and impact of genetic variables that contribute to phenotypic variation.

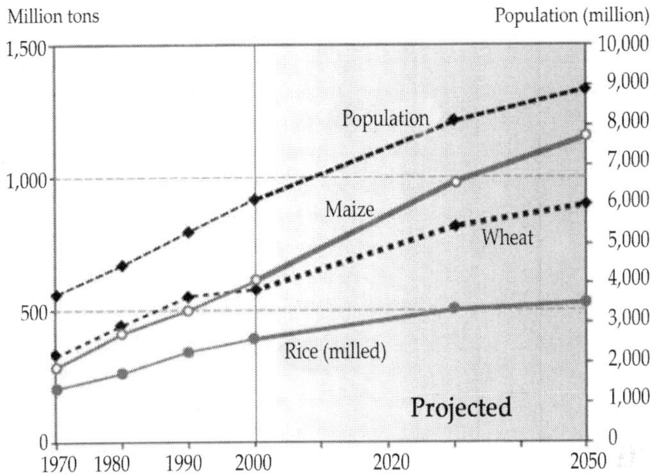

Figure 6.3. Demand for wheat.

Source:https://www.researchgate.net/profile/Petr-Kosina-2/publication/46472222/figure/fig2/AS:669368941809674@1536601400460/World-demand-for-wheat-maize-and-rice-1970-2050-Source-FAO-2006_W640.jpg

6.5 FOOD DEMAND

After rice and maize, wheat is the third most extensively farmed crop. It is a staple food and the primary source of calories for millions of people around the world since it is consumed by more than 40% of the world's population. Wheat demand is expected to rise by 60% by 2050 (Figure 6.3) in order to feed the world's growing population. Furthermore, as climate change scenarios deteriorate, wheat production is expected to decline by 29%. Drought, heat, salinity, metal toxicity, cold, and other abiotic stressors are expected to be key contributors. As a result, abiotic stress resistance is a prominent goal in wheat breeding projects around the world. Traditional breeding procedures have pushed annual genetic gains in grain yield up to 1%, but to deal with a 2% annual rise in world population under changing climatic conditions, more efforts in a short period of time are required to develop climate tolerant high yielding varieties(Raza et al., 2019). As a result of the introduction of new sequencing and genotyping platforms, such as next-generation sequencing, breeding approaches have experienced a paradigm shift, to the point that the use of genomics-assisted breeding is now commonplace in most breeding programs. Despite wheat's complicated genome, these modern NGS platforms have helped wheat by yielding a large number of markers for gene discovery, which had previously been a limitation.

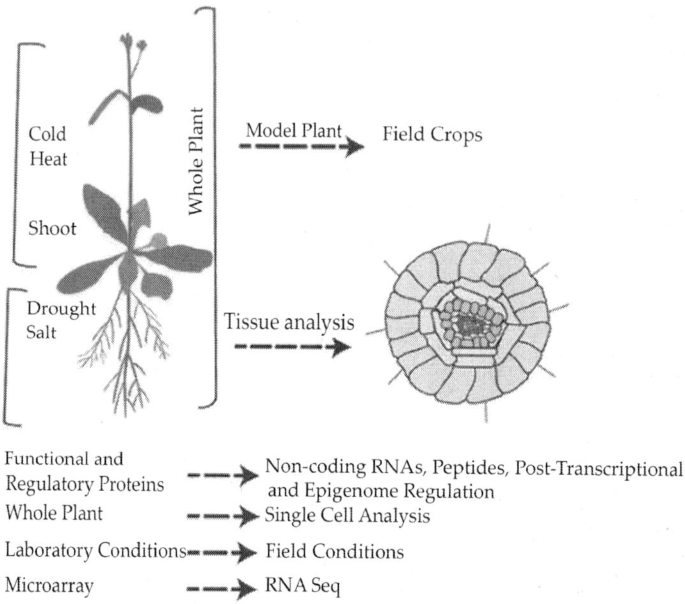

Figure 6.4. Transcriptomics.

Source:https://www.researchgate.net/publication/333094941/figure/fig1/AS:758627228995587@1557882234751/Transcriptomic-analysis-under-drought-salinity-heat-and-cold-stress-The-focus-of_W640.jpg

Transcriptomics (Figure 6.4) is the most explored of all the omics technologies, including metabolomics, proteomics, and ionomics, for identifying genes and understanding pathways for abiotic stress tolerance in various plant species. The transcriptome is the entire set of RNA transcripts found in a specific cell type or tissue at a given developmental stage and/or physiological environment. A plethora of information may be gained by analyzing the transcriptome of any plant under different treatments or at different stages of development, which aids in the identification of genes and related pathways. Microarrays, which quantify a set of pre-set sequences, and RNAseq, which uses high throughput sequencing to collect all sequences, are two significant contemporary tools in this discipline. Any external stimuli cause changes in the plant's constituent proteins and the accumulation of primary and secondary metabolites. Plant response to varied stimuli may be examined and potential genes can be discovered by evaluating the proteome and metabolome of plants using proteomics and metabolomics techniques, respectively. For example, a proteomic approach was recently used to identify proteins involved in wheat cold responses. The two-dimensional electrophoresis approach was used to analyze two cultivars that differed in cold tolerance, as well as their descendants who differed in cold tolerance. To analyze expression alterations at the RNA level, a real-time polymerase chain reaction was used to confirm a handful of these differentially expressed protein sites. The transcriptional expression patterns of 11 genes matched their protein expression models, indicating that they could be candidates for cold tolerance. Similarly, the proteomics approach was used to investigate abiotic stress-responsive proteins in wheat grain(Raza et al., 2019).

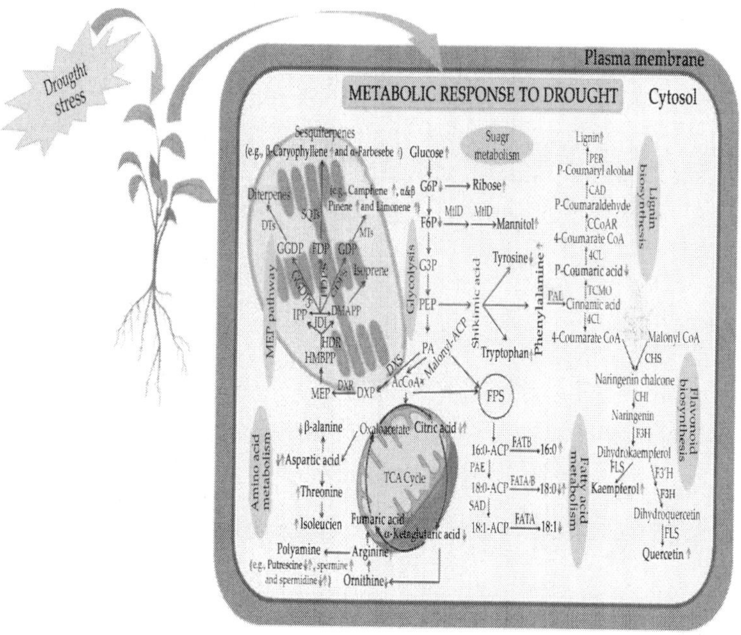

Figure 6.5. Metabolome.

Source: https://www.mdpi.com/ijms/ijms-22-09108/article_deploy/html/images/ijms-22-09108-g001.png

6.6 METABOLOME

Metabolome (Figure 6.5) changes during abiotic stress are also important to characterize because different metabolites represent different developmental changes. As a result, the characteristic metabolites identified under specific conditions such as developmental stage or stress could be used to identify metabolic markers diagnostic for plant stress. Under stress conditions, targeted or untargeted metabolic profiling can help us better understand plant metabolism and produce cultivars that are more tolerant and adaptable to stress. Metabolic profiling, on the other hand, is limited to the examination of pre-determined metabolites for specific pathways, whereas metabolic fingerprinting is a high-throughput technique. ATP binding, serine/threonine kinase activity, zinc ion binding, and metal ion binding were shown to be the most common differentially expressed genes (DEGs). RNAseq investigation of the st1 wheat mutant has revealed potential genes for salinity tolerance. In a screen for induced wheat mutants cultivated in high salinity hydroponics, the mutant st1 was discovered. Multiple genes involved in Na+ transport, as well as genes encoding arginine decarboxylase and polyamine oxidase, are involved in st1's salt tolerance, according to its transcriptome sequence variation study. Butonate metabolism was also discovered to be a new pathway for salinity tolerance. In terms of harvested area, trade value, cattle feed, and human nutrition, barley is one of the most significant cereal crops on the planet. It is one of the best-adapted plant species to abiotic conditions such as drought, heat, cold, low fertility, and salinity, which limit agricultural output. Barley is typically the only grain that can be produced in many countries in West Asia and North Africa due to its flexibility, making it a valuable resource for farmers and the primary feed for cattle in many regions. Selection for barley cultivars with steady and economic yield under a variety of environments is a major necessity of any breeding program to cope with these adverse situations.

New genomic and molecular methods have recently expanded the number of genes involved in abiotic stress tolerance and adaptation to unfavorable conditions that have been found in the barley gene pool. For future yield advances to meet global food, feed and industrial demand, as well as to cope with the consequences of climate change, traditional breeding procedures must be supplemented with innovative analytical selection methodologies. As a result, it's critical to use modern genomics and molecular-based breeding procedures to boost barley output and generate new varieties that are more adapted to abiotic stressors. Drought stress has always been a key selection factor in plant growth, development, and physiological evolution. Drought, temperature, and other associated factors have much greater combined effects on physiology, growth, water relations, and yield than their individual effects. As a result, identifying and utilizing novel genetic material with high elasticity to climate change is critical in order to meet these difficulties.

6.7 PLANT ADAPTABILITY

Plant adaptability is a critical factor that will decide how crop production systems respond to climate change in the future. Shifting planting dates or switching to crop varieties with short growing seasons may be the most effective method to mitigate the negative effects of climate change and associated pressures. Under arid circumstances, selecting drought-tolerant genotypes with shorter growing seasons is seen to be a viable evasion strategy that could boost crop yield. For many breeders in marginal locations, developing new crop types with early blooming and maturity, as well as better stress tolerance, is now a top priority. New varieties that can withstand pressures at the most vulnerable periods of crop growth, such as reproduction and grain filling, should be studied as an option to mitigate the negative effects of high temperatures and drought. By combining recent discoveries in genomics with existing breeding practices and the use of modern molecular techniques, better lines that are more acclimated to dry settings while remaining highly productive can be developed. Breeders in dry regions are developing programs based on physiological or developmental features or direct selection for grain yield. Due to a diversity of reasons, including the intricacy of the characteristics themselves, the complexity of the stress tolerance systems, and the annual fluctuation in the amount and timing of rainfall, direct selection for grain production and biomass under under water-limited environments is difficult. Breeders have targeted a variety of secondary traits in barley to select for stress tolerance, including growth habit, early growth vigor, maturity rate, plant height, peduncle length, and grain filling duration. Low heritability, epistasis, and high genotype by environmental interactions have limited direct selection for grain yield under stress situations.

Barley is the most salt-tolerant member of the Triticeae family, and its better tolerance is due to its rapid growth and phenological development, which leads to early maturity under less-than-ideal conditions. Salt tolerance is a physiologically complicated multigenic trait that necessitates adjustments in numerous biochemical pathways as well as all important functions like photosynthesis, protein synthesis, energy, and lipid metabolism. Plant responses to salt are separated into two phases: an osmotic phase that restricts young leaf growth and an ionic phase that accelerates adult leaf senescence. Salt tolerance at these two separate development phases is controlled by numerous genes located at various loci, according to a QTL analysis utilizing two independent mapping populations. Barley is grown in a variety of production systems that are suited to specific climatic and soil conditions, as well as end-user preferences. Many factors, particularly the incidence of abiotic stressors and their combinations, contribute to the considerable diversity in yield seen in different growing locations. Greater knowledge of the genetic mechanisms that influence genotypes' ability to cope with abiotic stresses and their interactions with the environment is critical, especially in light of projected climate change scenarios and food security concerns. Despite the barley scientific community's tireless efforts over the last two decades, the number of abiotic stress markers now used in MAS remains limited. In breeding for multigenic characteristics with substantial genotype by environment interaction, MAS has had a limited influence. The absence of proper type and number of molecular markers, as well as the ineffectiveness of how to employ markers in breeding programs, has been blamed for the limited influence of MAS on barley breeding in previous years.

Recent improvements in next-generation sequencing technologies and bioinformatics, along with high-throughput phenotyping methodologies, will improve our ability to detect and utilize natural variation at abiotic stress tolerance loci for breeding novel varieties through marker-assisted selection. Genomic selection (GS) has been used in animal breeding and also be used in plant breeding, albeit more research is needed to determine the optimum technique. Wheat breeding became systematic roughly 100 years ago, but farmers' efforts to improve wheat strains through selective breeding may be dated back almost 10,000 years to the dawn of agriculture(Venske, dos Santos, Busanello, Gustafson, & Costa de Oliveira, 2019). Wheat yields increased dramatically following the Green Revolution and subsequent breakthroughs in agricultural research, technology, and infrastructure. Wheat production, on the other hand, has struggled to satisfy global demand, and an increasingly changeable and uncertain climate is exacerbating the situation. Furthermore, in modern agricultural systems, popular wheat cultivars are spread over a large region in order to generate elite cultivars with desirable agronomic qualities such as high yield or higher quality, as well as adaptability to a variety of climates. Plants are inadvertently exposed to a number of challenges, including abiotic and biotic pressures, as a result of this practice.

Drought is the most damaging abiotic stress to agricultural yield, and it is characterized by a complex interaction of numerous elements such as limited water supply, decreased rainfall, changing precipitation patterns, decreased ground water level, and rising temperature. Drought stress is the most common constraint to wheat productivity and stability in rain-fed production systems, and it has become a major issue as a result of rising water scarcity and uneven rainfall distribution. As a result, several wheat improvement initiatives have focused on generating wheat cultivars with improved drought tolerance and high yield. The difficulty in identifying and precisely assessing the key morpho-physiological determinants of production under drought circumstances is the key reason for the limited success in developing drought resistance.

6.8 GENETICS AND GENOMICS

Advances in genetics and genomics have substantially improved our understanding of the structural and functional components of plant genomes, as well as our ability to improve crop plants, in recent years. Drought tolerance, for example, is a complex feature that can be studied using genomic techniques, and the genes underlying it can be cloned. Genomic techniques are assisting in the detection of quantitative trait loci and the identification of existing beneficial alleles of minor effect, which have often gone unrecognized and not been included in the breeding gene pool. We were able to discover QTLs for features that influence drought resistance and yield in wheat using linkage and association mapping based on high density markers. Cloning important genes and QTLs that affect yield under drought conditions gives a more direct path for mining and altering favorable alleles after they've been identified. While approaches based on resequencing, candidate gene identification, 'omics' platforms, and reverse genetics will accelerate the pace of gene/QTL discovery, QTL mapping and cloning addressing natural variation will continue to shed light on mechanisms of adaptation to drought and other adverse conditions. Without prior knowledge of the relevant QTLs, genomic selection that assesses marker effects throughout the entire genome is a valuable option for improving wheat performance under drought conditions.

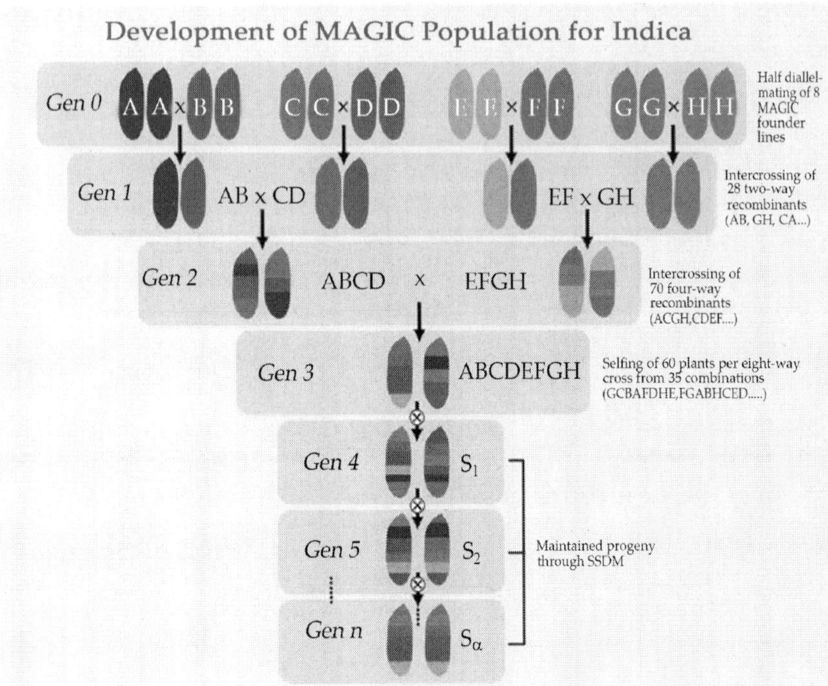

Figure 6.6. Multiparent advanced generation inter-cross (MAGIC) population.

Source: https://www.researchgate.net/profile/Rakesh-Singh-2/publication/257883368/figure/fig3/AS:267595920179249@1440811247429/Crossing-schemes-to-produce-four-multi-parent-advanced-generation-inter-cross-MAGIC_W640.jpg

One of the new generations of developing mapping resources in plant genetics is the multiparent advanced generation inter-cross (MAGIC) population (Figure 6.6). For the creation of a mapping population in a MAGIC population, many parents are used. In wheat, MAGIC populations have been developed and are being developed in a number of breeding initiatives. The selection of parents based on geographic, genetic, and phenotypic diversity is the first step in the formation of a MAGIC population. Hybridization in a predetermined pattern is used to combine parents. Individuals descended from the pioneer population are intercrossed for future generations. Selfing and double haploidization of individuals are used to generate inbred lines, either directly on the hybrid population or after advanced intercrossing. They are formed by the intercrossing of many lines over several generations. In a normal bi-parental mapping population, many parents allow for greater allelic variability to be captured than two parents. Furthermore, many intercrossing cycles provide more opportunities for recombination and, as a result, more precision in QTL placement. Double haploidization of individuals may be preferable at the advanced intercross stages of MAGIC population development to form inbred lines more quickly and easily.

Genomic Selection, which predicts phenotypes using information from genome-wide markers. GS is a ground-breaking method in which a breeder's choice is based on genomic estimated breeding values derived from genome-wide DNA marker data. It entails the creation of training populations for the purpose of simulating selection criteria for lines in a breeding program. Because doubled haploid populations are immortal, they may be used to provide data for modeling, which is required for the application of genomic selection. The improved resolution of genomic selection models should be attributed to the increased precision in phenotyping attributable to doubled haploid populations. Under controlled and osmotic stress treatments, doubled haploid mapping populations in hexaploid wheat were recently employed to compare genomic selection models to predict flowering time and spike grain number. After establishing a sufficient mapping population, the next stage in producing drought-tolerant crops is to analyze drought-related features at both the molecular and phenotypic levels. In recent years, omics have played a key role in crop development by allowing the identification and characterization of genes, proteins, and metabolites linked with drought tolerance. Furthermore, advances in genome mapping and functional genomics technology have brought new and sophisticated tools for studying drought tolerance at the molecular level.

6.9 MOLECULAR MARKERS

The discovery of molecular markers and candidate genes linked to drought tolerance or avoidance-related phenotypes contributes to improved knowledge of the genetic basis of drought tolerance, and once validated, they can be employed in molecular breeding. However, careful genotypic and phenotypic monitoring of the crop is essential for the optimal use of molecular markers or candidate genes. First and foremost, there should be enough heterogeneity in accessible germplasm for these features. To find the best and most stable lines, the appropriate germplasm can be examined and screened for all attributes in multiple locations trials. These best lines can be used to build mapping populations such as bi-parental mapping panels, association mapping panels, and so on. Following the creation of mapping populations, a variety of genetic and genomic technologies can be employed to discover the main genes behind these features and incorporate them into desired cultivars.

Several software applications for extracting data from digital photographs of roots, shoots, leaves, seeds, and grains have been developed. The current and future phenomics methods will allow for the scanning of thousands of plants in a single working day, analogous to high-throughput DNA sequencing in genomics. The precise and accurate data supplied by these facilities is crucial for the meaningful genetic separation of complex traits and has a wide range of uses in crop improvement projects. The platform can readily monitor nearly infinite sets of characteristics, perform complete screening, and provide dynamic data on many plant features. Plants are manually placed on the 3D Scan analyzer or carried directly from the greenhouses to the imaging chambers via conveyor belts, depending on the degree of automation. Plant height/width, biomass, and plant architecture can all be measured using such chambers, which enable top and side imaging of both shoot and root systems. Three approaches have emerged from research over the last three decades that have made major contributions to crop improvement efforts. The identification of the most relevant loci governing drought tolerance and their related phenotypic expression has been made possible by combining molecular biology and genetics with physiology. The problem for breeders, however, is to create drought-tolerant cultivars without sacrificing production potential. Drought tolerance traits should therefore be tested in both stressed and non-stressed conditions before being used in MAS breeding programs. Priority targets for MAS should be QTLs for drought-related variables that are closely linked to QTLs for yield potential. Once the target traits have been found and introduced into the MAS breeding program, advanced breeding techniques such as two-fold haploidy breeding should be used to integrate them stably into the wheat genetic background. This integrated approach will result in the development of stable drought-tolerant elite cultivars, ensuring the rapid evolution of agriculture in the direction of meeting the growing population's food supply.

Water deficits can be caused by drought, heat, cold, or salinity, and this is a fundamental factor in how plants are distributed geographically. Water stress has elicited a variety of responses from plants. Some plants, for example, are able to avoid drought by flowering before the commencement of the drought. Avoidance and tolerance are two types of resistance strategies. Plants that escape water shortages do so by conserving or spending water at various stages and times throughout their life cycle. Water-saving plants minimize stomatal opening throughout the day to prevent water loss, whereas water-demanding plants sustain high transpiration rates by accessing water resources via vast root systems. Water shortage tolerance can be achieved by maintaining osmotic potential by the production of osmolytes or other suitable solutes. Repair mechanisms, on the other hand, can be used to keep cells healthy. While the escape method is difficult to implement for perennial crops, forage grasses use the other strategies outlined above to deal with water shortages.

Figure 6.7. Chilling injury.

Source:https://www.canr.msu.edu/floriculture/uploads/images/ChillInjuryFigure%203.jpg

Chilling injury (Figure 6.7) is defined as a set of molecular, biochemical, and physiological changes in plants caused by low non-zero temperatures, as well as the morphological symptoms that result. The severity of chilling injury in crop fruits is largely determined by both the low temperature and the length of exposure to cold stress. Fruit ripening and development delays are the two most severe chilling injury signs seen in fruit crops.

Low temperatures during fruit storage can cause serious damage such as browning, reddening, etc., and reduce the fruit's storage life. Reduced leaf growth, loss of stiffness, loss of leaves, chlorosis, and necrosis are some of the other phenotypic signs described in fruit crops under cold stress.

6.10 RICE

Rice has been widely studied for drought resilience, and various potential contributing features have been discovered. Increased abscisic acid synthesis, leaf rolling, reduced respiration, root development promotion, and stomatal closure are among the features that aid to avoid water scarcity. Drought resistance is conferred by the root system's ability to sustain water and mineral intake in order to maintain nutritional homeostasis and ionic balance in the plant system. A dense and deep root system of rice variety, in particular, impacts its ability to tolerate adverse water shortages by obtaining water from deeper soil layers. When it comes to penetrating deeper into soil, having a thick or thin root system is vital. A hard soil pan may require bigger roots with penetration ability, whereas a coarse textured soil may require thin roots for penetration. Similarly, various shoot-oriented characteristics are important in conferring drought tolerance in rice. Osmotic adjustment and dehydration tolerance, high water use efficiency, membrane stability, and so on are among them. Drought stress affects rice plants severely during the flowering stage, resulting in significant yield loss, the magnitude of which is dependent on the degree and duration of water scarcity, as well as the time between drought spells. The cumulative loss due to all other abiotic stresses combined is predicted to be greater than the yield decline caused by water stress alone. In order to maintain hypothetical rice production in water-scarce settings, drought-tolerant rice cultivars would be required. The majority of popular rice cultivars used in upland rice fields, on the other hand, are lowland varieties created for irrigated habitats.

These types perform well in normal years with enough and evenly distributed rainfall, but they completely fail in drought years due to their extreme drought vulnerability and suffer severe yield losses. Rice is a self-pollinating, inbred crop that produces homogeneous and homozygous populations. Rice has a limited natural gene flow; therefore, any cross-breeding will eventually restore homozygosity. As a result, the establishment of discrete genetic groupings and subpopulations in rice is favoured by ecological and geographic isolation. Furthermore, nearly all of today's high-yielding rice varieties have low allelic diversity. These adaptation-specific genes, on the other hand, are maintained in local germplasm, such as landraces, traditional varieties, etc., which are suited to the local setting. Drought is a genetically complex trait to design a simple breeding solution for because many drought-related traits are damages rather than symptoms of stress, making them unsuitable for genotypic response evaluation. They are also highly unstable and weigh significant environmental influence, exhibiting very low heritability. As a result, when compared to breeding for other stresses, drought resistance breeding in rice lags behind utilizing traditional methods. Recent advances in genomic techniques and technology have aided breeding efforts by assisting in the identification of genes involved in drought adaptation and their deployment in elite cultivars lacking resistance.

6.11 COWPEA

Cowpea has made significant progress in terms of resistance to biotic stressors, including diseases such as bacterial blight, stem blight, macrophomina, parasitic weeds such as Striga and Alectra, and insects such as aphid, leaf and flower thrips, among others. Cowpea is a crop that is widely farmed in dry areas and is prone to seedling, midseason, and terminal droughts. The amount of rainfall in the dry savannah parts of SSA during the agricultural season has been decreasing in recent years. As a result of the late start or early end of the rainy season, the cropping season is becoming shorter. Farmers in the cowpea-producing areas of SSA typically do not have access to irrigation, therefore their crops are grown rainfed. Climate change is probable to increase the frequency of drought in the dry savannah region, hence efforts should be undertaken to develop drought-resistant cowpea types that farmers can grow. Efforts have been undertaken to improve drought tolerance in some improved cowpea types by conventional breeding, but progress has been modest. Drought tolerance is a complex trait that is inherited through a number of genes. As a result, it would be preferable to pyramid these genes in enhanced cultivars. Such pyramided gene variants are expected to maintain their performance over time and across multiple savannah locales.

Drought is a major block to food production in Sub-Saharan Africa, where irrigation infrastructure is inadequate. Water shortages are expected to affect 67 percent of the world's population by 2050, according to estimates(Islam & Karim, 2019). Terminal drought is one of the most prominent environmental challenges that farmers and plant breeders in Sub-

Saharan Africa continue to face. Drought tolerance is a multifaceted feature governed by polygenes whose expression is altered by external influences. As a result, figuring out its genetic foundation is critical for both breeding and basic research. Using traditional breeding methods to breed for drought resistance could be challenging and time-consuming. Geneticists, breeders, molecular biologists, physiologists, and agronomists, among others, must work together to achieve this. Traditional breeding processes may not be suitable in creating improved varieties that can mitigate the issues due to high population expansion, particularly in underdeveloped nations where climate change is anticipated to have a greater destructive impact. New molecular breeding tools have the potential to help accelerate the development of appropriate technology to fight the effects of climate change. With recent advances in genomics, it might be possible to create drought-tolerant cultivars in a shorter amount of time.

Drought resistance is assumed to be conferred to cultivated agricultural plants via root-related features such as deep, profuse, thick, and widespread root systems. Although root growth and morphology can increase water uptake in drought conditions, and root traits are used as surrogates for soil water extraction, the relationships between root traits and water acquisition, as well as their contribution to drought yield formation, are unclear across crop species. Few studies on cowpea roots have been conducted, with the majority focusing on root growth and morphological changes using a small number of test lines. Furthermore, none of this research looked into whether root characteristics affect soil water accessibility and usage during drought, particularly during critical pod and grain filling stages. To analyze the two-dimensional root system of a large number of cowpea plants and progenies, a "root-box pin-board" technique was devised, which allowed characterization of key changes in root system architecture in cowpea.

Using these resources, researchers discovered a number of possible contributors that can confer resistance to biotic and abiotic stressors. A number of cowpea lines that are resistant to abiotic and biotic stresses have been identified. These genetic resources will continue to be relied on by breeders as sources of genes for desired qualities in cowpea improvement. They have the potential to give genes for the development of novel types that will aid in tackling growing challenges that may come as a result of climate change, as well as human food and animal feed needs. It's also worth mentioning that wild cowpea relatives haven't been used much in the production of improved varieties. The likelihood of linkage drags from using cowpea wild relatives as parents has sparked a lack of enthusiasm in using them as parents. Wild cowpea relatives, for example, have relatively less seeds with smooth, unappealing seed coat colors. To regain the cowpea seed size preferred by consumers, several backcrosses to the recurrent parents would be required. However, current advances in genomics may lead to a shift in interest in crop wild relatives as sources of new genes, allowing for faster progress in breeding new varieties through marker-assisted selection.

Cowpea yields are extremely poor due to a number of factors, including a lack of access to high-quality seeds of improved kinds. Scientists at worldwide and national cowpea research organizations have made significant progress in developing new varieties through conventional breeding during the last three decades. However, additional efforts are needed to speed up variety development in order to meet the rising global demand for cowpea to feed the growing human population. With the availability of the most up-to-date genetic and genomic resources, as well as the establishment of high-throughput SNP genotyping platforms, modern molecular methods in cowpea can now be used to successfully and quickly develop and release improved varieties to farmers, bridging the current yield gap. However, a bottleneck for modern breeding is high throughput plant phenotyping for precise and accurate agronomical, morphological, and physiological data in huge numbers of genotypes and segregating populations.

Abiotic stressors represent a serious danger to global food security since they reduce agricultural yield and productivity by more than half. These abiotic pressures represent a severe challenge to ensuring sustainable food production for the world's rapidly growing population in the face of global climate change. To survive stress and/or maintain growth, plants respond to a wide range of adverse environmental situations by dynamically regulating numerous physiological, developmental, and biochemical pathways. To design adaptable crop types, a detailed understanding of such reactions to abiotic stressors is required. Adopting modern crop breeding tactics to create enhanced varieties for abiotic challenges employing novel biotechnological approaches, along with genetics, molecular biology, breeding, and physiology, has resulted in genetic breakthroughs in sugarcane. Transgenic approaches have recently emerged as adaptable tools for combating the negative effects of abiotic stresses on crop productivity, and they have proven to be one of the most promising methods for genetic modification. The expression of a certain set of stress-related or responsive genes is dependent on the use of modern molecular biology methods to determine the regulatory mechanisms for abiotic stress tolerance and developing stress-tolerant crops. As a result, multiple abiotic stress-responsive genes have been found, extracted, cloned, and used to help sensitive genotypes gain stress resistance. Agrobacterium-based approaches, as well as other means of gene transfer, were used to create transgenic sugarcane lines harboring genes for abiotic stress tolerance.

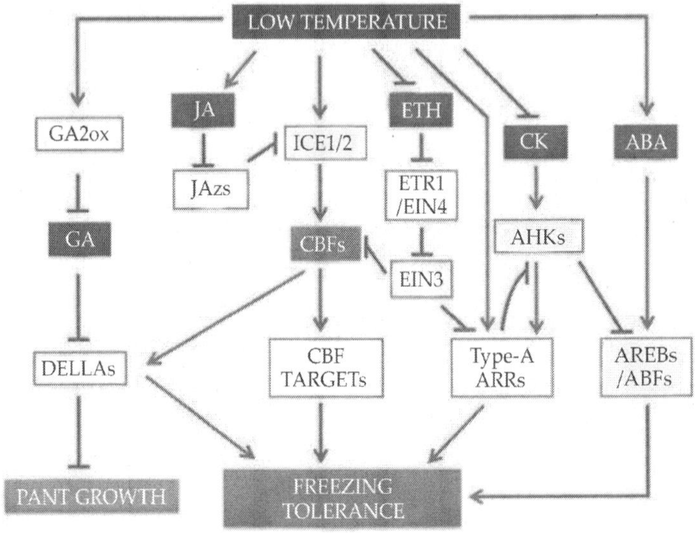

Figure 6.8. Hormone signaling.

Source: https://www.researchgate.net/profile/Yiting-Shi-2/publication/265394667/figure/fig2/AS:506966367797249@1497881609590/The-involvement-of-plant-hormone-signaling-in-plant-responses-to-cold-stress-CBFs-can-be_W640.jpg

6.12 HORMONE SIGNALING

Components of hormone signaling, (Figure 6.8) a reactive oxygen species scavenging system, changes in amino acid profile, and lipid peroxidation are all part of the defensive process. Plant genetic engineering provides a potent tool for up- or down-regulating certain metabolic steps in order to modify specific abiotic stress responses by over-expressing a broad-spectrum single gene. However, this method ignores the fact that stress tolerance is the result of the action of multiple stress responsive genes, and single gene tolerance is unlikely to be long-lasting. As a result, there has been a resurgence of interest in developing new ways to convert plants with regulatory genes. Sugarcane biochemical research has uncovered various elements of cellular metabolism under salinity stress, laying a firm foundation for our understanding of high salinity adaptation mechanisms. In vitro experiments, for example, demonstrated greater salt leaching from NaCl-treated calli compared to control calli, implying that sugarcane can be regarded as a Na+-excluder plant species. The control calli, on the other hand, contained a larger quantity of K^+ ions than the NaCl-treated calli. Increased Na^+ and a matching decrease in K^+ ion concentrations eventually resulted in growth inhibition and reduced cell viability, revealing sugarcane's usual glycophytic nature. Flavonoids' antioxidant qualities are thought to be vital for tolerance, and they were shown to be higher in tolerant sugarcane clones compared to susceptible clones. The effect of flavonoids in protecting sugarcane from ion-induced oxidative stress during salinity stress was established in this study. Furthermore, sugarcane priming with NaCl solutions could explain sugarcane seedling resistance and stress memory in response to NaCl treatments. Priming therapies have now been shown to modify a number of pathways, including the induction of an antioxidant pathway that ensures plant growth under adverse conditions.

6.13 SUGARCANE

Sugarcane expression profiles have also been evaluated using microarray technologies. Initially, they were employed to compare gene expression differences between mature and immature sugarcane stem tissues. SAGE analysis was used to further characterize the sugarcane leaf transcriptome, which revealed that 70% of transcripts matched to at least one sugarcane assembled sequence with a pre-assigned putative function(Hoang et al., 2017). The gene product was most frequently found in the photosystem (PS) I reaction center, as well as other locations such as PS I, PS II, and thylakoid complexes, according to gene ontology investigations of these transcripts. Genetic engineering has shown considerable promise and potential for incorporating foreign gene resources in sugarcane in order to generate transgenic plants with unique qualities such as strengthened built-in defense mechanisms, increased nutritional content, yield, and agronomic attributes. Initial efforts to improve abiotic stress tolerance were made by genetically modifying the rearrangement of genes that are either directly linked to encountering adverse environmental conditions; directly concerned in cell defense against water scarcity; or genes that encode proteins important for regulating signal transduction pathways in response to environmental stress. Plant and non-plant genes are being successfully transferred, either separately or in combination, to improve sugarcane abiotic stress tolerance.

CHAPTER 7

VAM IN DROUGHT TOLERANCE

CONTENTS

7.1 Introduction 128
7.2 Arbuscular Mycorrhizal Fungi (AMF) 129
7.3 VAM 131
7.4 Symbiotic Relationship 132
7.5 Interaction Between Fungi and Plants 133
7.6 Mycorrhizal Fungi 134
7.7 Bacteria 134
7.8 Leaf Adaptations 136
7.9 Soil Bacteria 138
7.10 Weathering 139
7.11 Symbiosis 143
7.12 Heavy Metal Stress 144
7.13 Biocontrol 147

7.1 INTRODUCTION

Climate change, as well as agricultural malpractices such as excessive fertilizer and pesticide use, have exacerbated the effects of abiotic pressures on crop productivity and deteriorated the environment. There is a pressing need for environmentally friendly crop management strategies, such as the utilization of arbuscular mycorrhizal fungi (AMF). Bio-fertilizers (Figure 7.1) are a term used to describe AMF. Furthermore, it is widely assumed that inoculating host plants with AMF confers tolerance to different stressful circumstances such as heat, salinity, drought, metals, and severe temperatures. AMF may help host plants both up-regulate tolerance mechanisms and prevent important metabolic pathways from being down-regulated. Because AMF are natural root symbionts, they deliver vital plant inorganic nutrients to host plants, enhancing growth and yield in both stressed and unstressed conditions. AMF's position as a bio-fertilizer has the potential to improve plants' ability to adapt to changing environments. AMF are soil-borne fungus that can boost plant nutrient uptake and resilience to a variety of abiotic stressors. They are obligate biotrophs who complete their life cycle by consuming plant photosynthetic products and lipids. Not only does AMF-mediated growth promotion improve water and mineral nutrient intake from the surrounding soil, but it also protects the plants from fungal infections. As a result, AMF are important endosymbionts that contribute to plant productivity and ecological function. They are crucial for crop enhancement in the long run(Begum et al., 2019).

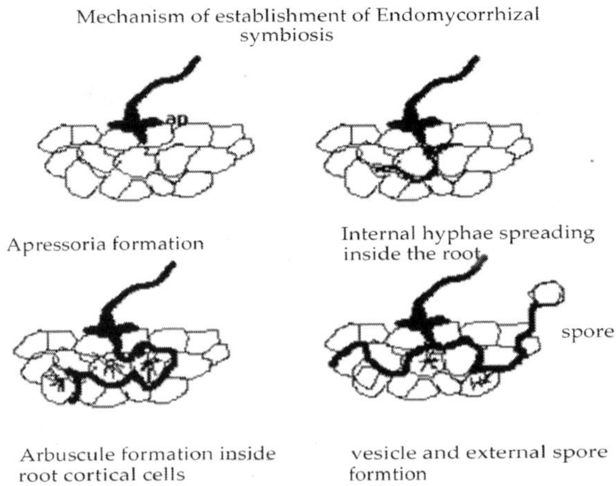

Figure 7.1. VAM.

Source: http://edugreen.teri.res.in/explore/bio/bioferti.gif

7.2 ARBUSCULAR MYCORRHIZAL FUNGI (AMF)

The formation of a hyphal network by the AMF with plant roots improves root access to a vast soil surface area, resulting in improved plant growth. AMF increases the availability and transport of different nutrients in plants, improving plant nutrition. AMF improves soil quality through altering the structure and texture of the soil, and thus plant health. The breakdown of soil organic materials can be accelerated by fungal hyphae. Furthermore, mycorrhizal fungi may influence ambient CO_2 fixation by host plants by boosting the sink effect and photo-assimilate migration from aerial portions to roots. Bio-fertilizers are a blend of naturally occurring compounds used to boost soil fertility.

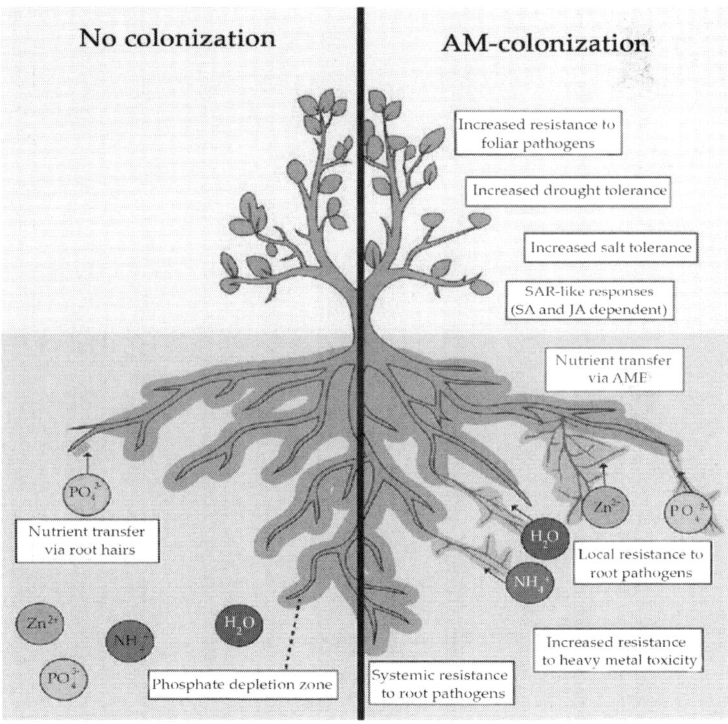

Figure 7.2. Symbiotic relationship.

Source: https://www.researchgate.net/publication/321113317/figure/fig1/AS:561481268056065@1510878974672/Positive-effects-of-arbuscular-mycorrhizal-AM-colonization-The-hyphal-network-of_W640.jpg

These fertilizers are extremely beneficial to both soil health and plant growth and development. AMF has been the subject of numerous research studies over the last two decades, all of which have demonstrated its numerous advantages to soil health and crop productivity. Because mycorrhizal treatment may significantly reduce the quantitative usage of chemical fertilizer input, especially phosphorus, it is assumed that AMF could be considered as an alternative for inorganic fertilizers in the near future. Inorganic fertilizers, herbicides, and fungicides have a negative impact on food quality, soil health, and air and water systems, resulting in a variety of difficulties for soil, plants, and human health.

AMF form symbiotic relationships (Figure 7.2) with roots to receive critical nutrients from the host plant and, in exchange, offer mineral nutrients such as N, P, K, Ca, Zn, and S. As a result, AMF provides nutritional support when the root cells are in an unfavorable state. AMF generates fungal structures called arbuscules, which aid in the interchange of inorganic minerals and carbon and phosphorus molecules, giving host plants a boost in vigor. As a result, they can considerably increase the concentration of phosphorus in both root and shoot systems. The mycorrhizal association promotes phosphorus delivery to infected roots of host plants in phosphorus-limited environments. Pi uptake rate, for example, was significantly improved in AMF-colonized maize plants. Augmented photosynthetic activities and other leaf functions are associated with improved AMF inoculation growth frequency, which is linked to the intake of N, P, and carbon, which travel towards roots and stimulate tuber development. AMF has been found to maintain P and N uptake, supporting plant development at both higher and lower P levels under various irrigation regimes. Under drought stress, for example, mycorrhizal symbiosis enhanced N, P, and Fe concentrations in *Pelargonium graveolens* L.

In response to AMF-mediated combined stress adaptations, tolerance mechanisms may resemble one another. AMF-mediated changes in phytohormone profile, mineral absorption and assimilation, build-up of suitable osmolytes and secondary metabolites, and up-regulation of the antioxidant system may be common processes induced during various stresses. Particular mechanisms such as damaging ion compartmentation and sequestration, phytochelatin synthesis, and protein expression, on the other hand, can be specific and vary significantly depending on the stress type and the AMF species involved. Changes in root properties such as hydraulic conductivities can significantly enhance osmotic stress tolerance. The aforementioned features of AMF may improve crop nutraceutical quality and may be of significant agronomic value for the production and management of many prospective crops. However, more research is needed to fully understand the role of AMF in reducing the consequences of coupled stressors.

7.3 VAM

VAM's effects on host plants may be reflected in changes in defensive adaptation. Biochemical parameters employed in plant protective adaptation research include soluble carbohydrate accumulation, amino and imino acids, and protective enzyme activity. Researchers have traditionally assumed that the differences in these biochemical parameters between VAM and NM plants reflect the VAM plants' increased drought resilience. Lower soluble sugar build-up, for example, indicates that plants are better at avoiding dryness and hence have less need to osmotically modify osmoprotectant enzymes, or that they are less stressed or injured. The VAM symbiosis was found to have a considerable impact on soluble carbohydrate storage and partitioning in tobacco plants during drought. In dryness, their VAM plants collected less glucose and fructose in their leaves and roots than NM plants. Following drought, similar results were found for rose and pepper. After dryness, VAM maize plants had higher foliar concentrations of soluble sugars than NM maize plants, implying that VAM plays a favorable function in increasing host plant drought tolerance. Plants with VAM symbiosis have been discovered to have higher amounts of amino and imino acids in drought conditions.

During drought, the activity of numerous protective enzymes was investigated in VAM and NM plants and found to be much higher in VAM plants. VAM lettuce, for example, exhibited stronger superoxide dismutase activity in the root and shoots than NM lettuce, and VAM alfalfa had higher acid phosphatase activity than NM alfalfa. Other studies have found glutamate-ammonia ligase activity in maize, as well as glutamine and glutamate synthase activity in soybean. Numerous investigations have found that VAM symbiosis increases nitrate reductase activity in leaves and roots. During drought, total protein concentrations in VAM plants were consistently greater than in NM plants. Almost every plant in a natural ecosystem forms a relationship with fungus, either intracellularly in the case of arbuscular mycorrhizal fungi (AMF) or extracellularly in the case of ectomycorrhizal fungi. The most common symbiosis with land plants is arbuscular mycorrhizas (AMs). The related fungus colonizes the plant roots and lives in the host plant's interior tissues. This mutualistic relationship not only helps plants grow by aiding the intake of water and necessary nutrients, but it also shields them from harmful soil conditions. Mycorrhizal fungus application is a viable alternative technique for long-term crop development under normal and biotic and abiotic stress situations. Mycorrhizal plants have a better ability to absorb nutrients and can withstand stress. The use of the AM fungus to improve plant growth and crop output is gaining popularity. The AM fungus also benefits crop growth by enhancing soil quality and minimizing soil erosion by increasing water infiltration and retention.

7.4 SYMBIOTIC RELATIONSHIP

A symbiotic relationship between a plant and a fungus in which the fungus penetrates the root cortical cells of a vascular plant is known as arbuscular mycorrhiza. Due to the production of specific vesicles and arbuscules after colonization of the root cortical cells, this association is known as vesicular-arbuscular mycorrhiza (VAM). The most prevalent type of symbiotic relationship is arbuscular mycorrhiza. According to some estimates, up to 90% of the world's plant species form mycorrhizal associations. Arbuscular mycorrhizal fungi can grow in a variety of soil types(van der Heijden, Martin, Selosse, & Sanders, 2015). Although a stressful environment with excess salts and oxygen deficit is detrimental to mycorrhizal fungal survival and efficiency, some fungi are able to endure and colonize the plant root under such conditions. In a semiarid environment, where aggregate stability is one of the most critical features for improved plant growth and development, the symbiotic relationship between plant and fungus improves aggregate stability even in stressful conditions. The fungus' synthesis of the glycoprotein glomalin plays an important role in enhancing aggregation stability. Although the processes by which the fungus improves the water relations of the host plant are yet unknown, arbuscular mycorrhizal fungus has been shown to boost water uptake by controlling stomatal conductance. In addition to delivering macro and micronutrients, the mycorrhizal association helps the plant survive in a water-stressed environment because of this water intake. Arbuscular mycorrhiza plays an important role in nutrient availability, enzyme production, increasing photosynthetic rate, and improving plant tolerance to stresses in general, making it an important component of sustainable agriculture.

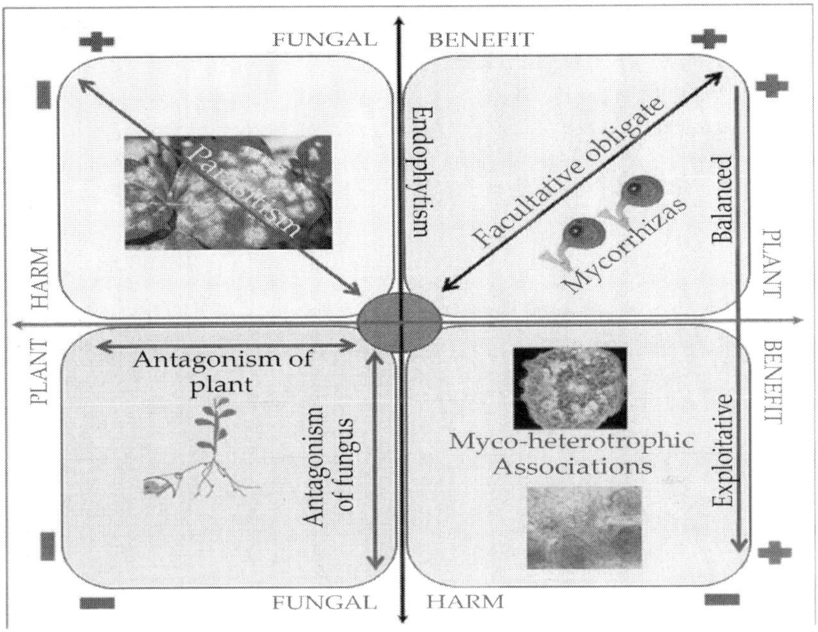

Figure 7.3. Interaction between fungi and plants.

Source:https://www.researchgate.net/profile/Susanne-Zeilinger/publication/284355767/figure/fig1/AS:299278670024706@1448365003217/Comparison-of-different-plant-fungus-interactions-Mutualistic-associations-occupy-the_W640.jpg

7.5 INTERACTION BETWEEN FUNGI AND PLANTS

The interaction between fungi and plants is diverse, ranging from pathogenic to symbiotic (Figure 7.3). Living together for mutual benefit is referred to as symbiosis. Molecular interactions occur during fungi-plant interactions, resulting in metabolic changes in the host, protection from external challenges, and favorable conditions for the symbiont. Mycorrhizal fungi are found predominantly in forest trees and live on or in the roots of plants. Over 90% of plant species have a symbiotic relationship with a mycorrhizal fungus. Mycorrhizal growth has been observed in fruits, vegetables, cereals, and ornamental plants. This sort of fungal connection improves the root's absorption surface area, which benefits the plant by increasing water and nutrient uptake while also protecting it from biotic and abiotic challenges. Because fungal hyphae are thinner than plant roots, they can easily penetrate even the smallest pores in the soil. Plants supply 10–20 percent sugar, which is generated by the plant via photosynthesis(Huey et al., 2020). Mycorrhizae are separated into two groups based on the fungi's hyphal arrangement inside cortical tissues: ectomycorrhiza and endomycorrhiza. Ectomycorrhiza, which are mostly found in temperate woods and belong to the basidiomycetes and some ascomycetes, are the most common. Between 10% of plant species and about 5000 fungal species, an ectomycorrhizal association exists. In comparison to non-mycorrhizal roots, ectomycorrhiza cause plant roots to swell and branch profusely(Bücking, Liepold, & Ambilwade, 2012). Around the feeder roots, the fungal hyphae form a fungus mantle. Fungi reach the roots but produce a hartig net around the cortical cells.

7.6 MYCORRHIZAL FUNGI

Mycorrhizal fungus promotes plant growth and development through a variety of methods. These mechanisms differ by species and are also affected by soil conditions. Synthesis of phytohormones and metabolites, such as vitamins and amino acids, solubilization of minerals, root colonization, improved water uptake, production of osmolytes, and improving soil structure are just a few of these mechanisms. Arbuscular mycorrhiza (AM) is a type of endomycorrhizal fungal association with vascular plant roots in which fungal hyphae enter the cortical cells and generate arbuscules, which can sometimes form vesicles. The morphology of roots does not alter significantly. Arbuscular mycorrhizal (AM) fungus first makes contact with and penetrates the epidermis of immature roots. After that, AM fungal hyphae expand into root tissues, forming a mycorrhizal network that allows plants and fungus to exchange nutrients. Sugars, organic acids, etc., secreted by plants in the rhizosphere cause soil fungi to colonize plant roots and interact with them on a molecular level. Endomycorrhiza have both arbuscules and vesicles in most situations, which is why it's named VAM.

7.7 BACTERIA

Bacteria linked with the AM fungus promote hyphal development and make root penetration easier for the fungus. It is usually caused by the bacteria producing particular substances that increase the rate of root exudates. The fungal hyphae are activated by increased root exudation, which enhances root colonization. Exopolysaccharide synthesis by specific bacterial strains contributes to soil aggregation and improved water retention in the rhizosphere.

Bacteria promote hyphal growth by increasing cell permeability, making it easier for the fungus to penetrate the roots. In a bacterial-fungal interaction, bacteria boost the activity of the AM fungus, while the mycorrhizal fungus aids the bacteria in nitrogen fixation and phosphorus solubilization from insoluble sources. Mycorrhizae co-inoculation with many strains has also been demonstrated to have a major impact on plant development. Plants inoculated with mycorrhizal fungus and rhizobium showed significant improvements in plant growth, survival index, physical and chemical soil characteristics, leaf nitrogen (N) and phosphorus (P), and soil P.

Arbuscular mycorrhizal fungi live in the soil and can colonize the roots of 80 percent of all land plants(Berruti, Lumini, Balestrini, & Bianciotto, 2016). Mutual benefits are characterized as mycorrhizal characteristics: Photosynthates are transferred into endosymbiotic AMF for development, and AMF provides the host plant with vital nutrients and water. Hyphae, which are formed by soil propagules of AMF, asexual spores, or mycorrhizal roots, colonize suitable roots. In general, mycorrhizal hyphae colonize roots via an appressorium and then penetrate into the root cortex, eventually generating intercellular and intracellular hyphae, coils, and arbuscules. Furthermore, colonized roots' extraradical mycorrhizal will continue to search the soil for mineral resources and may even colonize the roots of other nearby plants, forming a common mycorrhizal network (CMN). Plant development, nutrition absorption, root architecture, flowering, and stress tolerance have all been demonstrated to benefit from the AM symbiosis. The AM symbiosis has been shown to improve plant drought tolerance in studies, although the exact pathways remain unknown. AMF spores indeed exist in soils, and soil moisture has an impact on spore germination and development. Studies have shown that soil water quality has a significant impact on spore behavior. Spores are surrounded by soils, and one of the main elements influencing the spore-soil contact zone is soil water availability. Small variations in the matric potential of the soil surrounding a spore cause water to either retreat or migrate into bigger spores, resulting in changes in the degree of contact between spores and soil water, as well as the conductivity of water in the capillaries at the spore-soil interface. As a result, spore germination is promoted in soil with water potentials equal to or greater than field capacity, but it is reduced as soil water potentials fall below field capacity.

7.8 LEAF ADAPTATIONS

In addition to leaf morphological adaptation, mycorrhizae use root morphological adaptation as a drought stress response. Because of the improved root morphological adaptation generated by mycorrhization, the host plant can explore more of the soil volume to absorb water and nutrients, potentially improving drought resistance. The control of endogenous polyamine metabolism and phytohormone balance, particularly root putrescine synthetases by activation of arginine decarboxylase and ornithine decarboxylase, as well as root indole-3-acetic acid, may be responsible for AMF-induced root morphological alterations (IAA). Waterlogging is a natural occurrence that has a negative impact on plant survival, growth, and development. AMF has been found to have a symbiotic relationship with a variety of aquatic and wetland plants. AMF has been discovered in a variety of permanent and seasonal wetlands, including lake and stream flooding, mangrove, salt marsh, river, riparian and floodplain, peat swamp forest, and other forms of wetlands. Funneliformis mosseae, Rhizophagus fasciculatus, and F. geosporus are all members of the Glomeraceae family, and some of them are found in all sorts of wetlands. Many factors influence colonization, spore density, species richness, and diversity of AMF, including oxygen availability, seasonal fluctuations, P availability, water depth, AMF type, and other types of vegetation. Through increased nutritional status and possible adjustment, the presence of AMF can promote plant growth and biomass, as well as speed up the succession of early pioneer vegetation in specific types of aquatic habitat.

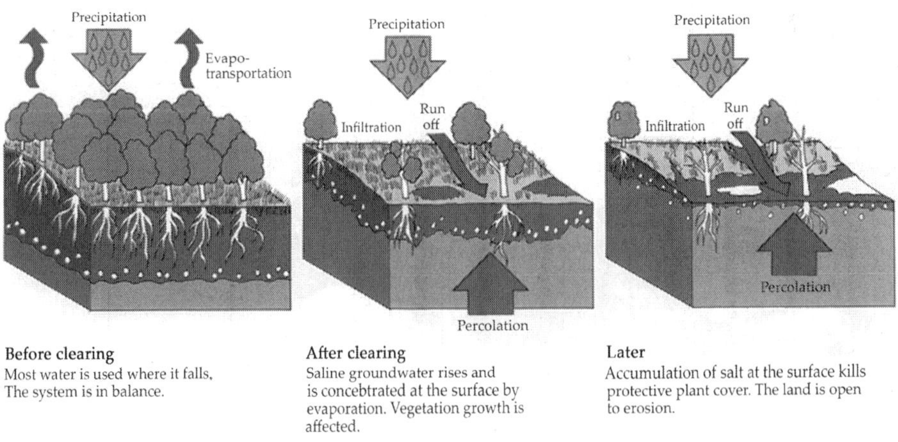

Figure 7.4. Salinization of arable land.

Source: https://9f416d62-a-62cb3a1a-ssites.googlegroups.com/site/102beetwo/salinity/causes/DLS222.jpg?attachauth=ANoY7cpwavYmNC LBaKnfdcUlgyeYtrQnaN47Sy5og_GqfAKb0C0o2Br_SnE7HZL9f8D-dBjqdQvf_zGT0RdQ7Xdts0l9jCwtxRXpe9KrjCzoYbaRq3otSkHCv50-OBST3sy96I_FBZyC5EecgqAaZ7J0FTEKAY82jEl3a8NKIEE6lTvTH8L-nM7LTjcKwR1DcnCGZa2rtqpqK6gyeHPy3ExvL1qJohBul84qiBO8lS60n-FzDNNpzJE%3D&attredirects=0

Increased salinization of arable land (Figure 7.4) is predicted to have catastrophic worldwide consequences, resulting in a 30 percent loss of land in the next 25 years and up to 50 percent by the middle of the twenty-first century(Machado & Serralheiro, 2017). Plants that develop in saline environments are subjected to three types of stresses: ionic, osmotic, and oxidative. Specific ions found in saline soils, such as Na^+ and Cl, disrupt enzyme and other macromolecule structures, damage cell organelles, disrupt general metabolic activities, limit protein synthesis, and cause ion deficit. Plants that are subjected to low water potential suffer from physiological drought. Furthermore, salinity causes nutritional imbalances in the plant as a result of reduced and differential nutrient uptake and translocation to the shoot and leaf tissues, making the plants weak and unproductive. Arbuscular mycorrhizal fungi (AMF) are soil microorganisms that form symbiotic relationships with the majority of higher plants, provide a direct physical link between soil and plant roots, are an important component of natural ecosystems, and are found mostly in saline environments. The extraradical hyphae of AMF extend several meters beyond the depletion zone, increasing root surface area and facilitating plant nutrient uptake. Indeed, AMF aids in the mitigation of physiological drought, which is frequently enforced in saline conditions, by improving physiological processes and general metabolic activities of the plant. As a result, the use of AMF could provide a less expensive and more cost-effective solution to the salinity problem.

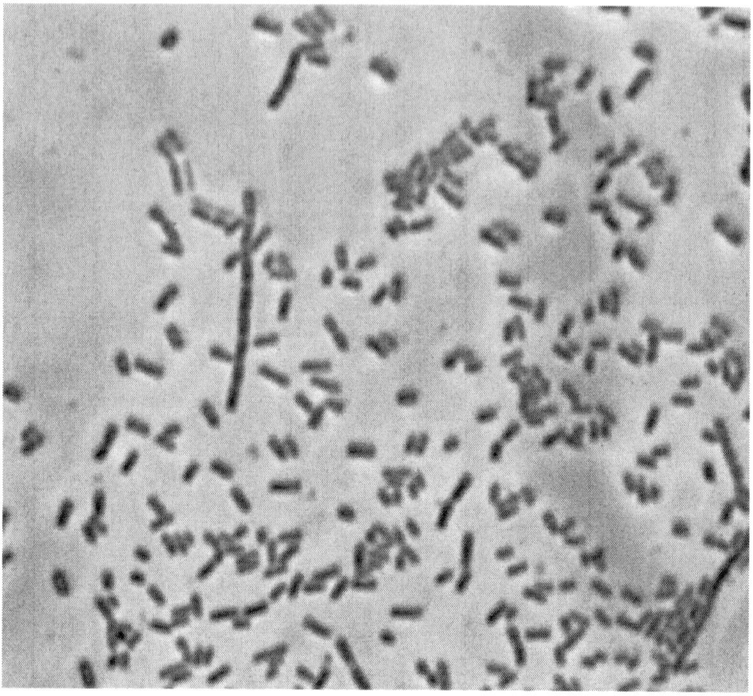

Figure 7.5. Soil bacteria.

Source:https://www.nrcs.usda.gov/Internet/FSE_MEDIA/nrcs142p2_049921.jpg

7.9 SOIL BACTERIA

Soil bacteria (Figure 7.5) have a major impact on the rhizosphere, which is the area immediately around the plant root. These bacteria can be harmful, such as plant pathogens that carry pathogenicity genes and cause biotic stress in plants, or they can be beneficial, reducing the negative impacts of biotic and abiotic pressures and encouraging plant growth and survival. They could be classified as stress relievers or potential bio-fertilizers. Arbuscular mycorrhizal fungi are one of the most common relationships among them, owing to their widespread distribution throughout terrestrial plant groups. They are important components of natural ecosystems and have been found in saline environments. AMF establishes a direct physical relationship between plant roots and soils, making it easier for the host plant to obtain mineral nutrients from soils, especially under nutrient stress conditions, and modifies the rhizosphere environment, reducing the negative effects of salt stress. Different plant species such as tomato, cucumber, maize, lettuce, clover, fenugreek, sesbania, and acacia have been reported to improve salt tolerance.

The use of AMF provides a less expensive and more cost-effective solution to the problem of salinity stress. Hundreds of research papers have shown that this association reduces the negative effects of salt stress and improves plant growth in saline conditions; however, the direct and exact physiological and biochemical mechanisms involved in mycorrhizal plants' increased salt tolerance have yet to be documented. Because phosphate ions precipitate with Ca^{2+}, Mg^{2+}, and Zn^{2+} ions and become immobile in a saline soil, excessive salt severely reduces the availability of phosphorus to plants. Under saline conditions, however, AMF has a significant impact on P uptake. AMF's role in saline conditions is not restricted to P uptake; in fact, AMF allows extraradical hyphae to explore more soil volume for higher nutrient absorption, especially in stressful situations. The fundamental mechanism by which mycorrhizal fungi ameliorate the negative effects of salt stress on plant growth could be improved nutrient uptake, notably P by AMF. However, greater nutrient uptake isn't the only way AM plants improve their tolerance to high soluble salt concentrations. It has also been claimed that non-mediated nutritive effects, which could play a more precise function than nutritional effects in improving salt tolerance in AM plants under salinity, may be involved. In reality, AMF regulates the expression of salt-related genes and modulates various physiological and biochemical processes to counteract the negative effects of damaging ions in a saline environment.

7.10 WEATHERING

Weathering of rocks,(Figure 7.6) deposition of oceanic salts, groundwater table fluctuation and flood water are all natural processes that contribute to soil salinization. Sodium and other salts are abundant in rocks. The presence of salts in the soil is thought to be caused mostly by the weathering of rocks. During this process, the parent rock loosens or breaks down into the salt component, releasing calcium, magnesium, and sodium salts, as well as carbonates and sulfates to a lesser amount.

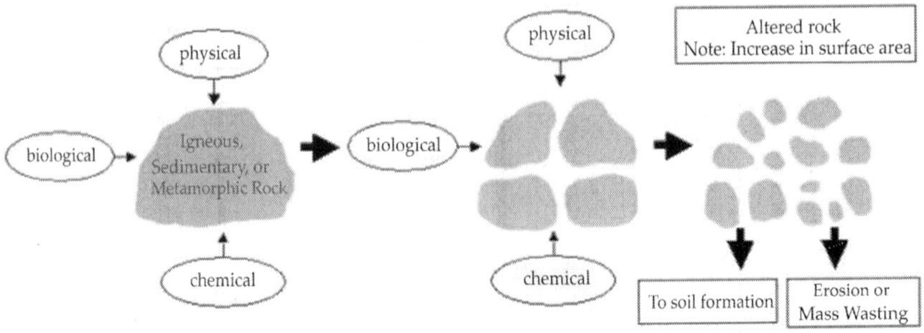

Figure 7.6. Weathering of rocks.

Source:https://passel2.unl.edu/image.php?uuid=8a61dac33566&extension=jpg&display=MEDIUM&v=1604681020

When the weather is humid, these salts seep down to the water table and are subsequently transferred to the sea. The soil fertility level and rate of soil formation are determined by the mineral composition of the parent rock and its porosity. Another aspect that contributes to salinization is flood irrigation. When water is irrigated into a crop area, it percolates lower and raises the water table, accumulating dissolved salts in the plant root zone. Because standing crops absorb root zone water, these salts move in the soil, adding to soil salinization. The problem is exacerbated by rising temperatures because evaporation of water from the soil causes the precipitation of specific ions such as Ca^{+2} and Mg^{+2} while leaving Na^{+} in the soil. To meet the world's food needs, the world's population is expanding, putting strain on agricultural lands. Continuous application of chemical fertilizers and pesticides to boost crop output has been identified as a possible source of rising salt levels in soil. NaCl is widely used to remove snow off the roadside in European countries. Plants are exposed to these salts by horizontal or vertical flow, and their health and survival are impaired. Increased urbanization, on the other hand, has resulted in deforestation in both developed and developing countries. Other issues that contribute to soil salinization include overgrazing-induced desertification of land and poor management of industrial wastewater discharge.

Numerous studies have shown that the genes extracted from these organisms may be utilized to successfully generate transgenic plants by either transferring them to susceptible plants or modifying the expression of existing genes. Osmolytes, oxidative stress, ion homeostasis, transcription factors, DNA helicases, and chaperons are among these genes. Because salt stress resistance is a multigenic trait, genes for drought and cold stress tolerance may be transmitted as well. Implementing such procedures could help to alleviate salt stress to a greater extent, but such techniques are expensive and not available to farmers. However, using helpful microbes like arbuscular mycorrhizal fungi to improve salt tolerance in crop plants could be an environmentally friendly, low-cost, and efficient way to solve the problem of salinity stress. The production of ROS and the plant's inability to prevent or detoxify the detrimental effects of free radicals by antioxidant neutralization causes oxidative damage. These reactive compounds are formed as by-products of electron transport in the chloroplast and mitochondria, and they have the potential to trigger a variety of negative effects.

In the absence of protective systems, free radicals have a tendency to interact with cell components such as DNA, proteins, and lipids and destabilize them. ROS play a role in cell signaling, including death, gene expression, and cell signaling cascade activation. When growth conditions are favorable, ROS production is modest; nevertheless, when salinity or drought stress is present, ROS production increases. Excessive salt accumulation may cause elevated levels of ROS in saline conditions, which affects not only different pathways in plant metabolism but also cell membrane integrity and permeability, resulting in a variety of ROS such as superoxide anion radical. Higher plants establish an excellent antioxidant system to detoxify ROS under normal and stress conditions by producing antioxidant molecules and enzymes to protect them from oxidative damage, but the production rates of these compounds substantially decreased under salinity stress. These changes are most likely caused by Na^{+} and Cl toxicity, as well as osmotic imbalance, in a saline environment. Excessive salts have a negative effect on cell membranes and organelles. Salinity stress damages ultra-structures in *Trigonella foenum-graecum* plants with *G. intraradices*, according to experiments; nevertheless, the level of salt-induced ultrastructural damage was smaller in mycorrhizal plants. The fact that AMF prevents membrane damage was confirmed by decreased lipid peroxidation and electrolyte leakage in mycorrhizal plants. The lower ultrastructural damage in mycorrhizal plants may be owing to greater osmolyte and polyamine accumulation, as well as more and larger plastoglobule production in mycorrhizal plants compared to nonmycorrhizal plants. Higher osmolytes and tocopherols ensure osmotic correction and greater free radical scavenging capacity of mycorrhizal plants under salinity stress, whereas lower Na^{+} and Cl accumulation ensures reduced ionic toxicity.

The development of plant endosymbiosis necessitates the expression of a collection of genes known as common SYM genes. Their protein products are part of a signal transduction cascade that is activated when fungal signals are detected by receptor-like kinases. Myc factor causes nuclear Ca^{2+} spiking, which is decoded by a nuclear-localized calcium–calmodulin kinase and results in CYCLOPS transcriptional activation of symbiosis-related genes. The reduction of common SYM signaling could be one explanation for the suppression of AM growth at high Pi. High Pi did not reduce the ability of *Medicago rhizodermis* cells to trigger nuclear $Ca2^{+}$ spiking in response to rare hyphopodia or germinating spore exudates, and the expression of downstream common SYM genes involved in the generation and of Ca^{2+} spiking was not affected. When plants come into contact with pathogens, fungal biocontrol agents, or nitrogen-fixing bacteria, changes in intracellular free Ca^{2+} content are one of the first steps in signaling pathways activation. After interacting with certain receptors, microorganism-secreted molecules cause transitory changes in cytosolic Ca^{2+} levels in plant cells due to ion inflow from the external environment and release from internal Ca^{2+} storage compartments. Legumes can form a symbiotic relationship with both rhizobia and AM fungus. Both symbioses share components of the Ca^{2+}-mediated signaling pathway. The role of Ca^{2+} in the mycorrhizal signal transduction pathway has long been suspected, based on observed similarities with symbiotic nitrogen fixation.

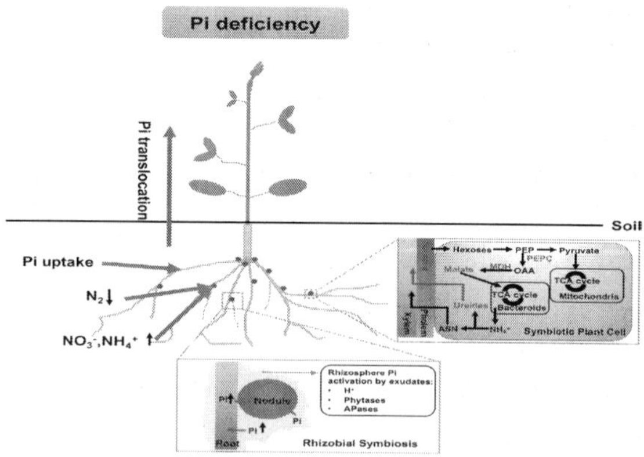

Figure 7.7. Symbiosis and Pi.

Source: https://www.researchgate.net/publication/329736387/figure/fig1/AS:705183944282115@1545140362631/Mechanisms-for-maintaining-Pi-homeostasis-under-Pi-deficiency-in-legumes-Legumes-use_W640.jpg

7.11 SYMBIOSIS

When the symbiosis (Figure 7.7) is due to Pi insufficiency, the corresponding signaling molecules of the host plant that govern AM growth become extremely complex. However, there was some indication that control occurs at various levels, including nutrition, miRNA, phytohormone, and sugar. Despite the fact that AM colonization grew more quickly when Pi was deficient, the signal transduction mechanism remained mostly unknown. P deficiency, which was attributable to Pi concentration variations, was normally detected in the root. The signaling molecules in the system, such as microRNA and plant hormones, were then stimulated, generating a systemic reaction across the plant. The systemic signal has an effect on the mycorrhizal root, causing AM to be promoted. Because of the crosstalk between these molecules and their complex consequences on AM development, defining the role of signaling molecules activated by Pi deficiency is difficult. Recently, significant links between ROS and Ca^{2+} have been discovered both within and between cells. These oscillations serve as a signal for the oxidative burst to create ROS and initiate signaling cascades that culminate in the production of mycorrhization-related chemicals.

Although research has shown that glomalin can sequester or bind Fe, nothing is known about the impact of GRSP on the effective Fe content of soil and the alleviation of Fe deficit. Metallic elements such as Fe can be sequestered by glomalin in general, but their efficiency in soil, particularly at high levels, will be considerably reduced by glomalin in order to reduce their toxic effects on mycorrhizas and plants. Approximately, glomalin-binding Fe on the surface of hyphae can act as an iron pool, supplying Fe for direct absorption or hyphae transfer to plants. As a result, glomalin may play a significant role in hyphae nutrient element uptake as a receiver of metallic elements. The chelating Fe content in glomalin was found to be negatively linked with readily extractable glomalin and total glomalin, which could be owing to the competitive interaction between Fe and Cu or Zn. Even so, if the Fe concentration of soils varies, the interaction between mycorrhiza-released glomalin and the binding capacity of glomalin to Fe will alter. Soil fungi known as arbuscular mycorrhizal fungi form a symbiotic relationship with most terrestrial plants. In this type of symbiosis, the fungi supply water and nutrients to the host plant in return for carbon. Research has shown that AM fungi have favorable effects on the growth of the host plant under stress conditions such as salt, drought, heavy metals, and so on. The fungi's ability to alleviate stress on plant growth and the environment is largely owing to their exceptional abilities in generating physiological and morphological mechanisms. Heavy metal stress is a significant factor impacting plant growth and the environment. Biological techniques, such as using soil microbes including AM fungi, plant growth-promoting rhizobacteria, and endophytic bacteria to eliminate the negative impacts of stress on plant growth and the environment, have been shown to be among the most effective.

7.12 HEAVY METAL STRESS

Under heavy metal stress, plants use a variety of processes to manage ion homeostasis, resulting in cellular detoxification. Heavy metal chelation by plant roots and binding by cellular walls, for example, can regulate heavy metal concentrations in plant cells. Plant cells can also synthesize metallothioneins and phytochelatins, both of which have a high affinity for heavy metal absorption. As a result of their transit through the tonoplast and sequestration in the vacuole, their cellular concentration is regulated.

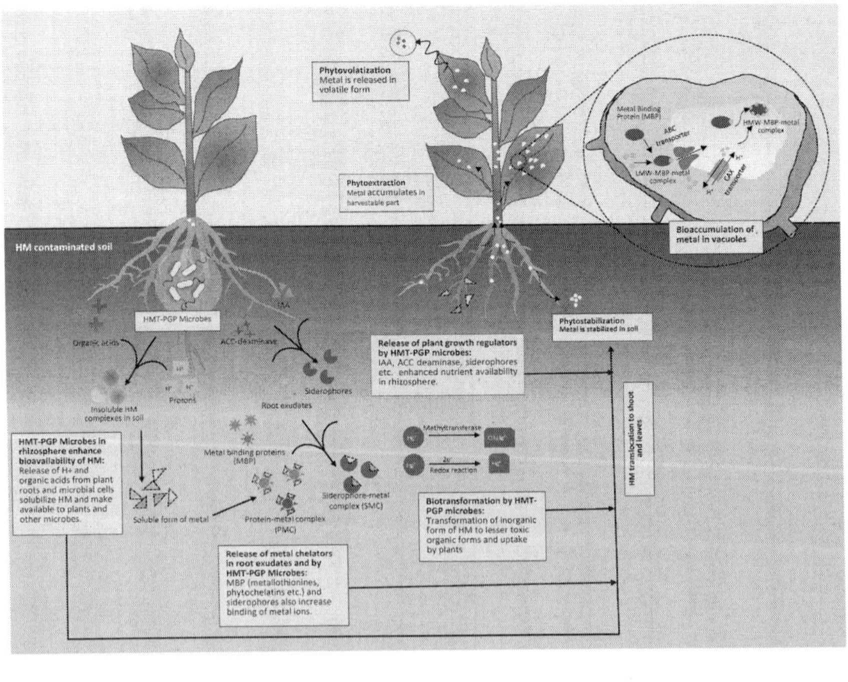

Figure 7.8. Heavy metal stress.

Source:https://www.frontiersin.org/files/Articles/295763/fmicb-08-01706-HTML/image_m/fmicb-08-01706-g001.jpg

Heavy metal stress (Figure 7.8) affects a variety of morphological and physiological activities in plants and microorganisms, including enzymatic activity, nutrient uptake, hormone function, and so on. Under various conditions, including stress, mycorrhizal fungi can boost the intake of certain micronutrients such as Fe, Zn, Mn, and Cu. For the treatment of heavy metal stress, various strategies have been used; however, biological methods such as soil microbes, particularly arbuscular mycorrhizal fungi, plant growth-promoting rhizobacteria (PGPR), and endophytic microbes have been found to be among the most efficient. AM fungi are soil fungi that form symbiotic relationships with most terrestrial plants. In this type of symbiosis, the fungi offer water and nutrients to the host plant in return for carbon. In the soil, fungal spores create a vast network of hyphae containing arbuscule and vesicles. The arbuscule is the structure that serves as the interface for the interchange of water and nutrients; the vesicle, on the other hand, has a large number of vacuoles and so has a high absorbing potential for elements not found in the environment. Under heavy metal stress, research has revealed an elevated concentration of Cu in the spore vacuoles and an accumulation of Cd in the hyphal vacuoles of *Glomus intraradices*. Such storage structures can help the fungi and the host plant cope with stress.

Because it includes hydroxyl, carboxyl, and free amino acids for heavy metal absorption, the fungal cell wall is an ideal location for binding heavy metals. The appearance and thickness of the cell wall determine the absorbing potential. Surprisingly, research has found that narrower hyphae can hold larger levels of heavy metals than thicker hyphae, owing to the thinner hyphae's higher surface area. As a result, *Rhizophagus clarus*, with an average hyphal diameter of 4 microns, absorbed more heavy metals like Pb, Cd, and Cu than *G. gigantean*, with an average hyphal diameter of 13 microns(Tchounwou, Yedjou, Patlolla, & Sutton, 2012). Heavy metals such as Ni, Fe, and Zn are found in considerably higher quantities in the root cortex of mycorrhizal plants, and different species of mycorrhizal fungi have different capacities for absorbing heavy metals. The enhanced formation of phytochelatins, glutathione-derived peptides, which are only produced by higher plants, is another mechanism that can make mycorrhizal plants absorb heavy metals. Plants will be able to chelate heavy metals and metalloids in this way. The species of mycorrhizal fungi and the kinds of heavy metals are two factors that influence the relieving effects of mycorrhizal fungi on heavy metal stress. AM fungi have varying degrees of efficacy in removing heavy metals from polluted soils, and some mycorrhizal fungi may be more efficient than other types of fungal species. Heavy metals are allocated differently by AM fungi to different areas of the host plant, with some heavy metals concentrating in the roots and others in the aerial part of the plant.

Metallothioneins and amino acids with a large number of cysteine residues are another key group of proteins that can retain heavy metals. When organisms are exposed to heavy metal stress, they generate proteins like this. Mycorrhizal fungi have the Mn-dependent superoxide dismutase (SOD) gene, which may convert the superoxide anion to hydrogen peroxide and detoxify the negative effects of reactive oxygen species under stressful situations. Other scavenging chemicals produced by mycorrhizal fungi, such as thioredoxins and glutaredoxins, have been shown to result in cellular redox homeostasis by modulating the redox potential of protein thiols and thereby protecting the fungi from oxidative damage. There are around 30 genes that can activate mycorrhizal fungi's antioxidant function. Vitamins B_6, C, and E, as well as glutathione, are other compounds that can act as antioxidants in AM fungus. The induction of glutaredoxins has been linked to heavy metal stress, and because glutathione can act as a redox donor for antioxidant enzymes, its scavenging role has been proposed. B_6 has a moderating effect on reactive oxygen species. Pollutants have been demonstrated to have negative effects on membrane lipids and they are required for the formation of mycorrhizal fungal symbiosis. Several genes have been discovered that can transfer heavy metals such as Zn, As, Cu, and Cd. By raising the rate of soil organic matter and promoting root growth and activity, mycorrhizal fungi are able to generate a suitable environment for the growth and activities of other soil bacteria. Although the use of fungi is a viable way for soil bioremediation, it is not the only option, and when used in conjunction with other bioremediation techniques, the process can be more effective. The most essential aspect of using mycorrhizal fungi in contaminated regions is that they can only be used when the host plant is not present. Although fungal spores can germinate in the absence of a host plant, they can only develop and form hyphae if the host plant is present.

Temperature is one of the most influential environmental elements affecting plant development and productivity around the world. Low and high temperature stressors have an impact on a variety of physiological and biochemical systems and functions. Plants with arbuscular mycorrhizal symbiosis have been demonstrated to be more resistant to temperature stress. The methods by which the AM symbiosis improves the host plant's temperature stress tolerance by increasing water and nutrient intake, improving photosynthetic capacity and efficiency, shielding the plant from oxidative damage, and increasing osmolyte accumulation. P is a vital macronutrient for plant development and crop production; it is a major structural element in nucleic acids, phospholipids, and enzymes, and it is also involved in matter and energy metabolism and signal transduction cascades. Temperature stress inhibits plant P uptake, transport, and metabolism. The hyphal network of the AM fungus and the expanded surface of plant roots, however, allow AM symbiosis to sustain plant P acquisition. Because AM plants improve the conversion of P from inorganic to organic forms in the leaves, increasing the P sink and thus promoting P absorption, the inorganic P gradient from root to leaves can be increased. AM hyphae can also store more absorbed P than roots, allowing for continuous P transport into the hyphae. Because of its stronger affinity for P ions or lower threshold concentration for uptake, the kinetics of P uptake into hyphae differ from those of roots. The majority of investigations have found that under low or high temperature stress, AM plants exhibited higher P content than non-AM plants, implying that the AM symbiosis increased P availability and absorption. Some studies have shown that the contribution of AM symbiosis to P uptake may be minor since AM plants have lower P content than non-AM plants.

7.13 BIOCONTROL

Biocontrol focuses on natural mechanisms that occur during interactions between microbes and host plants and is based on the use of living microorganisms or derived compounds to control plant infections and pests. Even while the presence of microorganisms in plants has long been thought to cause diseases, it is now accepted that the great majority of microorganisms encountered are not dangerous, with the majority of them being neutral or beneficial to plant health and growth. Plants are constantly interacting with microbes; as a result, the rhizosphere microbiome has gotten a lot of attention in recent years as a critical component of plant health and productivity. Plants can even influence their own rhizosphere microbial communities by releasing primary and secondary metabolites from their roots, which attract beneficial microbes. Obligate biotrophs from the phylum Glomeromycota can form symbioses with over 80% of terrestrial plant species, providing host plants with a variety of physiological benefits. They primarily offer mineral nutrients to the host plant, such as phosphorus or nitrogen, which boosts its development and fitness, in exchange for the plant providing carbohydrates to the fungus, accounting for up to 20% of its fixed carbon(Humphreys et al., 2010). AMF affects the plant host's ability to overcome abiotic and biotic stressors in addition to improving nutrition delivery.

CHAPTER 8

QUALITY IMPROVEMENT IN RICE

CONTENTS

8.1 Introduction ... 150

8.2 Abiotic Stressors ... 151

8.3 Salinity Stress ... 153

8.4 Evapotranspiration ... 154

8.5 Rice Quality ... 157

8.6 Breeding ... 158

8.7 Marker Assisted Selection ... 160

8.8 Gene Expression Analysis ... 162

8.9 Databases ... 162

8.10 Rice Genomics ... 163

8.1 INTRODUCTION

Rice is a diet of roughly half of the world's population. Rice was domesticated hundreds of years ago, and the two species currently produced worldwide are *Oryza glaberrima* (Figure 8.1) and *Oryza sativa*. Rice's domestication may be traced back to 10,000 years ago when *Oryza* sp. was just a wild grass, and artificial continual selection by humans led to the establishment of a stable and adaptable cultivar. Asian rice, or *O. sativa*, is descended from *O. rufipogon*, from which it was domesticated. *O. sativa* is divided into five subpopulations, with indica and japonica showing substantial genetic differences. Because the phenotypes of domesticated rice varieties differ greatly from those of their wild parent, organisms that interact with them are under pressure to adapt to these novel phenotypes. New insights into genetic variation, population structure, and the identification of domestication loci in rice species have been made possible by modern genomic methods (Doebley, Gaut, & Smith, 2006).

Figure 8.1. Oryza glaberrima.

Source:https://www.alamy.com/african-rice-oryza-glaberrima-plants-growing-in-an-agricultural-field-with-people-harvesting-the-crop-uganda-africa-image366217639.html

8.2 ABIOTIC STRESSORS

Rice is one of the key grains that is susceptible to a variety of abiotic stressors. Relevant physiological and biochemical parameters are inhibited by environmental restrictions, which determine grain production in rice. Heat stress causes considerable metabolic changes in leaves and developing seeds, lowering rice grain yield and quality. Salinity, dryness, and heavy metal stress are all factors that reduce rice grain production. Rice plants grown in an environment with both heat and drought stress have lower spikelet fertility. Furthermore, under stressful conditions, alterations in secondary metabolite build-up and altered ions homeostasis impair rice growth and output. Rice's defense mechanisms include osmotic adjustment, increased antioxidant enzyme activity, and increased generation of non-enzymatic antioxidant chemicals. Rice has been a prominent crop connected with Asians since the beginning of time. With a production of 668 million tonnes, Asia is the world's greatest rice producer, accounting for about 90% of global production(Meena, Bhusal, Kumar, Jain, & Jain, 2019). India and South China have been suggested as the initial places where domestication took place. There are two varieties presently known as indica and japonica subspecies, which differ in leaf and seed morphology and thus are classed as subspecies. Based on morphological, cytological, and genetic investigations, a third subpopulation was identified as javanica. Rice domestication began around 9000 years ago in Asia, mainly in India and China. Except for two qualities, certain traits related to domestication are related to the cultivated variety of rice, which is one of the primary issues associated with high diversity. The lack of secondary dormancy and the degree of spikelet rupturing are two fundamental characteristics of rice domestication. The ease with which different rice cultivars rupture varies.

In the Japonica variety, there is no abscission layer at the base of the seed; in the Indica species, the abscission layer is partially present; and in weedy rice, the abscission layer is totally present. Dormancy has a higher level of intricacy, with hull and seed dormancy. Secondary dormancy is significant in the Indica cultivar because seeds remain viable between planting seasons in humid and hot climates. It has been observed that the size of the organs that may be ingested has increased in domesticated varieties of various crops, although in the case of rice, the size of the seed in most cultivated varieties is slightly different from that of wild types. As a result, seed size cannot be utilized to distinguish between domesticated and wild types. Other key characteristics throughout the domestication process are synchronized maturity and the number of seeds per panicle. Domestication-related features have varied as a result of diversification after domestication.

The availability of fertile farming grounds, suitable irrigation facilities, seedling transplantation methods, and multiple cropping all contributed to a rise in human population settlement and consequently population growth. However, given the current situation, rice cultivation is unable to keep up with the rapidly increasing population. Feeding the continuously expanding population requires new farming practices. Due to biotic and abiotic limits, plant growth and development are hindered at physiological and biochemical levels, resulting in considerable yield losses in agriculture crops. Several environmental factors, including heavy metals, cold, heat, salt, and drought stress, significantly reduce rice growth and output. The developmental growth stage of rice plants determines how they respond to abiotic stimuli. A variety of crops show increased vulnerability to abiotic pressures during early growth stages, while later growth stages show increased tolerance to environmental perils. Drought is the most severe abiotic stress, resulting in significant reductions in growth and yield all around the world.

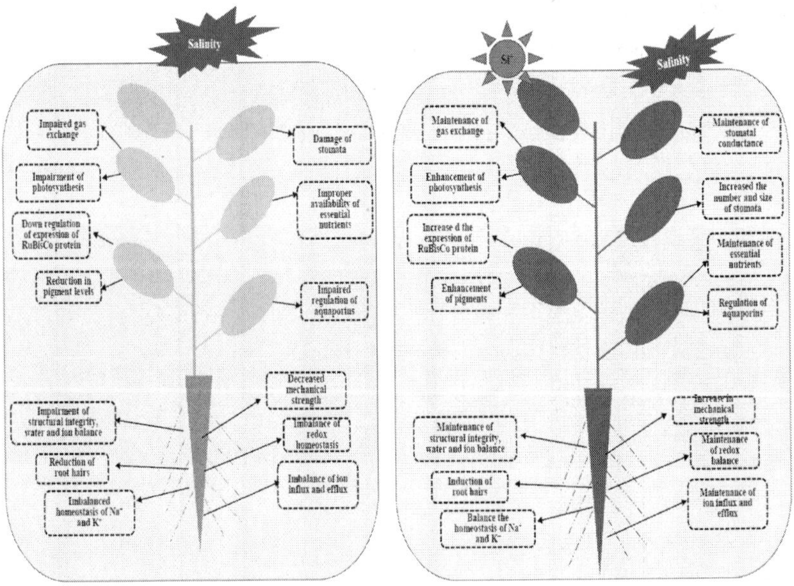

Figure 8.2. Salinity stress.

Source: https://www.mdpi.com/plants/plants-08-00307/article_deploy/html/images/plants-08-00307-g001.png

8.3 SALINITY STRESS

Salinity stress (Figure 8.2) is regarded as the second most severe abiotic stress on plant yield. Food insecurity is a growing issue, owing to large yield decreases in plants exposed to abiotic stressors. Due to biotic and abiotic environmental challenges, rice plant growth, development, and grain yield are significantly diminished. Abiotic stressors cause a variety of changes in plant morphology and physicochemical processes, which have a negative impact on plant productivity. Plant productivity is projected to be reduced by 50% as a result of these pressures, resulting in a major financial loss. Due to considerable yield losses in plants encountering environmental limits, the supply of food for the ever-increasing human population is jeopardized. Under salinity, the physiology of important processes like photosynthesis, respiration, and nutrient acquisition is hindered in rice. Salinity-mediated osmotic stress, which causes stomatal closure and reduced water content, disrupts the delicate balance in plant water content. Furthermore, salt causes considerable alterations in the expression pattern of proteins and genes that control rice growth and development. Salinity can come from a variety of places, and it can stifle crop development and output. It's crucial to understand what kind of salt generates salinity. Many other elements, such as crop and soil management, meteorological conditions, and soil types, contribute to the salt content of the soil. In a humid area, coarse-textured soil, for example, would have a higher infiltration rate and rainwater to leach down salt. Salt leaching does not occur in fine-textured soils in dry and semiarid environments. Salinity is a common issue that causes problems with germination, plant growth, and productivity. Reduced cell division and expansion, loss of membrane integrity, changes in metabolite functions, oxidative stress, nutritional imbalance, specific ion toxicity, and water stress are all factors that contribute to the negative effects of salinity on morphological and biochemical features.

Increased amounts of NaCl in the growth medium hampered growth due to the increased production of reactive oxygen species. Because ROS is highly reactive, it interacts with biological components, causing oxidative damage such as DNA fragmentation, protein oxidation, and lipid peroxidation. To reduce oxidative damage caused by salt, it is necessary to effectively scavenge ROS. Enzymatic antioxidants and non-enzymatic antioxidant molecules are found in plants' oxidative defense mechanisms. Superoxide dismutase, peroxidase, catalase, and ascorbate peroxidase are examples of enzymatic antioxidants. Tocopherols, glutathione, ascorbic acid, carotenoids, phenolics, and flavonoids are examples of non-enzymatic antioxidant substances. In rice, for example, the role of the antioxidant system in preventing oxidative damage caused by salt was investigated. Strong protein oxidation was also detected, as evidenced by decreased glutathione and ascorbate peroxidase. Antioxidant enzyme activity increased significantly. The degree of oxidative damage caused by Na^+ and Cl- toxicities was assessed in two rice genotypes in another investigation. The generation of H_2O_2 and $O_2^{\bullet-}$ radicals was boosted by an equimolar concentration of Na^+ and Cl-.

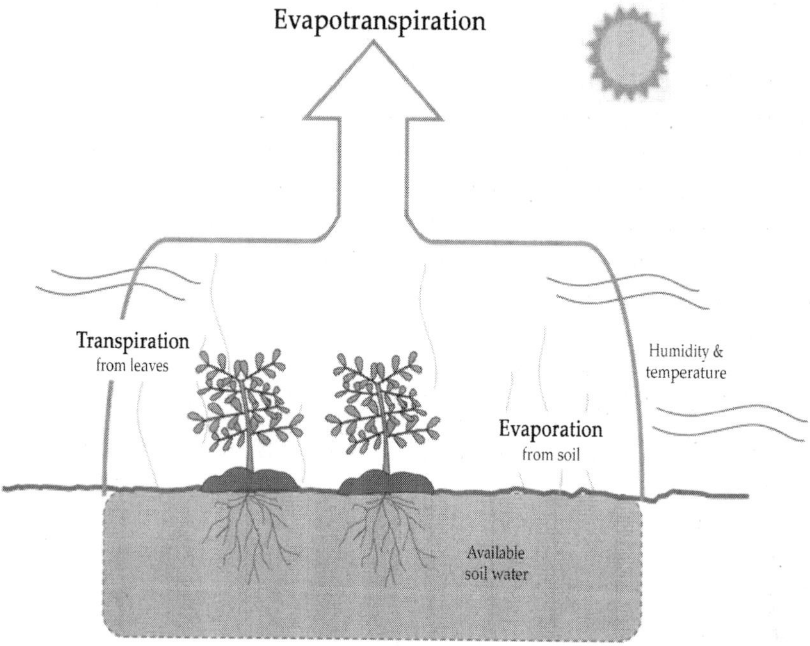

Figure 8.3. Evapotranspiration.

Source: https://d32ogoqmya1dw8.cloudfront.net/images/integrate/teaching_materials/food_supply/student_materials/schematic_evapotranspiration.png

8.4 EVAPOTRANSPIRATION

When evapotranspiration (Figure 8.3) is high or the availability of water to the roots is limited, plants experience drought. Under drought circumstances, rice growth and productivity are severely hampered. Growth inhibition is linked to inadequate root architecture as well as impaired leaf features such as leaf color, pubescence, cuticular wax composition, and leaf form and shape, all of which restrict plant growth. Rice spikelet sterility and grain reduction were both reduced when it was subjected to water-deficit circumstances during the anthesis stage. The direct effect of water stress on spikelet desiccation, spikelet opening, and panicle growth causes spikelet sterility. Low panicle water potentials had a substantial impact on the quantity of opened spikelets. The application of drought conditions during the early phases of rice growth resulted in a considerable reduction in vegetative growth. Root-shoot biomass and soil-plant water potentials, which are greatly affected by moisture stress, are the most commonly utilized criteria for estimating rice plant growth and vigor under water-deficit situations.

Drought stress was applied to rice lines like w-14 (drought-sensitive) and w-20 (drought-tolerant), and changes in physiological indicators and plant mineral nutrition were investigated(Hussain et al., 2018). Under water shortage, plant biomass, gas exchange characteristics, water potential, and PSII efficiency all decreased significantly. Rice's drought tolerance was improved by an exogenous Si spray, which resulted in increased photosynthesis, transpiration, and PSII efficiency. Under drought circumstances, endogenous levels of Fe, Mg, Ca, Na, and K were higher in plants, whereas Si treatment lowered nutritional content. Better drought resistance due to Si, according to the scientists, was the result of improved photochemical efficiency and regulation of ion homeostasis in rice. Drought caused a huge loss in plant fresh and dry biomass, as well as a significant decrease in leaf hydration status, water usage efficiency, and photosynthesis in rice plants. Under drought conditions, rice plants showed H_2O_2-mediated loss of membrane integrity. However, foliar polyamine treatment improved drought tolerance, as seen by increased photosynthetic and antioxidant defenses. Rice plants were tested for the effects of dryness on chlorophyll fluorescence indices, chlorophyll contents, photosynthetic light curve, lipid peroxidation, and ROS metabolism. Drought stress resulted in a considerable drop in photosynthesis, chlorophyll content, and PSII efficiency, as well as a significant increase in MDA levels. Antioxidant enzyme activities like SOD, POD, and CAT, were much higher under drought conditions, successfully reducing ROS and protecting plants from oxidative damage. Under drought conditions, rice had a considerable reduction in fresh and dry biomass, plant water relations, and photosynthesis. Exogenous administration of brassinosteroids resulted in an increase in plant fresh and dry biomass, as well as a higher rate of photosynthesis and a considerable reduction in H_2O_2 production and lipid peroxidation.

The complex and polygenic structure of rice genotypes has hindered breeding for stress responses and quality features using germplasms, as well as genetic diversity. Any breeding effort must have genetic knowledge about germplasm that can potentially be employed as a parent. Rice is a major grain crop grown all over the world. Moreover, half of the world's population eats it as their primary source of nutrition. Throughout the history of plant breeding, the primary goal has been to produce stress tolerance and high-quality rice cultivars. The need for biotic and abiotic stress tolerance, as well as higher grain quality, is increasing day by day in many rice-producing locations around the world, coupled with advances in the level of living. Drought, salinity, water logging/flooding, and temperature stress are the most important variables lowering agricultural productivity. Abiotic stressors cause plants to activate several systems in order to survive and maintain growth and production. As a result, it's critical to understand what systems exist to assist plants in surviving extreme stress. Rice is a salt and drought-sensitive crop, especially in its early phases of development. Landraces that are unique to specific crops provide a genetic resource that is distinct from another germplasm. Despite their established importance and potential uniqueness, patterns of genetic diversity among farmed species and their link to other germplasm pools are unknown. Local germplasm has been found to be genetically and morphologically unique, as well as a useful genetic resource for stress response and quality attributes enhancement in plants.

Rice species' sites of origin are characterized by topological heterogeneity and are thought to be the source of rice variety. With many rice growers switching to contemporary varieties, the variety of these centers is gradually dwindling. Farmers in many places adopted high-yielding contemporary varieties, and the cultivation of landrace variety decreased. Changes in land use, agricultural expansion into marginal areas, and deforestation are all putting wild species at risk of extinction. According to IRRI, there is still a lot of untapped genetic variation in rice's basic gene pool that can be used to improve the diversity of local germplasms and their performance under various agro-ecological conditions. Oryza species found in the wild could provide new alleles for rice cultivars to improve yield, quality, and stress tolerance. Improvements in the performance of valuable genes from wild germplasm into superior rice cultivars are critical. Crossing the wild species *Oryza rufipogon* with *Oryza sativa* cultivars resulted in lines with greater yields and tolerance to abiotic stressors than their progenitors. Low temperature, for example, is one of the most critical abiotic stresses impacting rice's early vegetative phases. Seedling emergence and early vegetative growth are inhibited or delayed at temperatures below 15°C, posing a threat to rice(Fahad et al., 2017).

Many academics from all around the world are interested in studying the genetic variety of different species. Throughout the history of plant breeding, the primary goal has been the cultivation of high-quality rice cultivars. The need for improved grain quality is increasing day by day in many rice-producing countries of the world, along with advances in the level of living. In the rice industry, grain quality is a major consideration. Rice consumers in different parts of the world, or even different sections of the same country, have distinct preferences for rice grain performance and cooking quality. Grain shape and size, and eating quality are all aspects that influence grain quality. A sieving mechanism separates milled rice into broken rice and whole grains, with the proportion of entire grains denoted as head rice recovery. Grain appearance is mostly affected by grain form and the quantity of chalkiness, which is defined as the translucent or opaque appearance of the grain. Chalky grains have an inferior density of starch granules than transparent grains, and because chalky grains are not as rigid as translucent grains, they are more prone to breaking during milling.

8.5 RICE QUALITY

Rice quality is a significant factor that determines its consumption rate, hence boosting output. As a result, after yield, grain quality improvement becomes the second most important concern in rice breeding. The milling, appearance, cooking, and nutritional characteristics of rice grains are all factors to consider. People are more concerned with the appearance and cooking quality of these. It has been shown to lower grain seed size and yield in rice, as well as alter endosperm quality. The use of semi-dwarf varieties has resulted in a significant increase in world rice production over the last half-century, requiring high inputs of fertilizers and pesticides, which indirectly cause environmental problems, disease outbreaks, and pest resistant insects, as well as affecting human health as a primary concern. The rapidly growing global population, the ongoing severity of global climate change effects, water scarcity due to declining water tables, a reduction in rice cultivable land, and an increase in the extreme weather events are all major concerns that have the potential to affect plant growth and rice yield, as well as the physical and chemical characteristics of the grains. Malnutrition has been a serious problem for developing countries; it is estimated that micronutrient malnutrition affects over three billion people and that 3.1 million people die each year as a result of malnutrition, with the number of deaths increasing day by day("Malnutrition in Children," n.d.).

8.6 BREEDING

The conventional breeding approach to improving grain quality and rice yield has had limited success because it requires sufficient genetic variation for a given trait in a species, is time-consuming, and occasionally exhibits linkage drag, in which undesirable genes are introduced in the same generation as desirable genes. As a result, finding fresh technologies such as molecular markers, genomics, and transgenic approaches to supplement rice breeding approaches that can aid with increased yield potential and grain quality enhancement is a must. To fulfill the requirement of rapid human population expansion and a parallel decrease in arable and fertile land, new varieties with high yield potential, good quality, and high tolerance to biotic and abiotic stressors are required. Size, shape, chalkiness, and translucency are all factors that influence grain appearance after milling. The clarity and translucency of the endosperm are commonly used as grain appearance parameters in the rice industry. Rice grain yield is influenced by three factors: the number of panicles per plant, the number of grains/panicles, and the grain weight. Grain weight, which is measured as 1000 grains, is the most trustworthy attribute among these(Li, Li, Ashraf, Liu, & Zhang, 2019). The physical dimensions of rice kernels are described by their length, breadth, and thickness, while the grain form is defined by the ratio of length to width. These four variables have a favorable relationship with grain weight. Aside from yield, grain shape is an important quality attribute that has a significant impact on rice grain product market values. People in Southern China, the United States, and South and Southeast Asian countries prefer long, slender rice grains, whereas consumers in Japan, Korea, and Northern China prefer short, spherical grains.

The nutritional content of food has a major impact on human health. Rice is a common staple food all throughout the world. Protein is the second most abundant ingredient in milled rice, after carbohydrates. Rice's nutritional value is governed by two main constituents: protein and lysine. Lysine is the first limiting essential amino acid in rice, according to human dietary requirements. People who consume rice require different nutritional qualities as the population grows and as society develops. Genetic research and biofortification of micronutrients through breeding are critical for improving rice nutritional quality, as micronutrient shortage is common in these areas. Whole grain rice is higher in minerals, vitamins, dietary fibers, and phenolics than milled rice for human health. Brown rice is becoming increasingly popular as a means of preventing chronic diseases, such as diabetes, among individuals with higher living standards. To provide high-nutritive rice to every segment of society, a lot of work is required. Biolistic transformation is one of the direct gene transfer technologies that allows microprojectile delivery. The biolistic transformation was first demonstrated in 1987 by Cornell University geneticist John C. Sanford and colleagues, who demonstrated the delivery of microscopic microprojectiles into a cell without destroying it(Lacroix & Citovsky, 2020). The microprojectiles are made of tungsten or gold and are coated with DNA before being accelerated and pushed into target cells. Acceleration is provided by the electric charge, CO_2, gun powder, and gasses, and DNA can be delivered into the tissue. A gene gun or particle bombardment technique is another name for it. A biolistic transfer technique has numerous advantages, including rapid gene transfer, efficiency, non-specificity to tissue, complicated cloning procedures with no biological constraints or host limitations, and simultaneous multiple gene transfer. Furthermore, the transport of lengthy fragment DNA is difficult, and biolistic transformation procedures are limited by intricate patterns of transgenic integration.

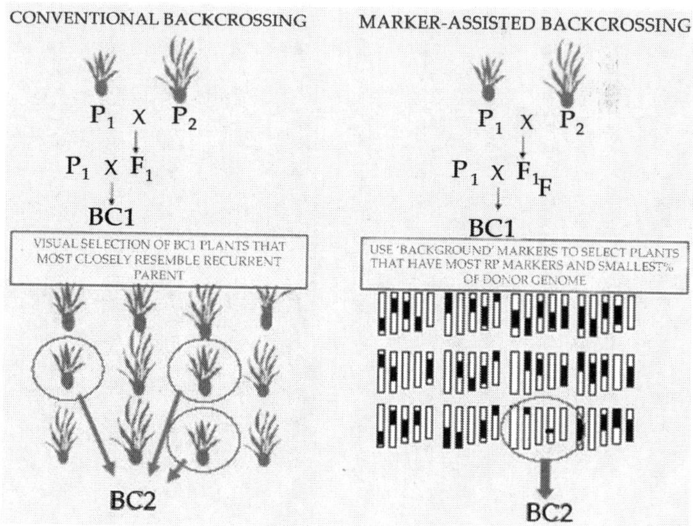

Figure 8.4. Marker-assisted selection.

Source: http://www.knowledgebank.irri.org/ricebreedingcourse/image99.jpg

8.7 MARKER ASSISTED SELECTION

Marker-assisted selection (MAS) (Figure 8.4) is an indirect selection strategy that uses molecular markers closely connected to a target gene as a molecular tag. Alleles at individual loci segregate and transmit according to Mendelian principles of inheritance from one generation to the next, and molecular markers are a sequence of nucleic acid that makes up a segment of DNA that codes for a phenotype. Amplified fragment length polymorphisms, restriction fragment length polymorphisms, random amplification of polymorphic DNAs, sequence tagged sites, simple sequence repeats or microsatellites, and single nucleotide polymorphisms are some of the molecular markers used for crop species. Molecular-assisted selection is more efficient and does not replace traditional breeding. It also does not transfer isolated gene sequences, but rather uses existing plant material as a breeding target. In a breeding program, there are two methods for doing MAS: foreground selection and background selection. Background selection is for yield potential and high resistance good quality rice improvement such as basmati rice, but foreground selection is for grain quality features, which is a time-consuming phenotyping screening technique. Grain quality markers are employed as a background selection to prevent the loss of good quality traits while the other qualities are introduced. Marker-assisted appraisal of breeding material, marker-assisted introgression, and marker-assisted pyramiding are the three main uses of molecular markers in rice breeding.

Starch is a form of carbohydrate storage, and it is found in nearly all plant organs during development. The functions of a number of key genes in the starch biosynthesis pathway have been evaluated in this way. The ability to customize granules or polymers with new functionality by altering starch production routes has immense potential. Until recently, much research on the different components of the starch biosynthesis pathway was done. Reduced phytic acid concentration in seeds is also an important breeding goal, since it would boost the availability of mineral nutrients while lowering the phosphorus load in the environment. If the rice genome's nucleotide sequence was accessible, the genome-wide physical-genetic maps might take advantage of genomics. The DNA sequence, in fact, can bridge the gap between the genome's structural organization and its functional dimensions. With advances in sequencing methodologies, automation, and computation, several genome sequencing projects of higher organisms, including *Saccharomyces cerevisiae*, *Caenorhabditis elegans*, *Drosophila melanogaster*, *Arabidopsis thaliana*, and *Homo sapiens*, were undertaken toward the end of the twentieth century. Rice was the natural choice for a genome sequencing effort on a cereal crop in that new era of sequence-based genomic research. Because of the decade-long significant research on rice genomics, the rice genome had been well mapped by that time, laying a solid foundation for the impending sequencing project. Rice has the shortest genome (430 Mb) among the major cereal crop genomes. Rice is also a good model because it shares a lot of synteny with other grass genomes like rye, wheat, maize, barley, and sorghum, and it's easy to change and edit genetically. Rice became the first crop and the first monocot genome to be sequenced when the International Rice Genome Sequencing Project was established in the second phase of RGP(Eckardt, 2000).

The fabrication of pseudomolecules, which was done using an in-house, database-based, semi-automated approach, was another significant step in the rice genome sequencing endeavor. These are virtual contigs created by choosing the best tiling path of BAC/PAC clones to represent all 12 chromosomes and resolving conflicts between overlapping BAC/PAC clones. Overlap sections were reduced based on similarity search results and tiling path information, while physical gaps were represented by a 1000-nucleotide stretch. By comparing the pseudomolecule sequence to the optical map for rice, the overlapping clones were also confirmed. Each rice chromosome's individual sequences were then joined together to make a contiguous sequence. On the basis of gaps and the presence of CentO repetitions, the location of centromeres was determined. The pseudomolecules were subjected to quality control using Nipponbare's whole-genome shotgun sequences, which included information on gene models and attributes associated with the BAC/PAC clones. For various reasons, not all BAC/PAC clones were incorporated into the pseudomolecules: some belonged to the subspecies indica, some clones were incomplete, redundant, or overlapping, some were released after the version of pseudomolecules, and some were difficult to map to their correct places. Sequences from BACs and Nipponbare's Syngenta whole-genome shotgun assembly that couldn't be aligned to the existing pseudomolecules were represented on two different pseudomolecules, the Unanchored and Syngenta chromosomes. Partially expressed genes, pseudogenes, and tiny gene models are not included in the annotation files for pseudomolecules.

8.8 GENE EXPRESSION ANALYSIS

Gene coexpression analysis identifies the patterns of connection between genes with known or unknown functions and biological pathways in microarray and other transcriptomic datasets. However, recent improvements in next-generation sequencing have allowed gene coexpression networks based on RNA-Seq data to improve gene function definitions and link noncoding genes and splice variants to biological roles. RiceFREND is a gene coexpression database based on vast amounts of rice microarray data gathered from diverse tissues at various developmental stages in natural field circumstances or under the effect of phytohormone therapies. RiceFREND is a platform that allows users to search for co-expressed genes or functionally relevant genes in diverse signaling or metabolic pathways using one or more guide genes. PlantPAN also provides features for gene coexpression analysis and the construction of regulatory networks based on co-occurrence TFs and protein-protein interactions.

8.9 DATABASES

Another integrated database, OryzaExpress, provides an interactive user interface for comparing rice gene expression networks to those of other plants. In the twenty-first century, having access to a sea of genetic sequence information of a cereal crop plant was a ground-breaking feat. There is a need to improve the plant's production, productivity, health, and acceptability, which has mostly been handled by traditional crop-breeding programs through generations. In the current global context, unmanageable population expansion, climate change, loss of cultivable land, greenhouse gas emissions, and excessive use of pesticides and fertilizers have all changed the nature of demands. The availability of the rice genome sequence has allowed for map-based cloning of QTLs, marker-based characterization of beneficial genes, functional analysis via functional knockouts with mutation or gene tagging, and exploration of the possibility of introgression of these genes or genomic regions to develop new agronomically competent hybrids. It also helped researchers better understand rice's effective ecogeographical adaptation, genetic heterogeneity across wild and domesticated species, and the molecular components that underpin agronomically favorable characteristics. The ability to map genes to genomes provides useful information on the rice genome's micro and macro level synteny with other cereal crops. The fundamental components of the toolset necessary to breed improvised crops are such extensive molecular linkage maps. By virtue of considerable collinearity in gene order and content conservation among members of the dicot and monocot families, a large amount of information can be extended to other plant species following gene functional assignments. Using rice as a reference genome, comparative mapping of grass genomes revealed significant macro-collinearity, which was represented using a graphical consensus map called the 'Circle Diagram.'

The introduction of various omics web-based technologies was another significant development that occurred concurrently. Genome browsers and databases were created, and high-throughput analysis tools for comparative and functional genomics, transcriptomics, metabolomics, phylogenomics, interactomics, and other fields were gradually added. Microarray, RNA-Seq, promoter analysis, and fluorescent protein localization, to name a few have now enabled researchers to comprehend the temporal and spatial expression patterns of a target gene, identify other targets that co-express with the gene of interest, predict functionality based on sequence and structure, and more. Several computational systems biology research projects have been conducted to better understand the processes of salt tolerance and regulatory crosstalks in hormone production pathways for modifying responses to biotic and abiotic stress. The diversity, user accessibility, and versatility of omics-driven tools, software, and programs are gradually removing the ambiguity around the rice genome.

8.10 RICE GENOMICS

Rice genomics has advanced tremendously in the last decade, with the primary goals of high-throughput gene identification, genomic analysis for pathways and gene networks that can be linked to important agronomic and biological traits, and the discovery of novel genes through the development of new genomic tools. More than 2000 genes have been identified and cloned that have been claimed to affect important agronomic traits; the molecular biological process underlying this is also partially understood. Understanding and establishing the basis for stress-tolerant and stress-sensitive types is a difficult task that necessitates the use of a variety of methodologies. The genomic tools enable successful introgression of these alleles into the genome for continued development; the genetic repertoire enables identification of the alleles responsible for the dramatic differences in stress responses. Deeper Rooting 1 (DRO1) is a significant QTL that has been demonstrated to improve root length and has been advantageous in drought tolerance experiments. It has been demonstrated to increase grain production significantly, especially in rice varieties with shallow roots. In contrast to traditional breeding, a trait is picked based on the marker associated with it in marker-assisted breeding. Identification of QTL and fine mapping investigations can lead to the progression of regions linked with tolerance at various phases of development, including early reproductive stages. The MAS is very useful since it allows you to screen for plants with the required genetic makeup without having to go through the time-consuming and expensive procedure of field testing. This aids in the selection of high-value agronomic features. Different combinations of QTLs that correspond to changes in crop yields show a synergistic link among the QTLs.

Drought stress can be effectively combated by using transgenics to develop superior crop kinds. It entails the transfer of desired genes that encode TFs, growth regulators, antioxidants, and enzymes associated with drought stress tolerance to desired characteristics. Stress-associated protein (SAP) was isolated from *Aeluropus littoralis* and overexpressed in rice, which reduced drought-induced leaf drying and rolling while also increasing biomass. For a long time, functional genomic methods have aided in the discovery of drought stress-responsive genes. Gene transformation techniques are being used to create drought-tolerant transgenic rice. Rice has exhibited drought resistance through the isolation, characterization, and overexpression of stress-related genes. The mutagenesis process was also crucial in the development of crop types that were more resistant to salinity stress. Approximately 15 mutant cultivars with improved salt tolerance have been created. In comparison to typical salt-tolerant controls, the rice mutant Shu-92 produced in Pakistan exhibited a significant increase in yield of 40–49 percent(Haque et al., 2021).

Cuba created another rice mutant, GINES, in 2007 that showed excellent tolerance to saline soils and higher producing potential. International Rice Research Institute (IRRI) and the Bangladesh Institute of Nuclear Agriculture (BINA) have developed two salt-tolerant rice varieties, BINA dhan 8 and BINA dhan 10, which have shown significant improvements in salt stress resistance. The IAEA and IRRI collaborated on mutagenesis research to improve rice genetics, with a particular focus on enhancing salt tolerance in the face of rising salt levels in rice fields. The creation of salt-tolerant variations of the IR29 type, which is highly susceptible to salt stress, was a big breakthrough. The improved salt-tolerant mutant lines developed from IR29 will play a critical role in resolving the scourge of salinity in Southeast Asian rice cultivation. The overall plant growth and development, as well as grain quality and quantity, of wild rice species, Pokkali, have received more attention.

Abiotic stressors have resulted in a significant reduction in rice productivity, and common transgenic techniques are being used to combat abiotic stress. The identification and description of many genes that influence stress tolerance are getting easier as our understanding of the mechanisms involved in abiotic stress tolerance grows. To counteract the impacts of abiotic conditions including salinity, drought, and heat, plants regulate the expression of genes that control the synthesis of low molecular weight molecules like glycine betaine, proline, trehalose, amphoteric quaternary amines, and LEA proteins. Increased GB synthesis conferred resilience to salinity and heat stress in transgenic rice overexpressing spinach choline monooxygenase was reported. Another gene in transgenic rice that contributes to abiotic stress tolerance is OsTPS1 (trehalose-6-phosphate synthase), which leads to a significant increase in the synthesis of osmoprotectant solutes such proline and trehalose. Several researchers have proposed various underlying mechanisms that impart resistance to salt stress, with one of them being the overexpression of the sodium proton pump gene, which resulted in quick and effective sodium ion extrusion across cell membranes. Another gene, superoxide dismutase, which was obtained from fission yeast and overexpressed in rice, resulted in a significant increase in photosynthetic rate, increased grain production, and improved salt tolerance.

Overexpressed PpENA1 isolated from the moss Physcomitrella patens was found to have increased salt tolerance in genetically altered rice. Similarly, overexpression of the OsECS gene in genetically modified rice resulted in improved seed germination and seedling survival in rice exposed to abiotic stress, which was linked to improved redox homeostasis. Aside from genes, numerous microRNAs, such as miR160, have been revealed to play a key role in reducing the negative impacts of drought stress in rice(Nadarajah & Kumar, 2019). Aquaporins, members of the main intrinsic protein (MIP) family, have been found to play a role in enhancing the ability to endure salt and drought stress, in addition to transcription factors and signaling pathways. Drought reduces the overall yield of rice crops and is thus seen as the biggest impediment to achieving the targeted rice production targets. As a result, plant breeding programs must be planned with the primary goal of enhancing drought resistance in crops. Isolation of QTLs that control drought tolerance in plants could help rice seedlings survive drought stress. As a result, MAB research is being conducted in order to get a better knowledge of drought tolerance processes and to accelerate the development of drought-tolerant cultivars. For example, root length and density (RLD) play a key role in reducing the negative consequences of drought stress, and hence RLD-based indirect selection for drought tolerance is a hot topic in stress research. Selection for RLD is cumbersome, time-consuming, and arduous due to severe disadvantages in traditional breeding; however, modern breeding procedures such as MAB may prove effective in enhancing drought tolerance.

CHAPTER 9

ABIOTIC STRESS TOLERANCE

CONTENTS

9.1 Introduction........168
9.2 Messengers........168
9.3 Abiotic Stress........169
9.4 Proteins........171
9.5 Microarray........173
9.6 Climate Change........174
9.7 Tropical Forests........175
9.8 Ultra-Violet Radiation........176
9.9 Carbon Dioxide........178

9.1 INTRODUCTION

When plants are exposed to abiotic stressors, one of their main responses is to activate their defense mechanisms. Many genes are affected by cold, salt, and drought stress. Maize and soybeans are two crops that are grown in the United States. It has been hypothesized that up to 30% of the genes in abiotic stress may have an impact on the Arabidopsis genome at the transcript level. Multiple stresses, as well as the stress itself, can activate some of these genes. abscisic acid is a hormone. In general, there are more upregulated genes than downregulated genes. This is likewise true of ABA-induced gene expression. Using microarray assay with Arabidopsis seedlings, using long-oligo chips Over 2000 genes were discovered to be upregulated by more than 50% after being treated with ABA. Around 500 genes were downregulated by 2-fold(Finkelstein, 2013). Some of these stress-inducible products could have a role in stress signaling and stress tolerance. Other gene products may also protect against stress-related damage, although their mechanisms of action are unknown. Late-embryogenesis-abundant protein (LEA)-like proteins are the most common. Some stress-regulated genes, on the other hand, may not play the main function in stress response. Their introduction could simply be a result of stress injuries. Stress may regulate genes physically connected with some essential stress-induced genes in a chromatin region, even if these genes are not otherwise related. The UFC gene is one such example. Long-term cold treatment reduces the transcript level of FLC, a flowering repressor. UFC is regulated by vernalization in a similar way to FLC, however, it is unrelated to FLC in sequence or function. They're just genes on the same chromosome that are next to each other. This shows that chromosome position may have a significant impact on gene induction.

9.2 MESSENGERS

It's frequent in plant cells for receptor activation to result in the production of second messengers, which are intracellular signals that are translated from the main external signal. Other signaling components interpret these intracellular messengers further, resulting in the activation of downstream pathways with numerous outputs. Reversible protein phosphorylation is frequently involved in these mechanisms. Protein phosphorylation may cause the activation of transcription factors that induce the expression of stress-responsive genes, among other things. Additional components that recruit and build signaling complexes, target signaling molecules, and regulate their lifespan are frequently required for signal transmission. In many circumstances, the signaling pathways that may have been triggered by the same stress signals govern these components as well.

The molecules that detect the stress stimuli and subsequently convey the signal to downstream molecules to start the signal transduction pathway are known as receptors. Finding these receptors, on the other hand, is a difficult task. The nature of many abiotic stress signals is complicated. It's possible that they're not just physical or chemical signals, but a combination of them. Low temperatures, for example, can cause both osmotic and mechanistic stress. Drought, which is characterized by a decrease in soil water potential, can cause osmotic stress, ionic stress, and heat stress in some cases. Each of these stress attributes is likely to have different weights depending on the plant's condition or the severity of the stress. It's also possible that a stress signal or one of its attributes will be perceived by many cellular sensors. The redundancy of abiotic stress perception machinery, as well as the complicated nature of abiotic stress as signals, pose significant challenges in identifying cellular machinery that perceives abiotic stress. Membrane lipids may play a direct role in the stress response by altering membrane fluidity or other physicochemical properties. These lipid components also play a crucial role in the production of intracellular signaling molecules. Lipids and their biosynthesis and enzymes perform a variety of functions in plant stress signaling and tolerance, both directly and indirectly.

9.3 ABIOTIC STRESS

Plants may vary their growth and development in response to abiotic stress. Cell expansion and division may be slowed, resulting in slower growth than under normal conditions. Plant phase transitions may be influenced by longer-term abiotic stress. Drought stress might encourage flowering. Abiotic stress may modify the homeostasis of growth regulators, based on these developmental changes. Despite the fact that plant hormones are not considered secondary messengers, abiotic stress activates ABA biosynthesis; ABA mediates numerous downstream pathways; and ABA can be transported long distances and play physiological roles at locations other than where it is synthesized. It has long been known that ABA has a function in plant stress responses. To reduce transpiration water loss, ABA controls ion channels and promotes stomatal closure in guard cells. Many stress-responsive genes are activated by ABA, either alone or synergistically with stress. Ethylene biosynthesis is inhibited by ABA, and plants' sensitivity to ethylene may be reduced as a result. ABA may also regulate the production of some aquaporin genes or the activity of these water channel proteins. Because ABA is involved in these processes, plants will respond to the stress by lowering their water potential and, as a result, slowing their growth rate.

SCaBP5 or PKS3 mutations result in ABA sensitivity phenotypes in the mutants. Furthermore, PKS3 was discovered to interact with ABI2, a type 2C protein phosphatase. ABI2 may inhibit the overactivation of downstream signaling pathways by negatively regulating the signals detected by SCaBP5-PKS3. There was no detectable dephosphorylation or phosphorylation of either as a result of the interaction between PKS3 and ABI2. It's possible that ABI2 is targeting another component connected with this complex. Researchers have validated this relationship and pinpointed the SOS2 protein domain that interacts with ABI2. Through interactions with CaM-binding proteins, calmodulins have been linked to a variety of cellular activities. Various environmental stimuli, such as mechanical stress, cold, salt, or drought stress, influence the expression of many plant CaM genes. These CaMs are likely to be involved in the transmission of these external stimuli. CaM3 was proposed as a negative regulator of COR gene expression since overexpression of this CaM resulted in lower transcript levels of the stress-responsive genes RD29A and KIN1. Experimental evidence suggests that a CaM binding protein called AtCaMBP25 may operate as a negative regulator of osmotic stress tolerance, which is consistent. Plants overexpressing AtCaMBP25 in transgenic Arabidopsis are more vulnerable to osmotic stress, whereas antisense plants are more salt tolerant. The involvement of CDPKs in stress signal transduction has also been suggested. In protoplasts, constitutively active CDPKs were shown to modulate stress-responsive reporter gene expression under ABA or stress treatments, in addition to their stress-inducibility. Increased resistance to cold and salt stress was imparted by overexpression of a CDPK in rice. However, no evidence has been shown that loss of CDPK function affects stress signal transduction or stress responses.

Some targets are likely to be transcription factors, which, when activated, bind to cis-elements in the promoters of stress-responsive genes and activate their transcription. Transcription factors may be regulated by upstream transcription factors at the transcription level. These upstream transcription factors are constitutively active, but they are kept physically separated from their target genes by repressors. Thus, controlling gene expression through the regulation of these transcription factors is crucial. Conformational changes caused by protein phosphorylation, ubiquitination-mediated degradation, and trafficking between subcellular localizations are all common ways for repression. Protein kinases may play a part in all of these activities, although there are few instances in plants. Nonetheless, much has been discovered about transcription factor gene activation during abiotic stress signal transduction. Transcription, post transcription, translation, and post-translation all have the potential to regulate genes. The majority of current research on stress gene regulation focuses on transcription. Aspects of mRNA processing are crucial for stress and ABA signal transduction, according to genetic studies of ABA and stress signal transduction.

mRNA processing factors or RNA-binding proteins were discovered during explorations for components that regulate the activation of the RD29A-LUC reporter gene. SAD1 is a small ribonucleoprotein (snRNP) that is essential for mRNA splicing and export. Seed germination, and vegetative growth, sad1 mutant plants are hypersensitive to ABA and osmotic stress. Structured RNA may influence FRY2/CPL1 activities, as FRY2/CPL1 comprises two dsRNA binding domains. An RNA helicase, LOS4, was discovered to be necessary for cold acclimatization and cold-regulated gene expression. All of these investigations suggest that RNA processing may play a role in stress and ABA signal transmission. The components indicated above play a direct role in stress signaling. Protein glycosylation, methylation, sulfation, and ubiquitination are all regulated by protein modifiers, which affect protein targeting, activity, and longevity. Abiotic stress signaling is known to be influenced by some of these processes. Protein lipidation makes it easier for changed proteins to find their way to the membrane. For signal transduction involving small GTPases, prenylation, including farnesylation, is very important. Although the precise plant pathways regulated by prenylation remain unknown, it is known that prenylation is required for some of the components of ABA signaling. Plants with mutations in subunits of the proteins farnesyl transferase or geranylgeranyl transferase were hypersensitive to ABA in seed germination and stomatal regulation. Nonetheless, the era1 mutant's control of the stress-responsive genes NCED3 and ABA1 does not appear to be hypersensitive to ABA or NaCl, implying that stress and ABA activation of these genes may not require ERA1.

9.4 PROTEINS

Denatured or unfolded proteins are also removed by protein ubiquitination. Under abiotic stress, these non-natural proteins may grow prevalent. Some of these proteins could be restored to their natural state by chaperone proteins. The expression of numerous heat shock proteins is induced by abiotic stress (Hsp). However, it is unclear what role these Hsp proteins and other chaperones play in stress signaling. Some of these chaperones play a key role in the formation of hormone receptor complexes in animal cells. Despite the presence of all chaperone proteins in plants, such as Hsp90, Hsp70, immunophilins, cyclophilins, and TPR domain adaptor proteins, there are no evident nuclear hormone receptors in Arabidopsis and rice genomes that have been entirely sequenced. Environmental pressures impair crop output, and salt stress is one of the most important issues in global agriculture. Both water and ion homeostasis are affected by salt stress. Many genetic engineering approaches have been tried in the past to improve salt tolerance in plants. It has been observed that increasing the ability to synthesize osmoprotectants or removing sodium ions from the cytosol improves salt tolerance in Arabidopsis or rice. Rice, a model crop plant, is an ideal target for combining salt tolerance with genetic information from closely related salt-tolerant species like barley, wheat, or other monocotyledonous plants. Salt-inducible genes in barley leaves and roots using differential display to find viable candidates for enhancing salt tolerance in plants, and the role of several genes was identified.

Methionine synthase produces one of the necessary amino acids in all organisms, methionine. Methionine and its derivatives play key roles in the one-carbon metabolic cycle as substrates, methyl group donors, and betaine and polyamine synthesis, and these functions are critical for stress tolerance as well as normal development. Although the Arabidopsis and maize methionine synthase genes have been cloned, their function is still limited under stress conditions. Differential display under salt stress identifies barley methionine synthase (HvMS). The salt sensitivity of a yeast mutant lacking methionine synthase was restored by overexpressing HvMS to the same level as the wild-type strain. This suggested that methionine synthase in yeast and plants is functionally similar. HvMS expression is stimulated by a variety of stressors, including salt, drought, cold, ABA, and H_2O_2, but it is regulated by the circadian cycle. The amount of HvMS protein is raised by salt stress, but not in rice, which is a significant observation. Diurnal fluctuations in methionine synthase transcript concentration had little effect on protein amount in potatoes. The cDNA microarray is a sophisticated tool for comprehensive transcriptome analysis that allows you to track the expression profile of thousands of cDNAs all at once. As a result, identifying genome-wide expression, transcription factor/regulator targets, or signal transduction crosstalk is effective.

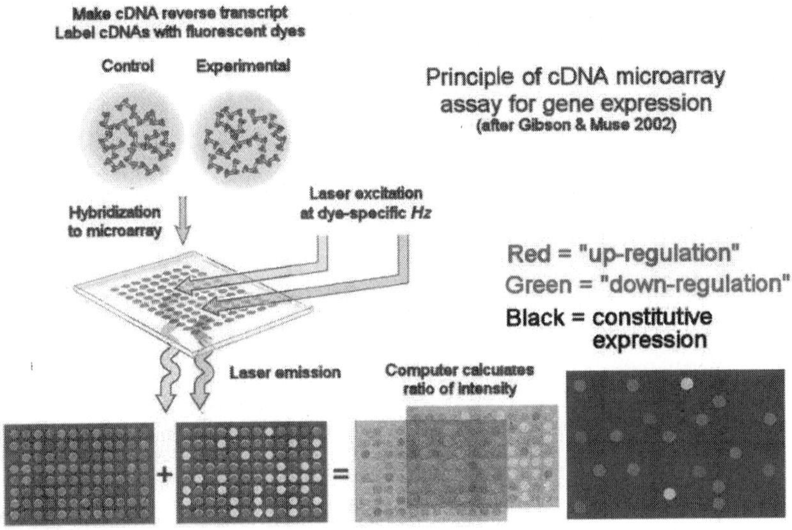

Figure 9.1. cDNA microarray.

Source:https://www.mun.ca/biology/scarr/cDNA_microarray_Principle.jpg

9.5 MICROARRAY

Many studies have used cDNA microarray (Figure 9.1) to analyze the transcriptome of model plants like Arabidopsis and rice. Other crops, such as barley, wheat, corn, and so on, have a limited bio-resource compared to the established bio-resource in model plants. As a result, reports on complete expression analysis employing such plants are currently rare. We can produce a cDNA microarray method using salt-responsive genes in barley and examine the transcriptome of the plant under salt stress. Furthermore, we used a barley cDNA microarray to examine the rice transcriptome. Barley and rice have highly conserved nucleotide sequences, allowing heterologous hybridization. At the molecular level, comparative expression analyses revealed disparities in early responses to salt stress. Under salt stress, the expression of a diverse collection of genes is regulated in barley and rice, according to comparative transcriptome analysis. This variance in sensitivity to salt stress could be attributable to evolutionary variation in the cis or trans-acting elements of such genes. Also, when exposed to salt, barley and rice have different physiological responses. During the early stages of salt stress, two distinct physiological responses were detected. Under salt stress, both plants' leaf water potential dropped. After 24 hours of salt stress, it recovered in barley but not in rice. This could be explained in part by the expression profiles of genes involved in osmotic adjustment. Water permeability across the plasma membrane would be reduced if water channel 1 and 2 genes were downregulated early in barley(Abdul Qados, 2011). Transgenic tobacco plants overexpressing a plasma membrane-localized water channel were more drought-susceptible, indicating that increased water permeability is an undesirable characteristic in a water-shortage situation. In barley, upregulation of osmoprotectant synthesis is triggered. The salt content of barley roots is higher than that of leaves. In rice, there is a contrastive pattern. Although the expression patterns of inorganic pyrophosphatase and PMP3 genes differ in barley and rice, comparative microarray research does not provide a clear frame to explain why.

9.6 CLIMATE CHANGE

Climate change is expected to occur on a worldwide scale in the twenty-first century when the mean air temperature rises due to rising carbon dioxide and other trace greenhouse-effect gases in the atmosphere. Because agriculture and forestry are businesses that rely on the natural environment for production, their productivity is especially vulnerable to climate change. In light of these circumstances, attempts are being made to find specific and practical techniques to improve plant tolerance to high-temperature environments, despite the fact that much study has been undertaken to analyze the consequences of global warming on agriculture. This efficacy is achieved without the use of bacterial or other extrinsic genes by controlling intrinsic gene expression directly. Trienoic fatty acids, in general, are the most abundant among the fatty acids found in plant membrane lipids. The concentration of trienoic fatty acids, which are polyunsaturated fatty acids with three double bonds, varies across plant species and the living environment. Through the activity of omega-3 fatty acid desaturase, trienoic fatty acids are produced from dienoic fatty acids. Due to the difficulty of characterizing the -3 fatty acid desaturase gene using conventional biochemical methods, genetic approaches, namely map-based cloning methods, were used to clone the gene using mutant Arabidopsis strains. The two types of omega-3 fatty acid desaturase genes that have been cloned so far are those found in chloroplasts (FAD7 and FAD8) and those found in the endoplasmic reticulum (FAD9 and FAD10) (FAD3). The FAD8 gene's expression is temperature-dependent, and it can be turned on and off by a few degrees Celsius, close to 25 degrees Celsius.

Nitrosative stress at an infection site probably contributes to the inhibition of potential invaders by combining antimicrobial activities that are activated in part by NO, such as phytoalexin release, crosslinking reactions, host cell death, and oxylipin formation. Soybean cell suspensions accumulate 10 M H_2O_2 and 2M NO within hours of exposure to an avirulent *Pseudomonas syringae* strain, values that are significantly higher than those caused by a virulent strain. NO concentrations of 2M activate the flavohemoglobin NO dioxygenase (NOD) in developing E. coli cells and can kill late stationary phase cells in rest. Under aerobic conditions, NOD expression reduces cellular NO toxicity by preventing NO from inactivating aconitase(Arai, Hayashi, Kuroi, Ishii, & Igarashi, 2005). *Erwinia chrysanthemi* strains with mutations in the HmpX flavohemoglobin gene are significantly less capable of causing sickness, suggesting that bacterial methods for removing NO could be phyto pathogenicity factors. Experiments in Arabidopsis and avirulent *P. syringae* using transgenic expression of bacterial Hmp have produced evidence that validates and further specifies the role of NO in starting plant responses to infection. Due to enhanced H_2O_2-degrading activity, the presence of Hmp, whether produced by a bacterium or a plant, produces a considerable reduction in tissue H_2O_2 build-up. After being exposed to avirulent *P. syringae*, Hmp plants developed hypersensitivity symptoms, which were exacerbated when the infecting bacterium also produced Hmp.

9.7 TROPICAL FORESTS

Tropical forests, despite accounting for over half of world production and a large proportion of total tree species, have gotten slight attention when it comes to the ozone depletion problem. Although there has been little or no ozone degradation in the tropics thus far, even a minor decline in ozone at these latitudes will result in a significant increase in UV-B radiation because solar UV-B radiation is already highly intense in these areas. Blocking existing solar UV-B light with filters can boost the growth of four broadleaf species. Global climate change has shifted the perception of increased UV-B radiation, yet the problem has not gone away. Even if ozone-depleting emissions were instantly ceased, the harmful substances currently in the stratosphere would take time to degrade. According to studies on agricultural species, roughly 60% are moderately vulnerable to high amounts of UV-B radiation. To some extent, increased carbon dioxide and global warming are thought to provide a barrier against UV-B damage. Increased carbon dioxide can help plants develop faster, but when combined with UV-B radiation's negative effect on photosynthetic enzymes in C_3 and C_4 plants, the outcome is uncertain. Physiologists are concerned about how increased UV-B radiation would affect tree carbon allocations. They may have to divert a large portion of their photosynthetic output into defensive systems, sacrificing growth. Growth is regulated by the content and activity of hormones and inhibitors, which utilize photosynthetic assimilation. In most situations, treating plants with UV causes the hormone inhibitor balance to move toward inhibitor dominance. There could be more severe direct repercussions as well, but given the importance of trees in controlling atmospheric carbon, even little changes could have a significant impact on climate change.

9.8 ULTRA-VIOLET RADIATION

UV-B (Figure 9.2) has an impact on trees through altering their biological and chemical environments. The nucleic acids are the most important, followed by proteins and other compounds to a lesser extent. Direct destruction of the genetic material - DNA, deactivation of enzymes, rupture of membranes and other cell structures, and the formation of extremely reactive chemical agents known as free radicals are all examples of damage. Elevated UV exposure can harm photosynthetic machinery in a number of ways, including pigment bleaching, loss of electron transport efficiency in cell division and growth regulatory mechanisms, and the composition and replication of genetic material. Reduced growth yield, variations in the amounts and effects of plant hormones, and changes in dormancy, flowering, and other times are among the consequences.

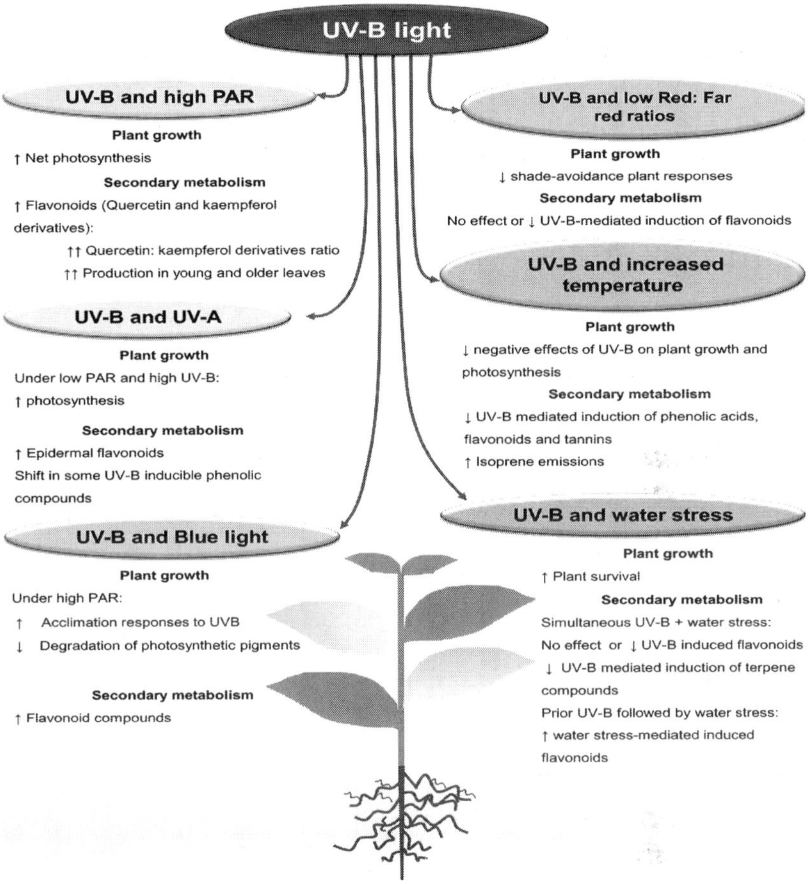

Figure 9.2. Effect of UV-B.

Source:https://www.frontiersin.org/files/Articles/246867/fpls-08-00278-HTML/image_m/fpls-08-00278-g001.jpg

Furthermore, strong UV exposures may inhibit the healing systems themselves. All of these mechanisms can interact to cause a range of negative consequences on plants. Many of the effects are minor, may interact with other factors, and are thus difficult to explain to UV alone. The ability of a tree to withstand the impacts of UV radiation is based on its ability to limit exposure through growth pattern optimization and the development of UV absorbing pigments as well as repair or replace damaged molecules. The quantity and quality of ambient light influence the expression of various

UV resistance systems. The effects of UV-B on plant height and rosette diameter growth are most pronounced in plants lacking the cyclobutane dimer photolyase enzyme, showing that the enzyme plays a significant role in UV resistance. A mutant deficient in the non-specific repair pathway nucleotide excision repair (NER) showed a significant, though reduced, sensitivity to solar UV-B, implying that either the transcription-coupled repair of UV-induced dimers, or the repair of another UV-induced lesion was also important for optimal growth. In nature, plants and other organisms are rarely influenced by a single stressor, such as UV-B radiation. Water stress, increased atmospheric CO_2, mineral nutrient availability, heavy metals, tropospheric air pollution, and temperature are all examples of variables that operate in tandem. As a result, it's crucial to remember that such conditions can considerably boost or decrease the efficiency of UV-B radiation. Because visible radiation is a significant ameliorating component, laboratory investigations should be used as close to natural levels as feasible to achieve realistic results.

9.9 CARBON DIOXIDE

CO_2 increases in the atmosphere are a certain component of global climate change, and CO_2 concentrations may get doubled soon. In semi-natural habitats, however, when nutrient or water availability may severely limit plant development, CO_2 responses are minimal. Furthermore, the majority of the ecosystem-level effects of increased CO_2 are mediated by changes in plant tissues. Plant growth responses to variations in UV-B radiation and atmospheric CO_2 concentration are assumed to be in opposite directions when evaluated separately. With a few exceptions, most tests involving increased CO_2 and UV-B light do not generate interactions between these components. For some species, increased CO_2 appears to provide some protection against increased UV-B radiation; nevertheless, increased UV-B radiation can impair their capacity to take use of increased CO_2 in photosynthesis. The allocation of biomass in plants can also shift in a sophisticated way when CO_2 and UV-B radiation treatments are used together. Temperature rise is also expected as a result of global climate change. The results of a study that combined two levels of UV-B radiation with two levels of CO_2 and two temperatures revealed that either increased CO_2 or a little higher temperature had equivalent benefits in minimizing the growth-inhibiting effects of sharp UV-B radiation.

Changes in plant shape are another route by which increasing UV-B might alter the competitive balance of plant species. Even if increased UV-B has no effect on plant productivity, changes in plant morphology can lead to changes in which species can compete for sunlight more successfully. A quantitative investigation of competition for sunlight in mixed stands with and without additional UV-B revealed that minor differences in plant morphology between the two species were enough to shift the balance of competition for photosynthesis-critical sunlight. As a result, one species can gain an advantage over another by capturing more sunlight for photosynthesis. Wheat benefited from higher UV-B in these studies, while weeds suffered. However, the condition could be reversed in different crop-weed mixes. Other changes in plant morphology, such as a larger allocation of biomass to roots, may alter specific species' competitive efficacy for soil moisture and nutrients. Similar shifts in species composition may occur in grasslands and forests that are not extensively maintained.

Plants and bacteria are used in phytoremediation to extract, sequester, and detoxify pollutants from soil and water. Because of their essentiality or chemical resemblance to an essential element, pollutants, particularly trace elements, are frequently carried into plant roots. The pollutants are transferred from the root to the shoot, where they can be extracted in the above-ground biomass, in phytoextraction. A subset of pollutants, notably selenium and mercury, can be converted to volatile forms in plants. This process, known as phytovolatilization, is particularly interesting since it has the ability to eliminate all contaminants from the surrounding ecosystem. Tolerating high amounts of trace metal pollutants in soil and water, plants have evolved three broad responses. The first is to stay away from metals entirely. This can be performed by releasing organic acid chelates through the roots or modifying the pH in the rhizosphere to convert trace metals to a less accessible form. Tolerance to high soil aluminum concentrations, for example, is linked to root exudation of organic acids in various plant species. The overproduction of chelating chemicals is a second approach for trapping toxicants in the roots. Because metals that exert their toxicity in the shoots are prevented from being delivered to the shoots, this technique is highly successful. Certain plants tolerate metals by detoxifying them to less hazardous forms, either by chelation, sequestration, or metabolism. Because the metals are collected in the easily harvestable plant shoots, only this technique is useful for phytoextraction.

Despite having a higher selenium content, the transgenic plants were just as tolerant of soil and field conditions as the natural variety. The GS transgenic line proved very resistant to contamination in the soil. The biomass of GS plants grown on the sediment-soil combination was 80 percent that of GS plants grown on control soil, showing that these plants were more tolerant of high levels of soil Se and salt, as well as other adverse growth conditions like heat or drought. In response to several and seemingly dissimilar stimuli, such as wounding, pathogen attack, drought, heat, and metal stress, elevated levels of metallothioneins, superoxide dismutases, other reducing agents, such as ascorbate, and organic acids, such as malate and citrate, have been found. This suggests that overexpression of certain oxidative stress response genes could provide plants used in phytoremediation with a broad range of protection against primary and secondary stresses.

Other technical issues confront researchers working on genetically altering plants for phytoremediation. The development of plants with truly superior levels of pollutant accumulation while maintaining near-normal biomass is perhaps the field's most pressing task. Transgenic plants, on average, have a several-fold higher accumulation of pollutants than wild-type plants. It is widely assumed that phytoremediation techniques will only be used when the accumulation exceeds 100 or 1000 times that of wild type. It's still unclear whether such concentrations are even possible(A. Yan et al., 2020). It is evident that we have a lot to learn if we are to break through this barrier. We can predict the impacts of gene overexpression in particular, as there are already multiple cases of oxidative response gene overexpression that did not result in plants with higher phytoremediation potential. Overexpression of the same gene in various species can have distinct effects, such as increasing tolerance in one species while increasing sensitivity in another. More sophisticated modeling tools, as well as the collection and analysis of genome, transcriptome, proteome, and metabolome data, could improve our capacity to forecast the influence of transgenes on contaminant absorption and tolerance. Betaine, as an osmolyte, protects cells, proteins, and enzymes from environmental stress. Betaine participates in the methionine cycle largely in the human liver and kidneys as a methyl donor.

Hypomethylation is caused by insufficient dietary intake of methyl groups in numerous key pathways, including altered hepatic methionine metabolism with higher plasma homocysteine and decreased S-AdoHcy concentrations, and inadequate hepatic fat metabolism with fatty build-up. This change in liver metabolism could lead to a variety of disorders, such as coronary, cerebral, hepatic, and vascular ones.

Rice is a crop that is known to be salt sensitive. As a result, many studies have been published on improving salt tolerance in rice using genetic engineering of only one gene in the majority of cases. If hydrogen peroxide acts as a signal molecule in the expression of genes involved in abiotic stress response or tolerance in rice, it may be able to boost salt tolerance by simply spraying the plants with the gas. Not only did the treatment increase the activity of active oxygen scavenging enzymes, but it also increased the expression of stress-related genes such as sucrose-phosphate synthase, pyrroline- 5-carboxylate synthase, and small heat shock protein 26. Hydrogen peroxide also enhanced the expression of many other genes and proteins. These findings imply that hydrogen peroxide, when given topically, can improve salt tolerance in rice seedlings by functioning as a signal molecule for the abiotic stress response. Furthermore, we discovered that the abiotic nitric oxide donor sodium nitroprusside improved rice seedling salt tolerance. Both hydrogen peroxide and nitric oxide improved rice plant heat tolerance. However, these effects only occurred at a certain dosage; excessive doses of hydrogen peroxide or sodium nitroprusside reduced rice seedlings' salt- or heat-stress tolerance as compared to untreated controls.

The knowledge of Na^+/H^+ antiporters has grown as the genome sequencing of higher plants has progressed. The Na^+/H^+ antiporters from barley, cotton, maize, and rice have been characterized in addition to the model plant Arabidopsis. The rice Na^+/H^+ antiporter OsNHX1 was overexpressed, which improved the salt tolerance of transgenic rice cells and plants. Numerous studies have concluded that these genes are crucial in the extrusion of Na^+ across the plasma membrane, the sequestration of Na^+ in the tonoplast, and the recirculation of Na^+ from shoot to root. Many Na^+/H^+ antiporters, on the other hand, have uncertain roles. The regulation of the expression of several Na^+/H^+ antiporters is poorly understood. Another crucial feature of salt tolerance is K^+ homeostasis. Salt stress affects the transcript levels of numerous K^+ transporter genes, which likely reflects the varying capacity of plants to maintain K^+ uptake under salt stress. Future molecular dissection of salt-tolerance processes in essential crop plants will be aided by the discovery of additional stress-associated gene resources from both crop plants and extremely salt and drought-tolerant model plants.

They accomplish this via controlling the expression and activity of K^+, Na^+, and H^+ transporters, as well as H^+ pumps. In many circumstances, H+ pumps provide the driving power for K^+ and Na^+ transport. Our understanding of K+ transporters has substantially increased in recent years as a result of molecular techniques combined with electrophysiological and structural research. The discovery of a large number of genes encoding K^+ transport systems reveals a high level of complexity. Functions including root K^+ uptake and secretion into the xylem sap, K^+ transport in guard cells, and K^+ inflow into developing pollen tubes are all being studied at the molecular level. At the molecular structural level, it appears to be more difficult to distinguish between the K^+ channel and K^+ pump.

CHAPTER 10

NANOTECHNOLOGY APPLICATIONS IN DROUGHT STRESS RESEARCH

CONTENTS

10.1 Introduction........184
10.2 Nanoclays........185
10.3 Food Packaging........186
10.4 Nanoparticles........187
10.5 Fertilizers........188
10.6 Nanofabrication........189
10.7 Nano-Biosensors........191
10.8 Advances in Nanotechnology........193
10.9 Applications of Nanotechnology........195

10.1 INTRODUCTION

Crop nutrition is now achieved through the addition of macronutrients and micronutrients to the field. NPK fertilizers are the most extensively utilized since they are directly linked to enhanced crop productivity. Excess fertilizer application causes soil physiology problems, resulting in altered soil dynamics, lower soil fertility, flora, and a significant increase in soil erosion; eventually, excess fertilizer flow into water resources causes eutrophication, posing substantial hazards to aquatic species. Furthermore, nutrient utilization efficiency is a critical aspect that refers to a plant's ability to obtain nutrients from the soil; it is estimated that only about half of standard chemical fertilizers are used by plants, with the residue converted into various insoluble forms. According to recent world population estimates, ensuring food security with present agricultural production systems is unfeasible. Extensive use of agrochemicals for maximum crop output and crop protection has a negative impact on the environment, resulting in a variety of health problems, some of which are even fatal to animals, including humans. Furthermore, current agrochemical-based farming has major losses in soil fertility and eutrophication. A novel technique that can help with increased productivity and crop protection is urgently needed. The key novel technologies that have been identified as having the potential in resolving these key restrictions are nanotechnology and biotechnology. Plant genetic engineering, nanofertilizers, and nanopesticides, for example, have a significant impact on agricultural systems and could significantly contribute to sustainable agriculture.

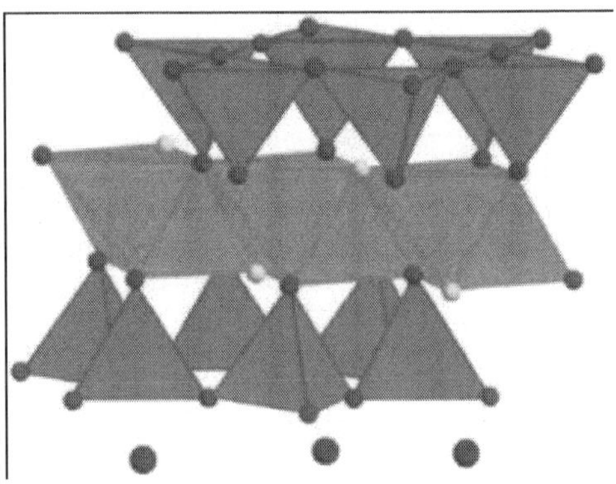

Figure 10.1. Nanoclays.

Source:https://www.google.com/url?sa=i&url=https%3A%2F%2Fwww.sciencedirect.com%2Ftopics%2Fearth-and-planetary-sciences%2Fnanocla

y&psig=AOvVaw2HPzXgYs2d--xOpPZzDUCC&ust=1643545173056000&source=images&cd=vfe&ved=0CAsQjRxqFwoTCPiR0rL51vUCFQA-AAAAdAAAAABBJ

10.2 NANOCLAYS

Nanoclays (Figure 10.1) are layered silicates with nanoscale dimensions that are divided into anionic and cationic nanoclays. Because of two factors, these nanomaterials are attractive candidates for carrying macro and micronutrients("Desert Control," 2021). First, their unique structural arrangement protects nutrients, and second, the formation of weak connections with nutrients allows for controlled and prolonged release. Some of the potential nanoclays employed as nutrient transporters are zeolites and kaolinite. In contrast to other nanomaterials, studies suggest that chitosan could be used as a nanocarrier. Chitosan is polymeric, biodegradable, and has better adsorption and bioactivities than other types of chitins. NPK fertilizer and plant growth hormones have been slowly released using chitosan nanoparticles. Controlled release of plant growth regulators and fertilizers has also been investigated using nanocomposites of various metals; more recently, potassium-controlled release has been claimed using very porous nanocomposites.

Foliar spraying, seed treatment, and direct soil application are all options for using nanofertilizers. Liquid-based formulation has been described in a number of investigations. Furthermore, nanofertilizers have been employed in a variety of farming systems, including hydroponics, aeroponics, and irrigation. However, certain severe issues have been raised in relation to their advantages. Recent reports reveal that high levels of exposure to these nanomaterials have negative effects on plant cells, including cell components, resulting in cytotoxicity and genotoxicity; nevertheless, other research shows that low levels of exposure do not cause toxicity. However, findings imply that larger nanomaterial concentrations can affect aggregation rates and cause instability. Furthermore, studies show that the surface charge of a substance is crucial in determining its hazardous effect. It has been suggested that the nanoparticle's positive charge is harmful. As a result, these considerations should be made prior to the use of nanofertilizers in agricultural production systems. Poor water solubility and oral bioavailability are major obstacles for these. To overcome these issues, new polyphenol nanocapsules have been designed. Dibenzoyl methane, a diketone derivative of curcumin, and curcumin nutritional supplement from the turmeric plant, for example, are both packaged in a homogenized oil-in-water emulsion capsule comprised of triacylglycerols and Tween 20 emulsifier. These are called curcumin nanocapsules, and they've been studied for their anti-inflammatory properties. To boost bioactivity, solubility, and nutritional absorption from food, nanoemulsions combining water and oil, solid-lipid nanoemulsions, and biopolymer nanoemulsions are being formed. In the structure-based design of tools to detect biological and chemical pollutants in water and food, nanocantilevers are utilized. These are gaining traction as potential instruments for improving food safety and quality.

10.3 FOOD PACKAGING

Food packaging makes use of bio nanocomposites. These have the ability to safeguard food and extend its shelf life. These help to safeguard the environment by reducing the use of plastic packaging materials. In the manufacturing of biodegradable plastic products, zein nanoparticles, also known as zein prolamin protein nanoparticles, are employed. In food processing, zein nanoparticles are used as edible carriers of flavor chemicals. Nanosensors are made up of thousands of protein-based nanoparticles that are placed in an array on a platform. When these nanoparticles come into contact with pathogens, they glow. Food pathogens are detected using these nanosensors. Food pathogens and food deterioration are detected using these nanosensors in food packaging. In the food processing industry, nanoemulsions, nano-bilayer emulsions, and surfactant micelles are employed as nanotextured nutrient and drug delivery systems. In the making of ice cream, nanostructured mayonnaise is utilized as a substitute for full fat creamy ingredients. Nanocochleates are soya bean-derived nanoparticles with a size of 50 nanometers. These are phosphatidylserine carrier systems that are used to keep antioxidants and micronutrients from degrading during the manufacturing and storage of food. Milk protein nanotubes and nano encapsulations are used as nutrient nanocarriers in drinks and foods. To improve gut function, live probiotic microbes are employed in food nanoencapsulation. This nanoencapsulation method is also utilized to make tasteless, colorless beverages that include nanoencapsulated nutrients. To treat selenium deficiency, this drink increases selenium bioavailability.

Typically, electrons are received by oxygen in the MFC cathode region. The oxygen reduction can be poor in some situations, affecting the operation of microbial fuel cells. Inside the cathode, some microorganisms aid in the catalysis of oxidation-reduction reactions. Because of their form and size,

CNT can maintain a good contact with the redox proteins of such bacteria, facilitating the oxygen reduction reaction and electron transfer. Apart from that, CNT helps other oxygen-reduction process electro catalysts, such as manganese dioxide. TiO_2, a titanium-based nanomaterial with gas sensing, photocatalytic, and semiconducting capabilities, is a type of titanium-based nanomaterial. They are employed in the elimination of organic contaminants due to their semiconducting properties. TiO_2 is used as a photocatalyst in wastewater treatment. The photodegradation of azodyes is achieved by combining TiO_2 nanoparticles with silver nanoparticles. Metals like Au and Pt are also utilized to improve photodegradation. The elimination of benzene is effective using a mixture of TiO_2 nanoparticles and carbonized bamboo powder. Butachlor is a soil and water contaminant, and titanium dioxide in its immobilized form is used to remove it.

10.4 NANOPARTICLES

Bimetallic nanoparticles are another form of nanoparticle that may be used to remove contaminants in a variety of ways. Pd/Au is a nanomaterial that can be used to reduce chlorinated contaminants in water. Lead or ferric nanoparticles are used to remove trichloroethane and tetrachloroethane, whereas nickel nanoparticles are used to remove carbon tetrachloride and trichloroethane, and copper nanoparticles are widely used to remove nitrate(Guerra, Attia, Whitehead, & Alexis, 2018). Degrading organic pollutants with nanoparticles generated from microbes such as fungi, bacteria, algae, and actinomycetes is also a part of nanotechnology. Certain Clostridia bacteria convert Palladium ions to generate Pd metallic nanoparticles, which can be utilized to clean wastewater and control groundwater pollution. Nanotechnology can also be used to reduce greenhouse gas emissions. Carbon dioxide may be extracted from flue gases produced after combustion using various membranes, including polymeric membranes that are thermally rearranged with CO_2 selective nanocavities. These polymeric membranes have a high gas permeability and good selectivity.

The most common methods of applying conventional fertilizers are spraying and broadcasting. The technique of application of conventional fertilizers is mostly determined by the actual final fertilizer concentration in the plants. The nutrients provided by conventional fertilizers are in a chemical form that is inaccessible to plants; additionally, leaching, drift, chemical runoff, hydrolysis, and evaporation by moisture in the soil, as well as microbial destruction. It is estimated that 40–70 percent of nitrogen,

80–90 percent of phosphorus, and 50–90 percent of potassium in applied fertilizers are lost in the environment before reaching the plants. All crops will see a decrease in production if they are not given the right fertilizer doses. Excessive fertilizer use can pollute groundwater, produce eutrophication in aquatic ecosystems, and pollute the air. As a result, creating smart approaches, such as nanofertilizers, is a superior choice for optimizing the usage of chemical fertilizers. Nutrients are contained within nanomaterials or protected with nanofilm and distributed as emulsions in nanofertilizers. Iron is an essential metal that serves as a cofactor for enzymes involved in a variety of physiological processes in plants. Soil contains a lot of iron; iron compounds account for around 5% of the earth's crust weight. Although present in vast proportions, the majority of the iron form is insoluble, rendering it unavailable to plants. Another benefit of employing iron-based nanofertilizers is that they contain no ethylene-based compounds, which prevents plants from aging and senescence prematurely. The application of iron nanofertilizers raises the ratio of ferrous to ferric iron in plants, which boosts chlorophyll synthesis.

10.5 FERTILIZERS

Fertilizer plays a critical role in agricultural production, accounting for 35–40% of all crop productivity. The practice of unbalanced fertilizing pollutes groundwater. Fertilizers are applied to the soil or to the leaves in a variety of ways, and these inorganic fertilizers are used to deliver the three primary components of nitrogen, phosphorus, and potassium. N, P, and K fertilizer use efficiency remained stable at 30–35 percent, 18–20 percent, and 35–40 percent, respectively. Eutrophication is caused by a large fraction of applied fertilizers remaining in the soil or entering the aquatic system. The way fertilizers are applied has an impact on their efficiency and environmental impact. The use of the right application method and the right amount of fertilizer increased nutrient uptake while lowering pollution. The kind of fertilizer, soil type, irrigation method, crop, and nutrient applied all influence fertilizer application methods. In agriculture, fertilizers are utilized to provide full-fledged macro and micronutrients that are deficient in the soil. Fertilizers have a direct impact on plant growth, accounting for 35–40% of crop productivity.

10.6 NANOFABRICATION

Nanofabrication is the design and construction of devices with dimensions measured in nanometers. Because it is such a dynamic subject, numerous new classes of materials with cutting-edge manufacturing techniques are predicted to emerge in the near future. Carbon-based materials, metal-based materials, dendrimers, and composites are the four types of engineered nanomaterials now available("New Technology Could Fight off Droughts," n.d.). Nanofabrication procedures are difficult to generalize with precision because they are produced using methods tailored to the needs of the materials and, in many cases, are protected by intellectual property rights. Traditionally, nanofabrication has been accomplished by scaling down integrated circuit fabrication, which involves removing one atom at a time to achieve the required structure, or by a more complex hypothetical scheme involving the atom-by-atom assembly of a device.

Nanotechnology is currently taught at both the undergraduate and doctoral levels in a number of engineering and conventional universities. Industry and industry-oriented institutions are served by their curricula and degree programs. Engineering and traditional universities' nanotechnology education programs do not prepare students to deal with difficulties that are vital to agriculture. For example, the intricate relationships that exist among the components of life, as well as the impact of nanomaterials on food chains, food webs, and farm wastes, are not adequately covered in technical institution courses. To successfully service nanotechnology applications in agriculture, it is critical to develop human resources who understand the complexity of the agricultural production system. Agricultural education has generally failed to attract a significant number of creative brains from throughout the world, and individuals from similar disciplines may lack a grasp of agricultural production systems. If started, agricultural nanotechnology instruction programs might fill this need by recruiting exceptional learners and producing a body of qualified farm-focused personnel.

Many scientists have been drawn to nanotechnology because of its distinct physical, chemical, and biological properties that differ from those of a large-scale model of the same substance. Nanomaterials have been produced for use in a variety of disciplines, including medical, drug delivery, electronics, fuel cells, solar cells, food, space, and more. Among these uses, nanoparticles have demonstrated numerous advantages in the remediation of various soil contaminants. In a variety of methods, nanomaterials can aid in the identification and remediation of soil contaminants. Soil contaminants are precipitated by nanomaterials, which also work as a solid waste stabilizer and limit soil erosion. The use of genetic engineering to manipulate genes in plants has become a powerful tool in modern molecular breeding. New advancements in plant transformation methods have been made in the recent decades. Agrobacterium and biolistic mediated transformation, in addition to a range of gene delivery techniques, has shown substantial results. Under climate change and an increasing global population, food productivity has grown through the genetic engineering of plants by transforming desired genetic features in agricultural crops. Plant genetic transformation remains hampered by the cell wall as a barrier to external biomolecular delivery. As a result of their ability to pass this barrier without the use of external force, nanoparticles are promising materials for the delivery of biomolecules. As a result, nanoparticles may be used to transfer biomolecules to plants via genetic engineering. The indiscriminate use of pesticides and fertilizers pollutes the environment and threatens biodiversity. In this setting, nanotechnology has a promising future in agrobiotechnology applications, where nanomaterials can be used to solve these difficulties.

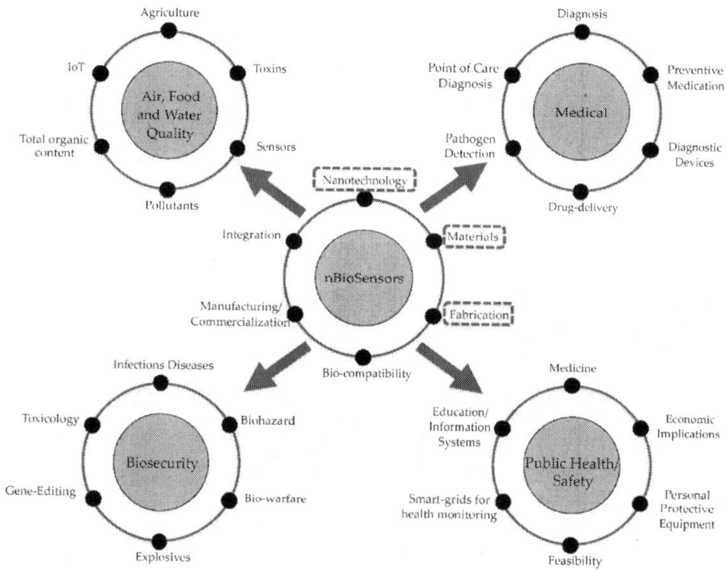

Figure 10.2. Nano-biosensors.

Source:https://www.mdpi.com/sensors/sensors-21-01253/article_deploy/html/images/sensors-21-01253-g002.png

around the world. Several studies are underway to document, analyze, and assess the potential difficulties and benefits, hazards, and known and unknown concerns associated with advances in nanoscience and technology. However, as the technology advances, health, environmental, and worker health risk studies, as well as a regulatory framework for nanotechnology and nanoscience, are some of the concerns that will need to be investigated further. Lack of regulatory harmonization also hinders the establishment of worldwide strategies for commercializing nanotechnology-derived products.

10.9 APPLICATIONS OF NANOTECHNOLOGY

Nanotechnology applications in agriculture are still in their infancy and are purely theoretical, but they have the potential to transform agriculture and food systems in the next few years. This novel scientific approach has the potential to improve agricultural productivity by involving genetic improvement of plants, delivery of genes and drug molecules to specific sites at the cellular level, nano-array based gene-technologies for gene expressions in plants and animals under stress conditions, protect the seed from diseases and pathogens, separate the unviable and infected seed, and maintain the genetics of the seed. In the food industry, small particle nanotechnology can be used to fortify foods with nutrients, antioxidants, vitamins, and nutritional supplements, as well as identify and detect the presence of any number and type of germs and pathogens quickly and accurately to keep food fresh for longer. Clay minerals, especially zeolite-based nanofertilizers, are becoming increasingly popular among researchers and the fertilizer industry. Zeolites are hydrated alkali and alkaline earth aluminosilicates with a three-dimensional structure furrowed by an inner network of holes and channels. Zeolites exhibit high cation exchange capacity and have been employed in a variety of applications as low-cost cation exchangers. This negative-charged natural super porous material can be used to retain positive cations like ammonium and potassium for slow and constant nutrient release throughout crop nutritional requirements. The surfactant hexadecyltrimethylammonium bromide (HDTMABr) can be used to modify the surface charge of zeolite to ensure the trapping of negatively charged components such as sulfate and phosphate. Slow-release fertilizers are a great alternative to soluble fertilizers since they release their nutrients gradually to match the plant's nutrient requirements, reducing application rate and frequency.

CHAPTER 11

ALTERNATIVE FARMING

CONTENTS

11.1 Introduction 198
11.2 Bacillus Thuringiensis 199
11.3 Dna Based Markers 199
11.4 RFLP 201
11.5 Blast Disease 202
11.6 RAPD 203
11.7 Phytoremediation 203
11.8 Agrochemical Contamination 204
11.9 DDT 205
11.10 Contaminated Soils 207
11.11 Pesticides 207
11.12 Phytoaccumulation 209
11.13 Phytoremediation 209
11.14 Wetlands 210
11.15 Soil Erosion 214
11.16 Environmental Issues 216
11.17 Ecological Fertilization 217

11.1 INTRODUCTION

The systematic design approach necessitates a closer examination of our current industrial model and its separation from a consumerist vision that focuses just on the product. Then we can offer a new paradigm that places humans at the center of an ecological context that understands social and natural structures' interconnection. In a sustainable production system, the indicators of sustainability are the complete lifecycle of the product and how it interacts with the context in which it is located, rather than economic growth, development, or competitive advantage. Our generation's objectives, as well as the opportunities and possibilities for future generations, are all dependent on these sustainability considerations. Sustainable production must correlate to sustainable consumption and pursue this correlation while creating products and services. The Systematic Design methodology can help us lessen the environmental impact of human activity: in fact, it can turn a cost into a benefit, and a waste product into a resource (Conrad, 2013).

To tackle these insects, many traditional pest management strategies have been devised, with various degrees of success. The commercial release and broad adoption of Bt transgenic hybrids in the mid-1990s virtually changed conventional maize pest management operations. The European and southwestern corn borers were the first insects to be targeted by Bt corn, which includes insecticidal protein producing genes from *Bacillus thuringiensis* (Bt). Bt hybrids for western and northern corn rootworms were released just a few years after Bt hybrids for stalk boring insects were introduced. Bt corn hybrids, on the other hand, have been the subject of intense study since their introduction. They've been shown to produce larger yields while also exposing humans, non-target organisms, and the environment to less pesticides. However, concerns have been raised about food-chain contamination, resistance development, the technology's overall viability, and, more lately, the high costs of Bt hybrids compared to non-Bt hybrids. Insect pest management in corn has traditionally relied on cultural and traditional chemical control methods. With the availability of *Bacillus thuringiensis*, transgenic corn hybrids over the last decade, pest management methods for economically significant insects have changed dramatically (Bt hybrids). Pest management efforts aimed at economically important insects have been reduced to merely planting Bt hybrids in some cases. However, some growers prefer not to plant Bt hybrids in places where pest pressures have been minimal. Crop rotation, administration of conventional insecticides, and asynchrony between crop vulnerability and pest infestation are among the methods used by growers.

11.2 BACILLUS THURINGIENSIS

Bacillus thuringiensis is a rod-shaped, gram-positive, spore-forming bacteria found in a variety of environments around the world. During sporulation, *B. thuringiensis* produces a proteinaceous parasporal crystalline inclusion body within the bacteria. The crystalline inclusion body contains one to many d-endotoxins that cause death in some insect species while being innocuous to humans and most non-target insects. Each holotype sequence is given a name based on the degree of evolutionary divergence as determined by phylogenetic tree methods. In diverse strains of *B. thuringiensis*, 204 holotype sequences for insecticidal proteins have been discovered. Cry endotoxins are currently categorized into 60 families based on holotype sequences, while Cyt endotoxins are grouped into two families. To prevent rootworm larvae and other soil pests, the corn rootworm refuge can be treated using soil-applied or seed-applied insecticides(Kathage & Qaim, 2012).

11.3 DNA BASED MARKERS

The use of DNA-based marker systems has the potential to boost crop output around the world. In phylogenetic investigations, genetic mapping, and comparative genomic analysis, a variety of DNA fingerprinting techniques, as well as more recently whole genome sequence information, have been extensively used to employ intrinsic genetic polymorphisms in the genomes of higher plants. DNA markers pave the way for genomic-based breeding strategies, which have significant advantages over phenotypic-based selection procedures in traditional breeding systems. Marker assisted introgression of novel genomic areas linked with anthesis-silking interval, marker-based diagnosis of plants carrying the opaque2 gene associated with quality, and marker-based prediction of hybrid vigor are examples of successful maize applications. To give resistance to both biotic and abiotic stressors, such as bacterial blight and blast, and to improve yield and quality, new rice varieties are produced utilizing DNA markers associated with genes and quantitative trait loci. The wheat variety 'Patwin' was produced using marker assisted selection for the resistance genes Yr17 and Lr37, which protect against stripe and leaf rust, respectively. The stay-green trait in sorghum that confers drought resistance has been extensively studied. Many genes conferring resistance to various biotic stresses have been incorporated from wild relatives into tomato, cotton, potato, soybean, and other crops using DNA markers. For many plants, the restricted genetic foundation of elite gene pools limits the scope of marker assisted breeding. Multiple studies have revealed that QTLs are conserved among various crop species, indicating that information from one crop can be used to improve others.

Recent advances in DNA marker methods have added a new dimension to traditional plant breeding by establishing associations between characteristics and genomic loci, paving the way for the rapid evolution of novel varieties. Polygenic features that were previously difficult to identify using traditional plant breeding approaches can now be easily identified by identifying tightly connected DNA markers. Marker-assisted selection is an indirect selection method for a trait that involves identifying the desired plants using a closely linked DNA marker. DNA markers facilitate not only the easy and reliable identification of breeding lines, hybrids, and cultivars, but also the monitoring of introgression from wild to elite cultivars, genetic diversity and relatedness assessment, gene pyramiding, genetic mapping, quantitative trait loci analysis, and MAS. DNA markers are useful for detecting and resolving problems like linkage drag, recombination suppression, and segregation distortion, which makes them essential for crop improvement.

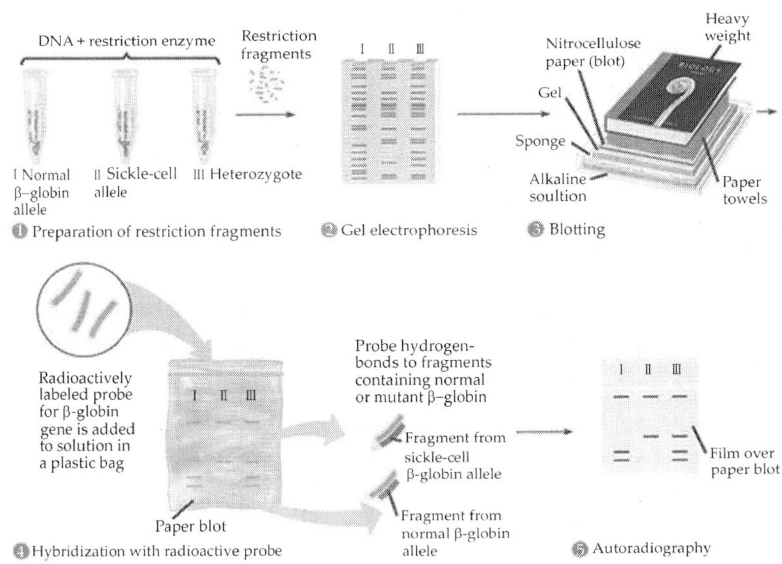

Figure 11.1. RFLP.

Source:https://lh6.ggpht.com/-qWr8WXDnwHI/USzrvwFSOcI/AAAAAAAAB-ZA/pkUZ9iYwg98/s1600/southern%25255B7%25255D.jpg

11.4 RFLP

Restriction fragment length polymorphism (Figure 11.1) markers are useful in linkage analysis and crop breeding, but they are time-consuming, expensive, and necessitate a large amount of DNA for restriction and hybridization analysis. RAPD analysis is much faster and less expensive than RFLP analysis, and it only requires a small amount of DNA. However, it is genotype-specific and difficult to replicate in different labs. Microsatellites are highly polymorphic, have a codominant expression pattern, and are generally durable. Amplified fragment length polymorphism is a reliable method that requires little prior knowledge, although it is technically challenging and costly. Single nucleotide polymorphisms (SNPs) are the most basic DNA markers, directly reflecting nucleotide differences between genotypes, and are increasingly being preferred over other marker systems as advances in DNA sequencing make them easier to use, as well as their high occurrence and codominance in various genomes. Aside from cost, major roadblocks to the development of MAS could include a lack of understanding of genetic phenomena such as gene networks, epistasis, and genotype-by-environment interactions, which complicate the genotype-phenotype relationship.

However, given the rapid advancement of marker technology, the employment of MAS techniques in crop development appears to be very promising. Breeders can benefit from categorizing genetic diversity to aid in parental line selection and breeding system design. At 20% of the single nucleotide polymorphism loci, long-term selection for hybrid performance has resulted in significant allele differentiation between heterotic groups. There was a lot of genetic variation between the breeding pools, which was reflected in the number of missing and unique alleles. There were two SNPs developed from the same candidate gene that were linked to the divergence between two Chinese heterotic populations. Two SNPs with associated allele frequency changes and their allele missing in Brazilian germplasm indicated a 142-kb linkage disequilibrium block. SNP markers have been shown to be effective for diversity analysis as well as a practical method for discovering and using unique alleles in maize breeding programmes.

Figure 11.2. Blast disease.

Source:http://www.knowledgebank.irri.org/images/stories/blast-leaf-4.jpg

11.5 BLAST DISEASE

Blast disease (Figure 11.2) is a fungal disease. Blast disease resistance that is race-specific has not shown to be a successful technique. It has been claimed that a previously unknown type of gene conferring non–race specific resistance has been cloned and successfully exploited in breeding. Pi21 induces a slowdown of the plant's defense reactions, which can help with defense mechanism optimization. Deletions in the proline-rich motif of the protein stop it from slowing down. Pi21 can be distinguished from a similarly related gene that causes bad taste. The resistant pi21 allele, which can be found in some japonica rice strains, has been shown to improve rice blast resistance. Early season cold tolerance is a quantitative trait, and numerous QTLs associated with microsatellite markers have been discovered. In multiple genetic backgrounds and situations, these microsatellite markers were verified for starting MAS for early-season cold tolerance.

11.6 RAPD

A RAPD marker OPJ011437 linked to Anthracnose disease resistance was mapped and transformed into an amplified region (SCJ01) that matched contig-3966 on chr-8 of the sorghum genome and could be used to identify resistant plants(Tullu, Buchwaldt, Warkentin, Taran, & Vandenberg, 2003). The shoot fly is one of the most devastating insect pests of sorghum seedlings. Some of the QTLs found corresponded to insect resistance QTLs/genes in syntenic maize genomic regions, implying that insect resistance loci are conserved between these crops. The QTLs will provide a foundation for MAS efforts aimed at enhancing shot fly resistance in sorghum. Synthetic insecticides, in particular, were utilized to protect crops and combat disease vectors. As a result, increased yields have made it possible to feed the majority of the world's population. Aside from the benefits to farmers, such as making their labor easier and minimizing harvest losses, and the benefits to humanity, such as supplying ample food of higher sanitary quality, the widespread use of pesticides has resulted in major health problems. Pesticides can be extremely harmful, and they are too responsible for agricultural diseases including cancer and neurological diseases. Pesticides are likewise becoming more and more expensive for farms as their efficiency and selectivity rise. However, in industrialized countries, the transition from subsistence farming to intensive farming, which can feed more people, is happening quickly.

11.7 PHYTOREMEDIATION

Phytoremediation can help to reduce pollution and pesticide damage. DDT is an organochlorine pesticide that has been shown to be highly effective against human disease vectors. Alternatively, some coleopterans, such as potato beetles, can be used to protect crops. It's time-consuming. POP, or persistent organic pollutant, is a pollutant that takes too long to degrade. DDT can be degraded by microbes, but its metabolites, dichlorodiphenyldichloroethylene, cannot. Plant uptake is very low, thus using plants to reduce DDT contamination is not an option. The second example is atrazine, an s-triazine group herbicide. Maize, for example, is heavily reliant on it. Atrazine and its metabolites are now the most commonly used pesticides. Plants can aid in the reduction of atrazine pollution by speeding up the microbial breakdown of the chemical, however, Deethylatrazine (DEA) or desisopropyl atrazine (DIPA) are two degradative chemicals. Plants' direct action on pesticides is determined by their ability to absorb, accumulate, or store chemicals.

Plants have an ability to reduce pesticide pollution and the utilization of buffer zones between fields and hydraulic networks. The effectiveness of vegetative filter strips (VFS) in protecting water from pesticide run-off contamination has led to its inclusion in good farming practices by the authorities. Plants could also be utilized to purify agricultural waste. Constructed wetlands with macrophyte plants are effective at purifying farm wastes, but their location is crucial. This means that the plant selection must be made from a large number of options. The wide range of plant tolerance makes selection more challenging. Gramineae in buffer zones, trees like poplar or willow in riparian zones or in phytoremediation processes because of their large evapotranspiration capacities, and aquatic plants for waste degradation processes are all useful. The latest genetic engineering technologies are among the few that have been created because they have the potential to expand the scope even further.

11.8 AGROCHEMICAL CONTAMINATION

Agrochemical contamination of farms and natural areas has resulted from the rapid increase and widespread usage of chemicals in crop protection and the control of numerous parasite diseases of people such as malaria and typhus. The positive impacts of these items for farmers and people were reflected in the enthusiasm for them. Their success aided farmers' work by removing the need for manual weeding and providing greater crop protection against insect pests and fungal diseases. This meant that people had a more consistent and plentiful food supply, as well as better sanitary conditions. The usage of these substances allows for a quick and effective reaction to a problem involving human health or agricultural sustainability. Even if there have been instances of misuse, agrochemicals have helped to alleviate some of the challenges associated with growing food crops. It seems likely that so-called organic farming systems will not be able to feed the entire world's population. As a result, pesticides are used in every developed agricultural country, regardless of farming management methods. As a result, during a time when the environment was just a secondary concern for our societies, the primary aim in farming was to produce sufficient quantities of food.

After leaching from the soil, atrazine, a more recent herbicide used extensively in maize growth is discovered in aquatic settings. It is suspected of causing endocrine disruptors, particularly in batrachians, and is therefore prohibited in several countries. Despite being outlawed for numerous years; it is still detected in waterways. As a result of the constant influx of pesticides through run-off, movement in aquatic habitats is the primary source of pesticide contamination. Contamination caused by them, on the other hand, must be remedied. Second, even though the present molecules are less persistent, there is still a risk of dispersion, and these new compounds should be safeguarded from the non-agricultural environment. The development of sustainable agriculture necessitates the restoration of environmental quality by eliminating pesticide contamination, and, second, the elimination of initial pollution by limiting use to what is strictly necessary and by erecting barriers between the application site, fields, and the surrounding environment, with no crop biotopes.

11.9 DDT

DDT was an insecticide that was used in large amounts to kill mosquitos, which are the carriers of diseases like malaria. It is now widely spread across the ecosystem. DDT's remanence is due to its slow breakdown by soil microorganisms, which has a half-life of several tens of years. Because the primary metabolites, DDE and DDD, have the same physicochemical features and breakdown resistance as DDT, metabolism does not effectively contribute to its elimination. As a result, natural degradation of these compounds, or natural remediation, by the bacterial flora already present in the environment is ineffective. As a result, they adsorb heavily on soil particles. Furthermore, alternate drying and wetting phases exacerbate the phenomenon, a weathering phenomenon that results in decreasing hydrophobic chemical bioavailability for plants and animals over time. DDT levels in the food chain increased, and these lipophilic chemicals were retained in body fat. The main effect of DDE in birds is eggshell thinning, but the mechanism is unknown. DDT is poisonous to insects as well as aquatic animals. Surprisingly, the isomer 2,4-DDE accumulates preferentially in courgette flowers, and a similar pattern is also seen in alfalfa. A preferred metabolic pathway is assumed to be the cause of this. Plants' DDT metabolism is nearly identical to that of soil microfauna, which results in the synthesis of DDD, DDE, and DDMU. These metabolic pathways, however, are ineffective at breaking down the drug. The parent molecule is not completely broken down, and the metabolites' environmental destiny and toxicological characteristics are similar to those of DDT.

DDT and other POPs have a long-term effect on the environment. In addition to a low availability due to soil adsorption, these chemicals' poor degradability can explain their long half-life. This demonstrates the need to preserve the ecosystem from these toxins, as well as the employment of alternative ways to natural attenuation to lessen the impacts of POPs. Degradation by rhizosphere bacteria will not be sufficient since it produces non-degradable compounds. Plants' inability to absorb POPs limits their application, with the exception of a few species that should be further researched. In the soil, atrazine is thought to be difficult to break down. By incorporating the rhizosphere, plant cover plays a crucial role. Apart from vegetated soil, atrazine decomposition is quite low in bare soil. Atrazine is broken down into deethylatrazine (DEA), deisopropylatrazine (DIA), and didealkylatrazine (DIDA) or hydroxyatrazine by microorganisms. However, repeated application of atrazine in some soils over several years caused bacterial communities to adapt to the disintegration, resulting in rapid pesticide degradation and complete mineralization. A metabolic mechanism in which bacteria employ atrazine as a single energy source is known as accelerated degradation. Dealkylation and other degradative routes are co-metabolism processes in which bacteria utilize soil organic matter as an energy source. As a result, in low-organic-matter soils, rapid deterioration is particularly effective. As a result, while full mineralization may be useful, it is usually ineffective in the field.

Figure 11.3. Contaminated soils.

Source:https://www.ehs.washington.edu/sites/default/files/site%20contamination.png

11.10 CONTAMINATED SOILS

Contaminated soils (Figure 11.3) are heated directly or indirectly for incineration and thermal desorption to vaporize hazardous contaminants, which are then burned, condensed, or trapped on granular activated carbon. Because soils must be excavated and a lot of energy is required, physico-chemical procedures are quite costly. These techniques, on the other hand, have proven to be successful as well as low-risk to the environment and human health. After treatment, the clean soil is usually not returned to the site and is considered waste. Except for certain attempts to detoxify soils using Fenton reactions, these approaches are mostly employed to treat water. They are expensive since they necessitate water pumping, particularly for groundwater. The effect of microorganisms and root exudates on the oxidoreduction state and pH of the soil in the rhizosphere leads in the modification of metal speciation, which could lead to their precipitation or binding to organic matter especially humic acids in a more stable, less water-soluble form. For organic molecules, this stabilizing mechanism does not appear to be important. Adsorption can cause a significant build-up on root surfaces, immobilizing the component at the root level. This process, like rhizo stabilization, is thought to reduce pollutants while posing few environmental issues. The procedure, on the other hand, could be transient. When the plants die, the substances may be released back into the environment as the roots decompose. The process should result in the co-degradation of chemicals and roots by soil microflora to be successful. However, it should be monitored to ensure that the stability does not result in new pollution of the site.

11.11 PESTICIDES

Plants absorb some pesticides outside of their range, even if the concentrations detected within them aren't of agronomic concern. Several experiments, undertaken in particular under hydroponic growing conditions or occasionally utilizing axenic plants, have revealed plants' ability to quickly remediate pesticide-laden solutions. This ability is determined by a number of factors, including the molecule's physico-chemical properties, particularly its lipophily, as well as the features of the plant in question, such as its water pumping capacity, structure, and root depth. The pesticide is fixed in the roots, transported to the aerial portions via a translocation process, or digested once it has been taken by the plant. Root accumulation is inefficient for remediation because it is difficult to gather roots to conclusively suppress soil pollution, even if the contaminant concentration in the soil lowers microbial degradation accounts for only a minor portion of the insecticide's evaporation. Both the leaves and the roots contribute to the build-up, but the root concentration is higher than the leaf concentration. Root accumulation in water hyacinth is an excellent phytoremediation procedure since the root system can make up to 50% of the biomass and full plants can be taken for subsequent chemical elimination. However, because shoots may be quickly harvested, accumulation in leaves is preferred.

A pesticide with a medium hydrophobicity can be transported to the xylem vessels and translocated to the shoots via the evapotranspiration stream, resulting in the build-up of the material in leaves. For example, sunflower shoots absorb more than 85 percent of imidacloprid, a genuine xylemien compound, which is then transported to leaves. The accumulation mechanism, also known as phytoextraction or phytoaccumulation, is effective at lowering concentrations in the soil. Shoot accumulation, like absorption, is significantly influenced by the hydrophobicity of the chemical, with an optimal log Kow of around two. Pesticide translocation in crop plants, which are being evaluated for use in phytoremediation due to their generally high growth rates, has received a lot of attention. For the purification of water, this remediation technique is frequently used with aquatic plants.

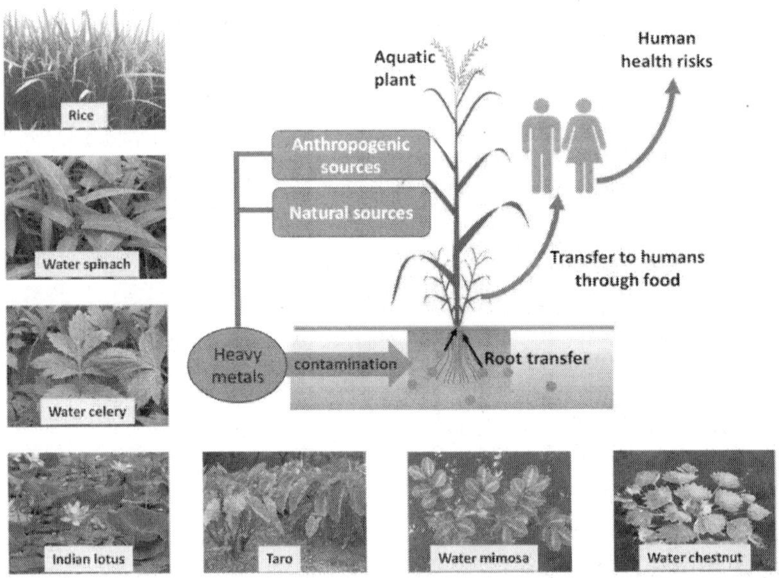

Figure 11.4. Phytoaccumulation.

Source: Uddin, M. M., Zakeel, M., Zavahir, J. S., Marikar, F., & Jahan, I. (2021). Heavy Metal Accumulation in Rice and Aquatic Plants Used as Human Food: A General Review. Toxics, 9(12), 360. https://doi.org/10.3390/toxics9120360

Typha latifolia, for example, is beneficial in lowering methyl parathion contamination of water and sediment. Sweet flag and pickerel weed both absorbed simazine, a triazine-based pesticide, and successfully translocated it into the leaves. Depending on the lipophily of a pesticide translocated to shoots, some of it could be absorbed in vessel macromolecules.

11.12 PHYTOACCUMULATION

Phytoaccumulation (Figure 11.4) is rarely a stand-alone process. Plants, like microflora, have a wide range of enzymes that may metabolize chemicals, and the prevalence of accumulation or degradation determines whether the pesticide is bio accumulated or not. Because perennial ryegrass does not metabolize lindane, it accumulates. Trifluralin, on the other hand, is rapidly metabolized by this plant and does not accumulate. Pesticides are usually metabolized by plants into more polar compounds, which are then compartmentalized into vacuoles or bound to cell walls as bound residues. The first is the fundamental metabolic processes, which convert physiologically active molecules into generally less poisonous or less effective compounds, primarily by oxidation and hydrolysis. The most common reactions are alkyl- or aryl hydroxylations, which are usually carried out by cytochrome P_{450} enzymes or peroxidases. Agrochemicals are sometimes converted to active molecules within plants using phase I processes, particularly hydrolysis. Pesticides are frequently esterified to increase penetration through the plant cuticle(Rasheed et al., 2020).

11.13 PHYTOREMEDIATION

Phytoremediation appears to be a good way to clean up pesticide-contaminated soil or water. There are, however, some limitations. Plants cannot tolerate high pesticide doses due to phytotoxicity. Furthermore, accumulation does not eliminate the need to harvest shoots at the end of the process and apply a follow-up treatment to eliminate pesticides. Because plant metabolism is unable to mineralize pesticides, plants collect pesticide residues as new compounds, pesticide degradation in plants typically necessitates plant harvest. In comparison to other procedures, phytoremediation processes have a number of advantages, including the preservation of soil cultivability and a stable ecosystem. As a result, this is a method that works effectively in some instances and should be carefully considered when a remediation strategy is required. The management of polluted wastes generated from agricultural spraying is also part of environmental protection. Previously, these effluents were dumped into aquatic ecosystems, where they were purified by the latter. In addition to a reduction of effluent, several approaches are now available for treating them utilizing physicochemical methods or biological techniques.

Figure 11.5. Wetlands.

Source:https://environment-review.yale.edu/sites/default/files/styles/flexslider_full/public/8070749192_72b1af084c_o.jpg?itok=qyqFEbdT

11.14 WETLANDS

Constructed wetlands (Figure 11.5) are an intriguing alternative that mimics natural environmental processes in a regulated and limited habitat. This method is already in use to reduce water contamination from mine wastes and in the treatment of urban sewage. Wastes flow into a macrophyte-planted artificial wetland. With pollutants adsorbed on periphyton, these plants were said to have a pumping effect, lowering the volume of effluent. Plants filter the water in the same way as VFS does, improving the sedimentation of impurities and providing a home for the microbes that disperse pesticides. Bulrush reduced simazine and metolachlor levels by 30 to 50 percent when compared to bare wetland, with degradation effectiveness of 90 percent(Kaur, Sharma, & Sodhi, 2022). The technology has proven its efficiency, but the size of the waste volume to be treated is critical. It's also important for the risk of phytotoxicity caused by herbicide contamination or the accumulation of certain elements like copper, which is often employed as a fungicide on a variety of crops.

Sustainable land use protects resources that will help to maintain the existing and future human populations in the area. However, in order to avoid resource depletion, demand for viable developmental processes and their possible collateral effects will continue to grow. Soil quality is an important indication of sustainable land use because agricultural soil provides the foundation for practically all land uses. Second, one of the key causes of soil degradation is land use and mismanagement of arable land by farmers and grazing land by cattle. Erosion, a reduction in fertility, changes in aeration and soil-water content, salinization, or a change in soil flora or fauna can all contribute to this. It is the measure of various potential uses since it reflects the basic working capacity of the soil. Agricultural management policies, on the other hand, will need to adapt to climate change by fostering flexibility in land use, crop output, and agricultural systems. It is vital to recognize agriculture's multifunctional role and strike a varied balance between economic, environmental, and social roles in various locations and sectors to do so.

Land is a living community on which all life on Earth depends, not an inert substance, a steady growing medium, or a mineral stockpile. It is alive in the sense that it is home to many millions of species both above and below ground. Because it is a part of a natural system, a complex web of biological and chemical interactions in which a change in one component causes changes in many others, maintaining this life is crucial for its own reason. As a result of reactions, particularly those connected to our own home environment, our management of the living earth is vital. Furthermore, land and the implications of its management have a significant impact on our society's food supply, water, air, employment, living environment quality, recreation, and, ultimately, our health and survival. Furthermore, the term sustainability refers to the ability of an agricultural land use system to maintain its productive potential throughout time. As a result, developing techniques that produce a sustainable and stable global system, one that minimizes the consumption of scarce resources and the development of waste and pollution, one that serves human needs while sustaining the natural environment, is an urgent problem. It is vital to build new land-management systems that, rather than addressing individual problems in isolation, effectively address the far-reaching impact they have on all of these critical sustainability challenges. As a result, one of the most important criteria for a healthy ecosystem is that it is sustainable, particularly in terms of preserving healthy soil.

Despite its shortcomings, the concept of payments for environmental services has certain advantages that make it a particularly effective measure. Payment schemes that are carefully developed and fit with ecological, economic, and social aspects of sustainability could be a significant tool for promoting sustainability. To put it another way, this goal necessitates the selection of environmental services based on ecological criteria, the development of an economic mechanism for efficient pricing of these services, and the establishment of a public framework to ensure transparency and implementation of these measures. Furthermore, the importance placed on sustainability reflects a convergence of forces reflecting, on the one hand, society's recognition of increasing demands placed on a finite resource base and rapid changes in natural resource quality, and, on the other hand, the political necessity to respond to these pressures and changes. Although there is no universal definition for sustainability, the common view is that it must be multifaceted, encompassing ecological, social, political, and economic considerations.

Land-use sustainability encompasses not only the long-term viability of a land-use model and biological production, but also the optimization of land-use patterns on a spatial scale. Traditional evaluations for sustainable land use, on the other hand, which focus on the social, economic, or ecological benefits of regional land use, can all be classified on the time scale. Landscape ecology, with its theoretical core of spatial heterogeneity and ecological holistic theory, can be a significant aid in creating a synthetic evaluation for sustainable land use on temporal and spatial scales, with a strong function in the analysis of regional land use patterns. Land use sustainability is determined not only by the stabilization of land-use patterns and the optimization of biological and non-biological land-use production, but also by human needs, which put pressure on regional land use. In general, the higher human needs for regional land use, the higher the goal of sustainable land use, and the lower the practicality of sustainable land use. As a result, the indexing system for evaluating regional sustainable land use may be developed from three aspects: landscape productivity, landscape threat, and landscape stability, using the Analytical Hierarchy Process approach and theories of landscape ecology. The index system will expose the gap between the goal of sustainable land use and the status quo of existing land usage, as well as the potential for reaching the goal of sustainability on a human time scale.

The capacity of land production, including biological productivity, economic benefits, and prospective yield of land usage, is reflected in landscape productivity. The bigger the land output and the greater the possibilities of achieving sustainable land use, the higher the landscape productivity. Only when research on sustainable land use is undertaken on a certain spatial and temporal scale is it effective, and the scale of human generation should take precedence over other temporal scales. Meanwhile, sustainable land use refers to the long-term viability of land use patterns as well as the optimization of patterns on a spatial scale. Traditional social, economic, and environmental research on sustainable land use only considered temporal studies and lacked spatial study of land-use patterns to some extent. Landscape ecology is useful for a synthetic investigation, evaluation, and management of sustainable land use on both spatial and temporal dimensions since it considers both spatial patterns and ecological connections. In a nutshell, landscape ecology is a new way of looking at sustainable land use research that emphasizes spatial elements. Although the popular basic definition of sustainable development encompasses practically all aspects of ecological, economic, and social development, it has been used to derive appropriate resource management guidelines, including multifunctional land development.

Figure 11.6. Soil erosion.

Source:https://assets.nrdc.org/sites/default/files/styles/full_content/public/media-uploads/nrcstn83005_2400.jpg?itok=MbalIUii

11.15 SOIL EROSION

Soil erosion (Figure 11.6) is growing, notwithstanding favorable results produced in areas covered by agri-environmental or afforestation techniques. Water erosion affects 115 million ha in Europe, while wind erosion affects 42 million ha. The Mediterranean region has its own set of issues. The contention between different types of land use is one of the causes of soil degradation and destruction. As a result, new perspectives and concepts for sustainable land use should be developed that are in line with nature's restrictions. In this perspective, sustainable land use and soil conservation can be described as the geographical and temporal harmonization of all major land and soil uses, with the least number of irreversible impacts. These harmful impacts are exacerbated by several agricultural practices. Industry, urbanization, road development, fire, other human activities, and, more broadly, demographic pressure and climate change are all key concerns to consider. Erosion, desertification, waterlogging, and compaction are the most major forms of physical soil deterioration caused by agriculture. Deforestation, overgrazing, some agricultural production practices, and the removal of vegetative cover are all examples of land-use activities that can worsen these problems. Water scarcity, as well as occasionally excessive automation and plowing, are contributing factors in this degradation.

Throughout the 1990s, the soil-quality concept evolved in response to a growing global emphasis on sustainable land use with a holistic approach, emphasizing that sustainable soil management necessitates more than only soil erosion control. They also developed a quantitative method for assessing soil quality, claiming that such assessments could aid in determining how soils responded to different management practices. To demonstrate the multiple functions that soils provide as the foundation for sustainable land management, assessment tools for indexing soil quality at various scales were pursued. These functions include nutrient and water cycling, contaminant filtering and buffering, decomposition of crop residues and other organic matter sources, and recycling of essential plant nutrients. Soil resources contain inherent features dictated by basic soil formation processes, as well as dynamic characteristics impacted by human decisions and management methods, as a result of global research and technology transfer activities. Multiple stakeholder groups are concerned about soil resources; society's interests and demands on soil resources are evolving; and soil resource and land-use decisions are made in a human or institutional setting. Due to inherent variances in soils, there is no single measure that can be used to evaluate soil quality all of the time.

Because of a number of barriers, few land managers have used soil quality as an indicator of sustainable land management. It is important to establish causal links between soil quality and ecosystem functions, such as biodiversity conservation, biomass production, and soil and water resource conservation. More than just comparing readings across management systems is required for genuine soil quality calibration. We need to improve the ability of soil quality indicators to predict disturbance response. Although there are numerous markers that represent a soil's current capacity to operate, only a few may forecast a soil's ability to continue to function under a variety of disturbance regimes. Both resistance and resilience must be taken into account. Many existing systems are either too complicated or too expensive. Soil quality is linked to various biophysical and socioeconomic factors. Not only will biophysical and socioeconomic indicators be included in effective early-warning monitoring systems, but models that incorporate feedback between soil quality and socioeconomic circumstances and trends will also be developed. The majority of ecosystem activities rely on long-term relationships between different sections of the environment. The area irrigated has grown, as has the usage of purchased inputs such as fertilizers, crop protection agents, and new technologies, resulting in higher productivity per hectare.

11.16 ENVIRONMENTAL ISSUES

High input levels cause nutrient and pesticide leaching, which causes a slew of environmental issues. Agro-ecosystem sustainability is clearly impacted by the mix of large inputs and new technologies. Overuse and misuse of agrochemicals have two effects: they pollute the land and water needed for production, and hurt human health directly and indirectly. For decades, the harmful environmental impact of fertilizers has been the subject of research as well as scientific and public discussion, focusing primarily on intensive agricultural systems in industrialized countries; systems that have just lately moved to tropical regions. Beginning with a better knowledge of the effect of the biophysical environment and the role of management at the plot and field stages, the analysis progressed to the farm and regional scales, incorporating socio-economic aspects of farm-level decision-making. Trade-offs and probable synergies of management and policy choices can be identified using this method.

Agriculture, on the other hand, contributes to biodiversity because biological diversity in agricultural crop species and varieties, as well as livestock species and breeds, is the result of adaptation to environmental conditions. The genetic base for increasing productivity is provided by the diversity of crop and livestock species, variations, and breeds. Changes in agricultural production, on the other hand, have led to the cultivation of only high-yielding types. The biodiversity mainstream focuses on so-called hotspots or locations that house significant numbers of species on the verge of extinction. Due to the low success rate of this strategy, subsequent efforts have been focused on the economic value of biodiversity. Climate change is currently one of the world's most critical development issues. Climate change has localized consequences that differ for different systems, industries, and areas. Climate change, on the other hand, has a far-reaching impact on development. In addition to the pressing need to reduce greenhouse gas emissions, emphasis must also be paid to adapting systems to changing environmental conditions.

Climate change and variability will undoubtedly have an impact on agricultural land use. Even a little increase in temperature might reduce yields. Changes in temperature and rainfall are particularly damaging to semi-arid and desert regions. Modifications in agro-ecological zones will need drastic changes in production techniques in some areas. Changes in water availability, as well as susceptibility to the occurrence of pests and diseases, will have an indirect impact on crop yield as a result of climate change. Unpredictable rainfall and insufficient water for crop and livestock production are often the cause of high intra- and inter-seasonal variability in food supplies. The concept of sustainable agriculture was born out of the environmental dangers of high-intensity agriculture. Plant development and fertilization in sustainable agriculture must be done in an ecologically sound manner, adjusting to ecological parameters and avoiding pollution, but also ensuring nutrient input to soil and hence food production. Sustainable agriculture refers to the sustainable management of agricultural resources to meet evolving human requirements.

Figure 11.7. Ecological fertilization.

Source:https://previews.123rf.com/images/miloszg/miloszg1611/miloszg161101260/66664861-man-working-in-garden-with-fork-fertilization-of-garden-with-ecological-compost-.jpg

11.17 ECOLOGICAL FERTILIZATION

Ecological fertilization (Figure 11.7) is a type of fertilizer that combines agricultural and environmental goals and is tailored to the local environment. Organic farming, ecologically friendly fertilizing, and sustainable agriculture all follow similar nutrient management concepts. In contrast to organic farming, which is based on principles and completely prohibits the use of mineral fertilizers, ecological fertilization is based on the principle that mineral fertilization should only be applied to the soil in the quantities and at the times required by the crop, thus avoiding environmental damage. It's critical to understand the dynamics of soil organic matter. If the organic matter in the soil decomposes, it releases mineral nitrogen and phosphate, which, providing K levels are appropriate, may be sufficient for the crop. Organic farming's main goal is to increase organic matter in the soil for this purpose, which is accomplished by the return of all organic wastes and the use of restorative crops. However, if these methods are used to provide nutrients to a crop, a proportional amount of organic matter must be destroyed, with the carbon transformed to CO_2. On the basis of general soil physical attributes such as soil structure, water-holding capacity, and erosion resistance, this loss can only be permitted to a certain level. In any event, because soil organic matter breakdown is governed by the weather and the soil organic matter level, and cannot be precisely adjusted, it can only produce a limited amount of mineral N and P in any given year(Tariq et al., 2017).

It's interesting to see how long the manure effect can last with a higher rate of release than in the original soil, but when the rate of release decreases, the crop production will decrease. In arid and semiarid environments, irrigation water is one of the most important and limited resources for agricultural output. In dry places, increasing productivity is largely dependent on the ability to expand water resources through better rainwater management and groundwater development. The scarcity of good water for irrigated agriculture is becoming a severe issue, requiring farmers to rely on low-quality water. Nonetheless, indiscriminate use of sodic and saline fluids endangers soil health by degrading the physical, chemical, and biological components of the soil. Problems with salinity, sodicity, and toxicity affect crop productivity while also limiting crop choice. To maintain crop output while minimizing salinization and deterioration of soil physical conditions over time, irrigation development plans must be properly planned and implemented. Alternatives have recently emerged for safely using waters that were previously deemed unsafe. Nonetheless, proper crop selection, improved water and fertility management, soil structure maintenance, and tailwater return systems are still required.

CHAPTER 12

DROUGHT MODELING

CONTENTS

12.1 Introduction........220
12.2 Indices........220
12.3 Drought Assessment........222
12.4 Adaptation........224
12.5 Early Warning System........225
12.6 Water Scarcity........227
12.7 Insurance........228
12.8 Climate Change and Groundwater........230

12.1 INTRODUCTION

Water is needed for the survival of all living things. It is not feasible to live a long and healthy life without it. When there is plenty of water, life is easier, but when there isn't enough, life is difficult. Humans rely on water for practically all of their activities in the environment. Humans migrate to better water resource areas, such as riverbanks, lakes, seashores, or shallow groundwater reserves, when water is scarce or not available in sufficient quantities in a given location. Water-related management activities are at the heart of any civilization's evolution and growth. Social gatherings, cultures, and civilizations all began with such activities. Groundwater resources played a major role in the history of civilizations, even in dry places, as evidenced by shallow wells or natural springs. Any society faces internal and external pressures, as well as demands for food security, which cannot be attained without water security. Internal and external demands have been and continue to be placed on water supplies. Practical indicators, in general, represent the degree, length, severity, or extent of drought coverage. Precipitation, temperature, streamflow, soil moisture, reservoir storage, and groundwater level are among the meteorological and hydrological variables used. Although drought indices can be simple to use, the scientific and operational meaning of an index value may raise questions about how each indicator is combined or weighted for an overall index derivation, as well as how an arbitrary index value relates to drought hydrometeorological and statistical characteristics(Şen, 2015).

12.2 INDICES

Snow accumulation, precipitation, runoff, and water reservoir storage are the four inputs for the SWSI. It is important for calculating snow accumulation in winter seasons, precipitation, and water storage in the reservoir, depending on the season. During the summer, runoff replaces snow accumulation in the SWSI equation. The SWSI is similar to the PDSI in that it incorporates hydrological and climatological characteristics into a single index value. Each component is given a monthly weight based on its normal contribution to the basin's surface water, and the weighted components are added together to provide an SWSI number that represents the entire basin. Full information regarding the number of wet and dry spells is becoming more important in many industries, such as water management projects, agricultural planning, flood studies, and numerous industrial activities, for a variety of reasons. Thus, the geographical distribution of these meteorological data for any location is useful in the creation and planning of new water resources, particularly in agricultural and environmental engineering projects. For example, if the total amount of water available exceeds demand, excess water is kept during wet seasons for use during dry spells, if necessary, either from previously stored water in the reservoir or, if practicable, from alternate water sources. As a result, the amount of water that has accumulated in the reservoir plays an essential role in reservoir design. Water pollution, sedimentation, erosion difficulties are all major concerns that necessitate the prediction of wet and dry times.

In general, any period is classified as wet or dry based on the relationship between water supply and demand; in other words, if the water supply is less than demand at any point in time, the period is classified as dry; otherwise, the period is classified as wet. The bulk of drought models, in actuality, are based on historical precipitation (meteorological drought) and streamflow (hydrologic drought) data. Preliminary drought prediction approaches only take into account temporal fluctuations because early models are based on a single area. However, its validity is tested in a small area around the original site using a basic and local approach. Early models are based on probabilistic, statistical, and stochastic modeling approaches and are mostly focused on frequency analysis methodologies. In the application and production of the essential formulations, basic assumptions such as structural independence, stationarity, Bernoulli trials, Markov processes, chains, and others are applied. Drought studies are primarily based on annual data, hence the smallest time period considered in their estimates is a year. The duration component is certainly important in frequency characterization and forecasting, but the severity aspect is as important.

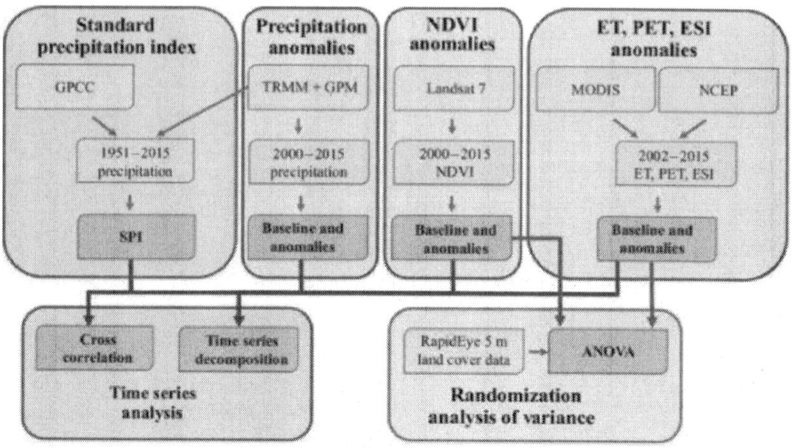

Figure 12.1. Drought assessment.

Source:https://www.researchgate.net/profile/Christopher-Williams-2/publication/329566537/figure/fig1/AS:727747232030721@1550519869290/Methodology-used-in-this-study-The-multiple-approaches-to-drought-assessment-used-in-the_W640.jpg

12.3 DROUGHT ASSESSMENT

Drought makes it difficult to assess its damage and may make disaster aid more difficult than for other natural disasters. These drought characteristics have hampered the creation of accurate, reliable, and timely assessments of drought severity and implications, i.e., drought early warning systems, and, eventually, drought preparedness strategies (Figure 12.1). In the context of uncertainty approaches, practically all objective drought quantities can be calculated probabilistically and statistically if a historical record is given. Average, standard deviation, correlation coefficient, and skewness values, as well as grouped data evaluation in terms of histograms and empirical PDFs, are among the statistical parameters. On the other hand, theoretical PDF statements can be calculated from the same record if the focus is on the frequency of drought occurrence. On the basis of competent modeling and forecasting work, rational decisions may be made that promote disaster vulnerability reduction and drought management with minor hazard and disaster inflictions. It is simple to combat future dangerous drought occurrences and put mitigation strategies into effect if the essential scientific and social knowledge and information sources are appropriately prepared for a region(Haied et al., 2017).

Droughts are difficult to detect, especially in the early phases of development. Any unusually dry period that leads to water stress and scarcity triggers an early drought warning. The initial drought warning is a lack of rain, which causes water shortages in the soil, rivers, and reservoirs, with the resulting impact appearing as a natural hazard. Drought catastrophes are best understood in practice based on their effects on human activities and resources, such as agricultural productivity, water supplies, and food availability, rather than rainfall insufficiency. Arid regions are those where rainfall is limited, yet drought is not the same as aridity. In dry locations, settlers adjust their everyday lives to the current circumstances, with little regard for the repercussions of drought. These places receive relatively little yearly rainfall; however, droughts are not limited to low-rainfall areas. The necessity for a new basis and framework for planning water resources structures, taking into account novel information and knowledge for necessary adaptation and mitigation efforts, is highlighted by the predicted climate change. Individual responsibility is paramount in raising public understanding about local climate change implications; otherwise, there may be serious consequences when dealing with severe weather. Experience exchange based on logical linguistic facts and a shared base of climate change impacts on future rainfall and subsequent runoff occurrences can increase public information.

Droughts are one of the slow-moving natural hazards that cover large areas and have a wide range of damage possibilities. Preparedness, vulnerability assessment, risk management, and appropriate mitigation efforts are the only ways to effectively mitigate drought hazard impact. Drought-effective factors such as meteorological, hydrological, agricultural, and social must be monitored for credible predictions and impact assessments with the least-expected hazard implications for proper drought preparedness and mitigation. Drought risks are divided into two categories: probabilistic risk based on an event such as precipitation in the case of meteorological droughts, runoff in the case of hydrological droughts, and harvesting risk in the case of agricultural droughts; and societal vulnerability to the event resulting in losses. Each drought-affected location has unique climatic, climatological, hydrological, and socioeconomic characteristics; therefore, no two droughts should be expected to have the same impact or risk. Vulnerability and risk are intertwined; therefore, a high level of vulnerability indicates that a population is vulnerable to unfavorable drought effects. As a result, in order to estimate the risk to society, the basics of vulnerability in the drought-stricken area must be properly understood, and methodical procedures must be put in place with the help of responsible authorities and the general public. Unfortunately, scientists disagree about the extent and severity of drought; as a result, the manager or decision-maker may be disappointed by such variances. This is also due to differences in perspectives between drought managers and scientists, as the former are more physically oriented and interested in drought modeling and predictions, whereas the latter are more physically oriented and interested in drought impact on society, the economy, and drought mitigation efforts in general. Encouragement of dialogue, mutual understanding, and integration between these two aims is important for successful outcomes. In most practical treatments, administrators, managers, and decision-makers are indifferent to scientific studies and their outcomes.

Increased storage capabilities in a region act as a buffer during drought periods, increasing drought resistance due to additional water storage. Water managers are required by law to maintain flows at specific levels under legal agreements between political jurisdictions regulating the amount of water to be provided from one jurisdiction to another. It's critical to avoid any potential friction or tension between different types of customers. The manager, decision-maker, and stakeholders should be aware of the various definitions of the disaster terminology that are also typical in any drought hazard case for a better understanding of any disastrous event. Acceptable risk is a level of injury or loss from a catastrophic circumstance that a society or authority considers reasonable in light of social, political, and economic cost-benefit calculations. In any uncertain investigation, there is no way to get a totally error-free result. As a result, a certain level of vulnerability is deemed acceptable while balancing issues such as cost, equality, public input, and the likelihood of drought are taken into account.

12.4 ADAPTATION

Adaptation is a physical or behavioral trait that has evolved in order for a civilization to better survive in its environment. It also refers to adjusting the social system to current and anticipated drought effects so that good opportunities can be managed through human intervention. Capacity is the ability to receive, hold, or absorb anything in general. It is the sum of all qualities, attributes, and resources available within a community, society, or organization that can be utilized to mitigate the effects of a disaster. It includes human experience, knowledge, skills, management methods, and administrative norms, in addition to physical capacities such as infrastructure and institutions. It deals with the handling of large-scale industrial and environmental disasters in this case. Drought responses and actions are dealt with by crisis management without any prior planning. Individual or governmental initiatives that are ineffective, poorly coordinated, and ultimately fail.

Enhancing drought preparedness and mitigation efforts in all parts of life makes it possible to better understand and improve drought hazard and subsequent mitigation research at micro and macro scales. Greater understanding and description of both the physical characteristics of the hazard and the social elements that determine societal susceptibility are required. Any civilization that does not have risk-based drought management plans and early warning systems is at risk of experiencing the greatest drought effects possible. Drought crisis management actions are the only way to address the drought's effects in such circumstances. Managers, decision-makers, and scientists should all be aware of drought risks and repercussions, but society as a whole should be as well. In the aftermath of a drought, a lack of pre drought preparedness is characterized by a delayed crisis response. This circumstance will have negative consequences and will necessitate a lengthy rehabilitation period. A drought-resilient society, a risk-based drought policy, preparedness plans, and proactive mitigation techniques are all necessary to avert such a worst-case situation. It's part of a long-term management plan aimed at minimizing societal drought vulnerability.

12.5 EARLY WARNING SYSTEM

An early warning system combines a variety of physical and social indicators, as well as their application. The information content of various drought-related alternative criteria can be crisp, probabilistic, or fuzzy. Each of these information sources must be analyzed together in a multicriteria method to find the appropriate mitigation option. Some of this data must be entered into a prediction model, but others may be used to linguistically support the conclusions of such models, leading to more refined solutions. Drought mitigation programmes can be divided into two categories: short-term and long-term. The utilization of relatively marginal water sources such as ponds, the relaxing of water-quality constraints, and the efficient use of water resources are also among the short-term alternatives.

In the event of a local drought, one of the short-term consequences is to restrict water supply at such a rate that the customer does not feel the loss physiologically. Municipal consumption cutbacks, annual crop restrictions, and the imposition of mandatory rationing can all help to reduce short-term demand("Early Warning Systems for Droughts | Climate Technology Centre & Network | Mon, 03/19/2018," n.d.).

To deal with drought, a proactive strategy to emergency management is required. Drought experience and watch systems have shown to be a successful instrument for implementing such a proactive approach over the years. Lessons learned from previous studies of short- and long-term drought mitigation measures in a water supply system show that a particular method can be used to describe multiple societal viewpoints and stakeholder interests in order to facilitate decision-making on drought vulnerability and mitigation tasks. Water supply augmentation is a primary objective in all drought mitigation efforts, but demand reduction is just as vital. Long-term strategic solutions include the construction of new impoundments, water transfer in addition to wastewater reuse, and the construction of desalination plants to boost water supply. Drought mitigation and combat necessitate constraints on a variety of social, economic, technological, and anthropogenic activities, and the costs, harm, and deterioration of human activities in every element of life are all consequences. It is vital to be on alert-based preparedness in light of mitigation measures and strategies in order to mitigate the actual consequences and losses during the next drought. Depending on the nature of drought, several mitigating methods might be used. Solutions to agricultural droughts, for example, can be found by expanding irrigation water infrastructure and, more importantly, by improving the efficiency of existing systems so that crops that use less water can be cultivated. In many irrigation networks, only about half of the water is used to irrigate crops, with the remainder being lost due to seepage from canals, evaporation and runoff from inadequate management that fails to provide water to crops at the right time and in sufficient quantities. The irrigation projects' future development is fraught with difficulties. If the essential precautions and steps are not adopted in a timely manner, such developments may result in other irrigation issues such as waterlogged and salinized soils, groundwater pollution, and thus reduced usage of water supplies and damage of aquatic habitats.

12.6 WATER SCARCITY

Water scarcity necessitates effective combat and mitigation strategies. Relevant drought consequences must be ranked according to their risk priority, taking into account the region's environmental, social, and economic factors. Such efforts give a critical chance to determine the best mitigation strategies for reducing short- and long-term drought hazards. As previously stated, an efficient mitigation strategy must take into account qualitative and quantitative qualities as well as repercussions based on prediction procedures. As a result, the essential emergency and preparedness measures can be devised and implemented in a timely way. Although there is a growing awareness of drought dangers and various initiative networks, effective drought combat requires risk assessment methodologies, monitoring-based prediction, and mitigation, as well as further advancements. Existing tactics, on the whole, are reactive rather than proactive when it comes to effective prediction work. Drought planning, monitoring, proper planning, and mitigation methods are always needed, not only through scientific methodologies, but also with the help of central and local government, public awareness, and stakeholder participation. A regional drought mitigation plan, based on factual data from the area, is required for success.

Unfortunately, risk-reduction measures are rarely considered as an integral part of water resource management because environmental, social, and economic factors are frequently overlooked. Multiple droughts, as well as a lack of political and institutional capacity, are among the major unknowns in developing an effective mitigation plan. However, the following major points are useful in the construction of effective drought mitigation planning. Modifying the water demand level allows for more water to be used in the event of future droughts. In both industrial and home operations, it is advised that water usage be kept to a minimum. Drought-resistant crop patterns are recommended to alleviate an agricultural drought situation and save water for drought combat and hazard reduction. In general, during droughts, required water rules and guidelines must be developed, publicized, and disseminated to ensure that water resources are used responsibly. As a pre-mitigation stage, all of these drought mitigation preparations and their combined effectiveness aid in the development of drought hazard reduction measures. Environmental, agricultural, social, and economic activities can all help to effectively incorporate and implement all of the aforementioned drought-reduction initiatives. In terms of environmental improvements, not only groundwater quantity measures, but also quality variations in both time and space must be examined. During drought periods, wildlife must be protected by implementing corrective remedy proposals.

12.7 INSURANCE

Insurance systems and relief funds are among the social and economic implications that must be expanded. Trade operations, in particular, can be directed toward the drought-affected region. Finally, in order to reduce the impact of agricultural droughts, it is possible to limit agricultural and irrigation areas based on the region's food requirement by selecting appropriate crop varieties. To lessen the impact of drought on agricultural lands, activities on such lands might be concentrated on areas where physical soil attributes are more convenient, while fertility enhancements are taken into account. Instream water demands may be affected by these changes, either directly or indirectly. Changes in streamflow, for example, would affect present and potential hydroelectric power generation. Changes in water amount have a direct or indirect impact on hydroelectric power output. As a result, changes in hydroelectric power generation may have a big impact on water demand in a watershed("Crop Insurance Systems | Knowledge Hub," n.d.).

The discharge of river confluences or lake water, as well as ice-free periods, are important components in any navigation system. Changes in aquatic system behavior, vulnerability to their benefits, and variations in hydrological regimes and cycles may all be caused by climate change. To ensure ecosystem sustainability, sufficient minimum water levels must be maintained in order to avoid jeopardizing the current scenario. When water supplies become scarce, especially during drought periods with the added influence of climate change, the water balance between instream and withdrawal may be jeopardized. Climate change is undoubtedly the most significant future effect factor on water demand and supply balance, with increased dry spells and widespread drought projected. Even if climate change has no effect, social, economic, cultural, industrial, technological, agricultural, land use, and many other human activities will have an impact on maintaining a sustainable water supply to meet continuously increasing needs in any corner of the world. All of these activities have an impact on the hydrological cycle, and human-induced climate change will have an impact as well, which is impossible to quantify without appropriate forecasts in general, and drought forecasts in particular.

Unfortunately, existing water supply and distribution systems practically everywhere in the globe are based on previous weather and climate impacts that may be extracted from public meteorological and hydrological records. Return periods, on the other hand, are taken into account for future planning, design, management, and operation. The return timeframes and accompanying risk levels are assessed for floods rather than droughts in practical applications. Adaptation studies conducted without reliable data may exacerbate the current problem. It is vital to have an integrated management of existing supplies and infrastructure at drainage basin levels in order to accomplish cost-effective adaptation, mitigation, and solutions, which leads to an increase in reliable supplies and efforts to address water conflicts in many places. In order to conserve and protect resources and opportunities in response to changing conditions, appropriate incentives might be provided to local residents. Climate change adds to the unpredictability of matching future supplies to future wants; nonetheless, the logical phases of any management and operation thought process are unaffected by such changes.

Water availability, quality, distribution, the complex infrastructure and mechanisms in place to manage it, and existing climate variability will all be affected by climate change. Unfortunately, global climate change is being portrayed solely as a sequence of calamities, with the positive parts being overlooked. For example, although some places are beginning to suffer from desertification, a lack of precipitation, or flooding, certain arid zones may be changing into more humid zones with considerable groundwater recharges. There is limited research on how various climate changes may affect the infrastructure and sophisticated systems designed to manage water supplies. There are still significant knowledge gaps, and much more research is required. Water managers and planners, as well as the more academic and scientific research community, should set the priorities and directions for future investigations. Increased atmospheric moisture can lead to higher relative humidity and more clouds, which can restrict solar radiation and the amount of energy available for evaporation at the surface. These feedbacks are incorporated into climate models and affect the amount of surface heat available for evaporation in various models.

In order to comprehend what is going on with precipitation locally, rainfall rates and frequency, as well as accumulations, are crucial. The accumulations are strongly influenced by the frequency, size, and duration of individual storms, as well as their rate, which is influenced by atmospheric static stability and other factors. Convection and extratropical weather systems, in particular, both of which operate to stabilize the atmosphere, require vertical heat absorption transfer at the surface. Increased greenhouse gas levels also help to keep the atmosphere stable. Changes in precipitation and evaporation, both of which are heavily impacted by temperature, cause changes in runoff and the source of a region's renewable water supply. There are a lot of unknowns regarding how a region's climate and hydrology may change as a result of global warming. However, locations where precipitation is currently primarily in the form of winter snowfall and streamflow is primarily derived from spring and summer snowmelt are one of the more likely impacts. A change in the relative amounts of snow and rainfall, as well as the timing of snowmelt and runoff, would almost certainly come from global warming. A change from snow to rain might raise the risk of floods early in the year and restrict water supply during high demand periods, particularly for irrigation.

12.8 CLIMATE CHANGE AND GROUNDWATER

Climate change must be detected by demonstrating that the observed change is greater than what would be expected from naturally occurring oscillations alone. In a statistical sense, a climatic fluctuation observed is exceedingly rare. Attributing change to human activity necessitates demonstrating that the observed change cannot be explained by natural forces, either forced or unforced. Groundwater and climate are inextricably intertwined. Climate change's effects on groundwater are poorly known, and the relationship between climatic factors and groundwater is more intricate than the relationship between climate variables and surface water. Climate change requires more attention from groundwater hydrologists. Drought monitoring and assessments, climate change and variability studies, water quality analysis, and assessments of the availability and sustainability of groundwater resources should all pay more attention to the linkages between groundwater and climate. The reaction of groundwater levels to climate change is delayed.

Groundwater is a key supply of freshwater, particularly in arid and semiarid regions, but there has been little focus on the possible consequences of climate change on these freshwater supplies. The majority of the research is focused on humid areas. Floods replenish aquifers in dry and semiarid locations through cracked and fissured rocks, solution voids in dolomite or limestone geological settings, and main stream channels of quaternary alluvium deposits. For groundwater recharge augmentation, engineering facilities such as levees, dikes, and sequential small-scale groundwater recharge dams may be created at convenient locations along the main channel. After each flood or at regular intervals, the groundwater recharge zones must be cleaned to keep fine silt from accumulating. Furthermore, in dry and semiarid regions, river inundation sites are among the most important groundwater recharge locations. As a result, their scope must be defined in light of potential climate change implications. Estimating recharge rates and determining the importance of these elements in controlling aquifer vulnerability to contamination require an understanding of the relative importance of climate, vegetation, and soils in controlling groundwater recharge.

REFERENCES

1. Abdul Qados, A. M. S. (2011). Effect of salt stress on plant growth and metabolism of bean plant Vicia faba (L.). Journal of the Saudi Society of Agricultural Sciences, 10(1), 7–15. https://doi.org/10.1016/j.jssas.2010.06.002
2. Abid, M., Ali, S., Qi, L. K., Zahoor, R., Tian, Z., Jiang, D., … Dai, T. (2018). Physiological and biochemical changes during drought and recovery periods at tillering and jointing stages in wheat (Triticum aestivum L.). Scientific Reports, 8(1), 4615. https://doi.org/10.1038/s41598-018-21441-7
3. Ahanger, M. A., Siddique, K. H. M., & Ahmad, P. (2021). Understanding drought tolerance in plants. Physiologia Plantarum, 172(2), 286–288. https://doi.org/10.1111/ppl.13442
4. Ahmad, Z., Anjum, S., Waraich, E. A., Ayub, M. A., Ahmad, T., Tariq, R. M. S., … Iqbal, M. A. (2018). Growth, physiology, and biochemical activities of plant responses with foliar potassium application under drought stress – a review. Journal of Plant Nutrition, 41(13), 1734–1743. https://doi.org/10.1080/01904167.2018.1459688
5. Ahmed, U., Mumtaz, R., Anwar, H., Shah, A. A., Irfan, R., & García-Nieto, J. (2019). Efficient water quality prediction using supervised machine learning. Water, 11. https://doi.org/10.3390/w11112210
6. Arai, H., Hayashi, M., Kuroi, A., Ishii, M., & Igarashi, Y. (2005). Transcriptional Regulation of the Flavohemoglobin Gene for Aerobic Nitric Oxide Detoxification by the Second Nitric Oxide-Responsive

Regulator of Pseudomonas aeruginosa. Journal of Bacteriology, 187(12), 3960–3968. https://doi.org/10.1128/JB.187.12.3960-3968.2005

7. Augustine, S. M. (2016). Function of Heat-Shock Proteins in Drought Tolerance Regulation of Plants. In M. A. Hossain, S. H. Wani, S. Bhattacharjee, D. J. Burritt, & L.-S. P. Tran (Eds.), Drought Stress Tolerance in Plants, Vol 1: Physiology and Biochemistry (pp. 163–185). Cham: Springer International Publishing. https://doi.org/10.1007/978-3-319-28899-4_7
8. Basu, S., Ramegowda, V., Kumar, A., & Pereira, A. (2016). Plant adaptation to drought stress. F1000Research, 5, F1000 Faculty Rev-1554. https://doi.org/10.12688/f1000research.7678.1
9. Begum, N., Qin, C., Ahanger, M. A., Raza, S., Khan, M. I., Ashraf, M., … Zhang, L. (2019). Role of Arbuscular Mycorrhizal Fungi in Plant Growth Regulation: Implications in Abiotic Stress Tolerance. Frontiers in Plant Science, 10. Retrieved from https://www.frontiersin.org/article/10.3389/fpls.2019.01068
10. Berruti, A., Lumini, E., Balestrini, R., & Bianciotto, V. (2016). Arbuscular Mycorrhizal Fungi as Natural Biofertilizers: Let's Benefit from Past Successes. Frontiers in Microbiology, 6, 1559. https://doi.org/10.3389/fmicb.2015.01559
11. Bücking, H., Liepold, E., & Ambilwade, P. (2012). The Role of the Mycorrhizal Symbiosis in Nutrient Uptake of Plants and the Regulatory Mechanisms Underlying These Transport Processes. In Plant Science. IntechOpen. https://doi.org/10.5772/52570
12. Burgess, P., & Huang, B. (2016). Mechanisms of Hormone Regulation for Drought Tolerance in Plants. In M. A. Hossain, S. H. Wani, S. Bhattacharjee, D. J. Burritt, & L.-S. P. Tran (Eds.), Drought Stress Tolerance in Plants, Vol 1: Physiology and Biochemistry (pp. 45–75). Cham: Springer International Publishing. https://doi.org/10.1007/978-3-319-28899-4_3
13. Butt, Y. N. (2017). Drought Tolerance In Plants: A Review. Research & Reviews: Journal of Ecology and Environmental Sciences, 5(4), 1–9. Retrieved from https://www.rroij.com/peer-reviewed/drought-tolerance-in-plants-a-review-86318.html
14. Chakrabarty, A., Aditya, M., Dey, N., Banik, N., & Bhattacharjee, S. (2016). Antioxidant Signaling and Redox Regulation in Drought- and Salinity-Stressed Plants. In M. A. Hossain, S. H. Wani, S. Bhattacharjee,

D. J. Burritt, & L.-S. P. Tran (Eds.), Drought Stress Tolerance in Plants, Vol 1: Physiology and Biochemistry (pp. 465–498). Cham: Springer International Publishing. https://doi.org/10.1007/978-3-319-28899-4_20

15. Chattopadhyay, S. (2016). Interplay Among Glutathione, Salicylic Acid, and Ethylene to Combat Environmental Stress. In M. A. Hossain, S. H. Wani, S. Bhattacharjee, D. J. Burritt, & L.-S. P. Tran (Eds.), Drought Stress Tolerance in Plants, Vol 1: Physiology and Biochemistry (pp. 145–161). Cham: Springer International Publishing. https://doi.org/10.1007/978-3-319-28899-4_6
16. Chave, J., Réjou-Méchain, M., Búrquez, A., Chidumayo, E., Colgan, M. S., Delitti, W. B., … Vieilledent, G. (2014). Improved allometric models to estimate the aboveground biomass of tropical trees. Glob Chang Biol, 20. https://doi.org/10.1111/gcb.12629
17. Chaves, M. M., Flexas, J., & Pinheiro, C. (2009). Photosynthesis under drought and salt stress: Regulation mechanisms from whole plant to cell. Annals of Botany, 103(4), 551–560. https://doi.org/10.1093/aob/mcn125
18. Coles, N. A., & Eslamian, S. (2017). Definition of Drought. In Handbook of Drought and Water Scarcity. CRC Press.
19. Comas, L., Becker, S., Cruz, V. M., Byrne, P. F., & Dierig, D. A. (2013). Root traits contributing to plant productivity under drought. Frontiers in Plant Science, 4. Retrieved from https://www.frontiersin.org/article/10.3389/fpls.2013.00442
20. Conrad, A. (2013, April 23). The benefits of alternative farming methods. The Guardian. Retrieved from https://www.theguardian.com/global-development-professionals-network/2013/apr/23/farming-methods-agroecology-permaculture
21. Crop insurance systems | Knowledge Hub. (n.d.). Retrieved January 28, 2022, from https://knowledge.unccd.int/drought-toolbox-action/crop-insurance-systems
22. Dalezios, N. R., Dunkel, Z., & Eslamian, S. (2017). Meteorological Drought Indices: Definitions. In Handbook of Drought and Water Scarcity. CRC Press.
23. Dastogeer, K. M. G., Tumpa, F. H., Sultana, A., Akter, M. A., & Chakraborty, A. (2020). Plant microbiome–an account of the factors that shape community composition and diversity. Current Plant Biology, 23, 100161. https://doi.org/10.1016/j.cpb.2020.100161

24. Desert Control: Tackling Desertification with Clay and Water. (2021, February 16). Retrieved January 28, 2022, from AZoCleantech.com website: https://www.azocleantech.com/article.aspx?ArticleID=1172
25. Deva, C. R., Urban, M. O., Challinor, A. J., Falloon, P., & Svitákova, L. (2020). Enhanced Leaf Cooling Is a Pathway to Heat Tolerance in Common Bean. Frontiers in Plant Science, 11. Retrieved from https://www.frontiersin.org/article/10.3389/fpls.2020.00019
26. Doebley, J. F., Gaut, B. S., & Smith, B. D. (2006). The Molecular Genetics of Crop Domestication. Cell, 127(7), 1309–1321. https://doi.org/10.1016/j.cell.2006.12.006
27. Early warning systems for droughts | Climate Technology Centre & Network | Mon, 03/19/2018. (n.d.). Retrieved January 28, 2022, from https://www.ctc-n.org/technologies/early-warning-systems-droughts
28. Eckardt, N. A. (2000). Sequencing the Rice Genome. The Plant Cell, 12(11), 2011–2018. Retrieved from https://www.ncbi.nlm.nih.gov/pmc/articles/PMC526008/
29. Fahad, S., Bajwa, A. A., Nazir, U., Anjum, S. A., Farooq, A., Zohaib, A., … Huang, J. (2017). Crop Production under Drought and Heat Stress: Plant Responses and Management Options. Frontiers in Plant Science, 8, 1147. https://doi.org/10.3389/fpls.2017.01147
30. Finkelstein, R. (2013). Abscisic Acid Synthesis and Response. The Arabidopsis Book / American Society of Plant Biologists, 11, e0166. https://doi.org/10.1199/tab.0166
31. Furlan, A., Llanes, A., Luna, V., & Castro, S. (2012). Physiological and Biochemical Responses to Drought Stress and Subsequent Rehydration in the Symbiotic Association Peanut-Bradyrhizobium sp. ISRN Agronomy, 2012, e318083. https://doi.org/10.5402/2012/318083
32. Giri, J. (2011). Glycinebetaine and abiotic stress tolerance in plants. Plant Signaling & Behavior, 6(11), 1746–1751. https://doi.org/10.4161/psb.6.11.17801
33. Goyal, K., Walton, L. J., & Tunnacliffe, A. (2005). LEA proteins prevent protein aggregation due to water stress. Biochemical Journal, 388(Pt 1), 151–157. https://doi.org/10.1042/BJ20041931
34. Goyal, M. K., Gupta, V., & Eslamian, S. (2017). Hydrological Drought: Water Surface and Duration Curve Indices. In Handbook of Drought and Water Scarcity. CRC Press.

35. Grime, J. P., & Curtis, A. V. (1976). The Interaction of Drought and Mineral Nutrient Stress in Calcareous Grassland. Journal of Ecology, 64(3), 975–988. https://doi.org/10.2307/2258819

36. Guerra, F. D., Attia, M. F., Whitehead, D. C., & Alexis, F. (2018). Nanotechnology for Environmental Remediation: Materials and Applications. Molecules : A Journal of Synthetic Chemistry and Natural Product Chemistry, 23(7), 1760. https://doi.org/10.3390/molecules23071760

37. Guleria, P., Mahajan, M., Bhardwaj, J., & Yadav, S. K. (2011). Plant Small RNAs: Biogenesis, Mode of Action and Their Roles in Abiotic Stresses. Genomics, Proteomics & Bioinformatics, 9(6), 183–199. https://doi.org/10.1016/S1672-0229(11)60022-3

38. Haied, N., Foufou, A., Chaab, S., Azlaoui, M., Khadri, S., Benzahia, K., & Benzahia, I. (2017). Drought assessment and monitoring using meteorological indices in a semi-arid region. Energy Procedia, 119, 518–529. https://doi.org/10.1016/j.egypro.2017.07.064

39. Haque, M. A., Rafii, M. Y., Yusoff, M. M., Ali, N. S., Yusuff, O., Datta, D. R., … Ikbal, M. F. (2021). Advanced Breeding Strategies and Future Perspectives of Salinity Tolerance in Rice. Agronomy, 11(8), 1631. https://doi.org/10.3390/agronomy11081631

40. Hoang, N. V., Furtado, A., Mason, P. J., Marquardt, A., Kasirajan, L., Thirugnanasambandam, P. P., … Henry, R. J. (2017). A survey of the complex transcriptome from the highly polyploid sugarcane genome using full-length isoform sequencing and de novo assembly from short read sequencing. BMC Genomics, 18, 395. https://doi.org/10.1186/s12864-017-3757-8

41. Huey, C. J., Gopinath, S. C. B., Uda, M. N. A., Zulhaimi, H. I., Jaafar, M. N., Kasim, F. H., & Yaakub, A. R. W. (2020). Mycorrhiza: A natural resource assists plant growth under varied soil conditions. 3 Biotech, 10(5), 204. https://doi.org/10.1007/s13205-020-02188-3

42. Humphreys, C. P., Franks, P. J., Rees, M., Bidartondo, M. I., Leake, J. R., & Beerling, D. J. (2010). Mutualistic mycorrhiza-like symbiosis in the most ancient group of land plants. Nature Communications, 1(1), 103. https://doi.org/10.1038/ncomms1105

43. Hussain, H. A., Hussain, S., Khaliq, A., Ashraf, U., Anjum, S. A., Men, S., & Wang, L. (2018). Chilling and Drought Stresses in Crop Plants: Implications, Cross Talk, and Potential Management Opportunities.

Frontiers in Plant Science, 9. Retrieved from https://www.frontiersin.org/article/10.3389/fpls.2018.00393

44. Ilyas, M., Nisar, M., Khan, N., Hazrat, A., Khan, A. H., Hayat, K., … Ullah, A. (2021). Drought Tolerance Strategies in Plants: A Mechanistic Approach. Journal of Plant Growth Regulation, 40(3), 926–944. https://doi.org/10.1007/s00344-020-10174-5
45. Islam, S. M. F., & Karim, Z. (2019). World's Demand for Food and Water: The Consequences of Climate Change. In Desalination—Challenges and Opportunities. IntechOpen. https://doi.org/10.5772/intechopen.85919
46. Kabir, M. M., Akter, S., Ahmed, F. T., Mohinuzzaman, M., Didar-ul-Alam, M., Mostofa, K. M. G., … Niloy, N. M. (2021). Salinity-induced fluorescent dissolved organic matter influence co-contamination, quality and risk to human health of tube well water, southeast coastal Bangladesh. Chemosphere, 275. https://doi.org/10.1016/j.chemosphere.2020.130053
47. Kathage, J., & Qaim, M. (2012). Economic impacts and impact dynamics of Bt (Bacillus thuringiensis) cotton in India. Proceedings of the National Academy of Sciences, 109(29), 11652–11656. https://doi.org/10.1073/pnas.1203647109
48. Kaur, M., Sharma, S., & Sodhi, H. S. (2022). Chapter 16—Phytoremediation of pesticides. In R. A. Bhat, F. M. P. Tonelli, G. H. Dar, & K. Hakeem (Eds.), Phytoremediation (pp. 323–350). Academic Press. https://doi.org/10.1016/B978-0-323-89874-4.00026-1
49. Khan, R., Ma, X., Shah, S., Wu, X., Shaheen, A., Xiao, L., … Wang, S. (2020). Drought-hardening improves drought tolerance in Nicotiana tabacum at physiological, biochemical, and molecular levels. BMC Plant Biology, 20(1), 486. https://doi.org/10.1186/s12870-020-02688-7
50. Khanna-Chopra, R., & Singh, K. (2015). Drought Resistance in Crops: Physiological and Genetic Basis of Traits for Crop Productivity. In B. N. Tripathi & M. Müller (Eds.), Stress Responses in Plants: Mechanisms of Toxicity and Tolerance (pp. 267–292). Cham: Springer International Publishing. https://doi.org/10.1007/978-3-319-13368-3_11
51. Kim, S. L., Kim, N., Lee, H., Lee, E., Cheon, K.-S., Kim, M., … Kim, K.-H. (2020). High-throughput phenotyping platform for analyzing drought tolerance in rice. Planta, 252(3), 38. https://doi.org/10.1007/s00425-020-03436-9

52. Kole, C., Muthamilarasan, M., Henry, R., Edwards, D., Sharma, R., Abberton, M., … Prasad, M. (2015). Application of genomics-assisted breeding for generation of climate resilient crops: Progress and prospects. Frontiers in Plant Science, 6. Retrieved from https://www.frontiersin.org/article/10.3389/fpls.2015.00563

53. Kumar, R., Aadil, K. R., Ranjan, S., & Kumar, V. B. (2020). Advances in nanotechnology and nanomaterials based strategies for neural tissue engineering. Journal of Drug Delivery Science and Technology, 57, 101617. https://doi.org/10.1016/j.jddst.2020.101617

54. Kumar, V., & Arora, K. (2020). Trends in nano-inspired biosensors for plants. Materials Science for Energy Technologies, 3, 255–273. https://doi.org/10.1016/j.mset.2019.10.004

55. Lacroix, B., & Citovsky, V. (2020). Biolistic Approach for Transient Gene Expression Studies in Plants. Methods in Molecular Biology (Clifton, N.J.), 2124, 125–139. https://doi.org/10.1007/978-1-0716-0356-7_6

56. Lamaoui, M., Jemo, M., Datla, R., & Bekkaoui, F. (2018). Heat and Drought Stresses in Crops and Approaches for Their Mitigation. Frontiers in Chemistry, 6. Retrieved from https://www.frontiersin.org/article/10.3389/fchem.2018.00026

57. Li, R., Li, M., Ashraf, U., Liu, S., & Zhang, J. (2019). Exploring the Relationships Between Yield and Yield-Related Traits for Rice Varieties Released in China From 1978 to 2017. Frontiers in Plant Science, 10. Retrieved from https://www.frontiersin.org/article/10.3389/fpls.2019.00543

58. Liu, N. T. (2014). Development and validation of a machine learning algorithm and hybrid system to predict the need for life-saving interventions in trauma patients. Medical & Biological Engineering & Computing, 52. https://doi.org/10.1007/s11517-013-1130-x

59. Ma, Y., Dias, M. C., & Freitas, H. (2020). Drought and Salinity Stress Responses and Microbe-Induced Tolerance in Plants. Frontiers in Plant Science, 11, 1750. https://doi.org/10.3389/fpls.2020.591911

60. Machado, R. M. A., & Serralheiro, R. P. (2017). Soil Salinity: Effect on Vegetable Crop Growth. Management Practices to Prevent and Mitigate Soil Salinization. Horticulturae, 3(2), 30. https://doi.org/10.3390/horticulturae3020030

61. Malnutrition in Children. (n.d.). Retrieved January 28, 2022, from UNICEF DATA website: https://data.unicef.org/topic/nutrition/malnutrition/

62. Meena, R. K., Bhusal, N., Kumar, K., Jain, R., & Jain, S. (2019). Intervention of molecular breeding in water saving rice production system: Aerobic rice. 3 Biotech, 9(4), 133. https://doi.org/10.1007/s13205-019-1657-0

63. Moloi, M. J., & van der Merwe, R. (2021). Drought Tolerance Responses in Vegetable-Type Soybean Involve a Network of Biochemical Mechanisms at Flowering and Pod-Filling Stages. Plants, 10(8), 1502. https://doi.org/10.3390/plants10081502

64. Muller, B., Pantin, F., Génard, M., Turc, O., Freixes, S., Piques, M., & Gibon, Y. (2011). Water deficits uncouple growth from photosynthesis, increase C content, and modify the relationships between C and growth in sink organs. Journal of Experimental Botany, 62(6), 1715–1729. https://doi.org/10.1093/jxb/erq438

65. Nadarajah, K., & Kumar, I. S. (2019). Drought Response in Rice: The miRNA Story. International Journal of Molecular Sciences, 20(15), 3766. https://doi.org/10.3390/ijms20153766

66. Nadeem, F., & Farooq, M. (2019). Application of Micronutrients in Rice-Wheat Cropping System of South Asia. Rice Science, 26(6), 356–371. https://doi.org/10.1016/j.rsci.2019.02.002

67. New technology could fight off droughts. (n.d.). Retrieved January 28, 2022, from Nano Magazine—Latest Nanotechnology News website: https://nano-magazine.com/news/2019/2/27/new-technology-could-fight-off-droughts

68. Nezhadahmadi, A., Prodhan, Z. H., & Faruq, G. (2013). Drought Tolerance in Wheat. The Scientific World Journal, 2013, e610721. https://doi.org/10.1155/2013/610721

69. Ngumbi, E., & Kloepper, J. (2016). Bacterial-mediated drought tolerance: Current and future prospects. Applied Soil Ecology, 105, 109–125. https://doi.org/10.1016/j.apsoil.2016.04.009

70. Patel, J., & Mishra, A. (2021). Plant aquaporins alleviate drought tolerance in plants by modulating cellular biochemistry, root-architecture, and photosynthesis. Physiologia Plantarum, 172(2), 1030–1044. https://doi.org/10.1111/ppl.13324

71. Peltier, G., & Cournac, L. (2002). Chlororespiration. Annual Review of Plant Biology, 53, 523–550. https://doi.org/10.1146/annurev.arplant.53.100301.135242

72. Ranjith, P., & Rao, M. S. (2021). Breeding for Drought Resistance. In Plant Breeding—Current and Future Views. IntechOpen. https://doi.org/10.5772/intechopen.97276

73. Rasheed, F., Zafar, Z., Waseem, Z. A., Rafay, M., Abdullah, M., Salam, M. M. A., … Khan, W. R. (2020). Phytoaccumulation of Zn, Pb, and Cd in Conocarpus lancifolius irrigated with wastewater: Does physiological response influence heavy metal uptake? International Journal of Phytoremediation, 22(3), 287–294. https://doi.org/10.1080/15226514.2019.1658711

74. Raza, A., Razzaq, A., Mehmood, S. S., Zou, X., Zhang, X., Lv, Y., & Xu, J. (2019). Impact of Climate Change on Crops Adaptation and Strategies to Tackle Its Outcome: A Review. Plants, 8(2), 34. https://doi.org/10.3390/plants8020034

75. Rosero, A., Granda, L., Berdugo-Cely, J. A., Šamajová, O., Šamaj, J., & Cerkal, R. (2020). A Dual Strategy of Breeding for Drought Tolerance and Introducing Drought-Tolerant, Underutilized Crops into Production Systems to Enhance Their Resilience to Water Deficiency. Plants, 9(10), 1263. https://doi.org/10.3390/plants9101263

76. Sah, S. K., Reddy, K. R., & Li, J. (2016). Abscisic Acid and Abiotic Stress Tolerance in Crop Plants. Frontiers in Plant Science, 7. Retrieved from https://www.frontiersin.org/article/10.3389/fpls.2016.00571

77. Salehi-Lisar, S. Y., & Bakhshayeshan-Agdam, H. (2016). Drought Stress in Plants: Causes, Consequences, and Tolerance. In M. A. Hossain, S. H. Wani, S. Bhattacharjee, D. J. Burritt, & L.-S. P. Tran (Eds.), Drought Stress Tolerance in Plants, Vol 1: Physiology and Biochemistry (pp. 1–16). Cham: Springer International Publishing. https://doi.org/10.1007/978-3-319-28899-4_1

78. Şen, Z. (2015). Chapter One—Introduction. In Z. Şen (Ed.), Applied Drought Modeling, Prediction, and Mitigation (pp. 1–41). Boston: Elsevier. https://doi.org/10.1016/B978-0-12-802176-7.00001-8

79. Senapati, N., Stratonovitch, P., Paul, M. J., & Semenov, M. A. (2019). Drought tolerance during reproductive development is important for increasing wheat yield potential under climate change in Europe. Journal of Experimental Botany, 70(9), 2549–2560. https://doi.org/10.1093/jxb/ery226

80. Shaji, E., Santosh, M., Sarath, K. V., Prakash, P., Deepchand, V., & Divya, B. V. (2021). Arsenic contamination of groundwater: A global synopsis with focus on the Indian Peninsula. Geoscience Frontiers, 12(3), 101079. https://doi.org/10.1016/j.gsf.2020.08.015

81. Squires, V. R. (2017). Desertification and Drought. In Handbook of Drought and Water Scarcity. CRC Press.

82. Steudle, E. (2000). Water uptake by plant roots: An integration of views. PlantandSoil,226(1),45–56.https://doi.org/10.1023/A:1026439226716

83. Tariq, A., Pan, K., Olatunji, O. A., Graciano, C., Li, Z., Sun, F., … Liu, C. (2017). Phosphorous Application Improves Drought Tolerance of Phoebe zhennan. Frontiers in Plant Science, 8. Retrieved from https://www.frontiersin.org/article/10.3389/fpls.2017.01561

84. Tchounwou, P. B., Yedjou, C. G., Patlolla, A. K., & Sutton, D. J. (2012). Heavy Metals Toxicity and the Environment. EXS, 101, 133–164. https://doi.org/10.1007/978-3-7643-8340-4_6

85. The Rhizosphere—Roots, Soil and Everything In Between | Learn Science at Scitable. (n.d.). Retrieved January 28, 2022, from https://www.nature.com/scitable/knowledge/library/the-rhizosphere-roots-soil-and-67500617/

86. Tiodar, E. D., Văcar, C. L., & Podar, D. (2021). Phytoremediation and Microorganisms-Assisted Phytoremediation of Mercury-Contaminated Soils: Challenges and Perspectives. International Journal of Environmental Research and Public Health, 18(5), 2435. https://doi.org/10.3390/ijerph18052435

87. Tracy, S. R., Nagel, K. A., Postma, J. A., Fassbender, H., Wasson, A., & Watt, M. (2020). Crop Improvement from Phenotyping Roots: Highlights Reveal Expanding Opportunities. Trends in Plant Science, 25(1), 105–118. https://doi.org/10.1016/j.tplants.2019.10.015

88. Tullu, A., Buchwaldt, L., Warkentin, T., Taran, B., & Vandenberg, A. (2003). Genetics of resistance to anthracnose and identification of AFLP and RAPD markers linked to the resistance gene in PI 320937 germplasm of lentil (Lens culinaris Medikus). TAG. Theoretical and Applied Genetics. Theoretische Und Angewandte Genetik, 106(3), 428–434. https://doi.org/10.1007/s00122-002-1042-x

89. Tuteja, N., & Mahajan, S. (2007). Calcium Signaling Network in Plants. Plant Signaling & Behavior, 2(2), 79–85. Retrieved from https://www.ncbi.nlm.nih.gov/pmc/articles/PMC2633903/

90. Ullah, A., Manghwar, H., Shaban, M., Khan, A. H., Akbar, A., Ali, U., … Fahad, S. (2018). Phytohormones enhanced drought tolerance in plants: A coping strategy. Environmental Science and Pollution Research, 25(33), 33103–33118. https://doi.org/10.1007/s11356-018-3364-5

91. van der Heijden, M. G. A., Martin, F. M., Selosse, M.-A., & Sanders, I. R. (2015). Mycorrhizal ecology and evolution: The past, the present, and the future. New Phytologist, 205(4), 1406–1423. https://doi.org/10.1111/nph.13288

92. Venske, E., dos Santos, R. S., Busanello, C., Gustafson, P., & Costa de Oliveira, A. (2019). Bread wheat: A role model for plant domestication and breeding. Hereditas, 156(1), 16. https://doi.org/10.1186/s41065-019-0093-9

93. Wani, S. H., Kumar, V., Shriram, V., & Sah, S. K. (2016). Phytohormones and their metabolic engineering for abiotic stress tolerance in crop plants. The Crop Journal, 4(3), 162–176. https://doi.org/10.1016/j.cj.2016.01.010

94. Yan, A., Wang, Y., Tan, S. N., Mohd Yusof, M. L., Ghosh, S., & Chen, Z. (2020). Phytoremediation: A Promising Approach for Revegetation of Heavy Metal-Polluted Land. Frontiers in Plant Science, 11. Retrieved from https://www.frontiersin.org/article/10.3389/fpls.2020.00359

95. Yan, J., Aznar, A., Chalvin, C., Birdseye, D. S., Baidoo, E. E. K., Eudes, A., … Scheller, H. V. (2018). Increased drought tolerance in plants engineered for low lignin and low xylan content. Biotechnology for Biofuels, 11(1), 195. https://doi.org/10.1186/s13068-018-1196-7

96. Zahran, H. H. (1999). Rhizobium-Legume Symbiosis and Nitrogen Fixation under Severe Conditions and in an Arid Climate. Microbiology and Molecular Biology Reviews, 63(4), 968–989. Retrieved from https://www.ncbi.nlm.nih.gov/pmc/articles/PMC98982/

97. Zhu, Y., & Gong, H. (2014). Beneficial effects of silicon on salt and drought tolerance in plants. Agronomy for Sustainable Development, 34(2), 455–472. https://doi.org/10.1007/s13593-013-0194-1

98. Zivcak, M., Brestic, M., & Sytar, O. (2016). Osmotic Adjustment and Plant Adaptation to Drought Stress. In M. A. Hossain, S. H. Wani, S. Bhattacharjee, D. J. Burritt, & L.-S. P. Tran (Eds.), Drought Stress Tolerance in Plants, Vol 1: Physiology and Biochemistry (pp. 105–143). Cham: Springer International Publishing. https://doi.org/10.1007/978-3-319-28899-4_5

INDEX

A

Adaptation 64
agricultural drought 5
agricultural malpractices 128
agricultural production 62
agricultural production systems 4, 184, 185, 189
agrochemical-based farming 184
agrochemicals 184, 192
alternate water sources 221
Aquaporins 86
Arabidopsis genome 168
Arabidopsis seedlings 64
arbuscular mycorrhizal fungi (AMF) 128, 131
arsenic (As) 108
Arsenic poisoning 108
ascorbate peroxidase 153
ascorbic acid 153
asynchrony 198
auxins 23

B

Bacillus thuringiensis (Bt) 198
barley 44, 50, 54
Bio-fertilizers 128, 129
biology 64
biotechnology 184
booting 44

C

Capacity 224
carotenoids 153
catalase 153
cell development 84
Cell expansion 169
cell homeostasis 22
cereals 104, 105, 106
civilizations 220
Climate change 128
competitive advantage 198
conventional insecticides 198
cowpea 44
crop enhancement 128
Crop nutrition 184
crop productivity 128, 130
Crop rotation 198
crop tolerance 62

crop vulnerability 198
cytokinins 23, 37

D

Direct selection 104
diversification 152
Dormancy 151
drought avoidance 22
Drought crisis management 225
Drought mitigation programmes 225
Drought planning 227
drought resistance mechanisms 22
Drought stress 22, 23, 36, 40, 41
drought tolerance 22, 23, 24, 30, 31, 34, 35, 36, 38, 39, 40
Drought vulnerability 4

E

economic growth 198
environmental stress 22, 24, 41
enzymatic antioxidants 153
ethylene 23, 24, 29, 38, 39
eutrophication 184, 188

F

fertilizer 128, 130
flavonoids 153
food chain 107
food-chain contamination 198
Food insecurity 153
food quality 130
food security 184, 194
fungal hyphae 129, 133, 134
fungal infections 128

G

genes 168, 169, 170, 171, 172, 173, 174, 180, 181, 182
germination 44, 50
germplasm 45, 52, 58, 59
gibberellins 23
global climate changes 4
glutathione 153
glycine betaine 87, 101
Grain yield 104
Grapevine stomatal regulation 86
groundwater level 220
Groundwater resources 220
growth reduction 62

H

habitats 64, 79
heat 128
heat shock proteins (HSPs) 45
Heat stress 151
Heavy metal accumulation 108
Heavy metals 107
high light intensity 22
Hormones 23
human activity 198
human consumption 4
human population grows 4
hurricanes 4
hydraulic conductivity 85, 88, 102
hyphal network 129, 147

I

inadequate rainfall 22
irrigation 44, 58

J

Jasmonates 23, 38

L

Late-embryogenesis-abundant protein (LEA) 168
Leaves 84, 86

Lipids 169

M

macronutrients 184
Maize 168
metals 128, 143, 145, 146
meteorological drought 4
methylation 171
micronutrients 184, 185, 186, 188
morphological traits 64
mycorrhizal fungi 128, 129, 131, 132, 135, 137, 138, 139, 141, 143, 145, 146

N

nanofertilizers 184, 185, 188, 192, 194, 195
nanopesticides 184, 192, 194
nanotechnology 184, 187, 189, 192, 194, 195
natural hazard 4
next-generation sequencing (NGS) 107

O

Oryza glaberrima 150
Oryza sativa 150, 156
osmotic adjustment 151

P

peroxidase 153
pesticide 128
pest infestation 198
pest management 198
phenolics 153, 159
phenology 85, 99
phenotypes 150
photosynthesis 85, 89, 90, 91, 92, 95, 96
photosynthetic metabolism 85
phytochelatin synthesis 130
plant cell 64
Plant genetic engineering 184
plant growth 129, 132, 134, 135, 136, 138, 139, 143, 145
Plants 45, 49, 50, 51, 53, 54, 59, 169, 170, 171, 179
plasma membrane 62, 71
Precipitation 220
Proline 87
Protein glycosylation 171
Protein kinases 171
Protein lipidation 171
Protein phosphorylation 168

Q

quantitative trait loci (QTL) 104

R

rainfall 4, 15
reserve storage 85
reservoir storage 220
RNA-sequencing data 107
root growth 63
runoff 220, 223, 226, 230

S

salicylates 23
salinity 22, 24, 25, 26, 38, 41
sanitation 4
seed dormancy 151
seedling 44, 51
seed morphology 151
Single nucleotide polymorphisms (SNPs) 201

small ribonucleoprotein (snRNP) 171
Snow accumulation 220
soil erosion 184, 190
soil health 129
soil moisture 44, 220
Soil salinity 84
soybeans 168
Stomatal function 106, 107
stomatal morphology 104
stomatal regulation 63
stomatal traits 104, 105
stomata management 104
streamflow 220, 221, 228, 230
stress hormone 63, 68
stress memory 22, 33, 35, 36
sulfation 171
Superoxide dismutase 153
sustainability 198, 204, 212, 216
sustainable agriculture 184
sustainable consumption 198
sustainable production system 198

T

thunderstorms 4
Tocopherols 153
tornadoes 4
triticale 44
Turgor 86

U

ubiquitination 170, 171

V

vesicular-arbuscular mycorrhiza (VAM) 132

W

water resources 4, 9, 15, 17, 184
Water scarcity 227
water security 220
water systems 130
Wheat 44
Wheat cultivars 105

Dedication

Kevin O'Connor

Barbara Evans

The Most Outstanding Editors in the World

Acknowledgements

We express our deep appreciation to Kevin O'Connor and Barbara Evans for their outstanding editorial work. An author could not have better editors to work with. Special thanks goes to Jimmy C. Chang, MBA, a system administrator at the Rand Corporation, for his valuable work on Chapters 4 and 5. Mr. Chang manages over 2,000 systems worldwide and provides support on 100 UNIX and Windows servers.

Contents

The Authors		vii
What This Book Will Do For You		ix
1	Introduction	1
2	Capacity Design and Management	36
3	TCP/IP Basics	49
4	Wiring Networks	70
5	Fault Tolerant Networks	119
6	LAN Switching	147
7	Data Storage and Management	162
8	Java Applications and Applets	173
9	Client Server Technologies	197
10	Network and Internet Security	230
11	Wireless Networks	270
	Glossary	286
	Index	293

The Authors

Anique Qureshi, Ph.D., CPA, CIA, is Professor of Accounting and Information Systems at the University of Tampa, and a computer consultant in network applications to business. He is a certified Java programmer and a web master. Dr Qureshi has authored books for Glenlake Publishing, American Management Association, Southwestern, and Prentice-Hall. He has also contributed chapters to books published by McGraw-Hill. His articles have appeared in *Accounting Technology*, *CPA Journal*, *Management Accounting*, *National Public Accountant*, *Journal of Business Research*, and *Internal Auditing*.

Marc H. Levine, Ph.D., CPA, is a computer consultant and Professor of Accounting and Information Systems at Queens College. Dr Levine was previously associated with Deloitte and Touche, CPAs. He has authored books for Glenlake Publishing, American Management Association, Southwestern, Prentice-Hall, and Warren, Gorharn and Lamont. He has authored many professional articles including in the *CPA Journal*, *Practical Accountant*, *National Public Accountant*, *Michigan CPA*, *Journal of Corporate Accounting*, *Accountants Record*, *Massachusetts CPA Review*, *Virginia Accountant Quarterly*, *Cost and Management*, *Management Accountant*, and *The Accountant*.

Jae K. Shim, Ph.D. is Professor of Business Administration at California State University, Long Beach. Dr Shim received his MBA and PhD degrees from the University of California at Berkeley. For over 20 years a consultant on information systems and computer applications, he is now president of the National Business Review Foundation, a management and computer consulting firm. Dr Shim has more than 50 books to his credit and published some 50 articles in professional journals, including *Journal of Systems*

Management, Financial Management, Journal of Operational Research, Omega, Data Management, Management Accounting, Simulation and Games, Long Range Planning, Journal of Business Forecasting, Decision Sciences, Management Science and *Econometrica*. Dr Shim has received a Ford Foundation Award, a Mellon Research Fellowship, and an Arthur Andersen Research Grant.

Joel G. Siegel, Ph.D., CPA, is a consultant to businesses on network applications, and a Professor of Accounting and Information Systems at Queens College. He was previously employed by Coopers and Lybrand, CPAs and Arthur Andersen, CPAs. He has served as a consultant to numerous organizations including Citicorp, International Telephone and Telegraph, and the American Institute of CPAs. Dr Siegel is the author of 65 books published by Glenlake Publishing, American Management Association, Prentice-Hall, Richard Irwin, McGraw-Hill, HarperCollins, John Wiley, Macmillan, Probus, International Publishing, Barron's, Southwestern, and the A1CPA. He has written over 250 articles on business topics, many on network applications to companies. His articles have appeared in such journals as *Computers in Accounting, International Journal of Computer Applications in Technology, Financial Executive, Financial Analysts Journal, CPA Journal, National Public Accountant*, and *Practical Accountant*. In 1972 he received the Outstanding Educator of America Award. Dr Siegel is listed in *Who's Who Among Writers* and *Who's Who in the World*. He formerly chaired the National Oversight Board.

Jeffrey Brauchler, CPA, is a computer consultant to business. He is the author of books published by Glenlake Publishing, Southwestern, and the American Management Association. He has contributed chapters to several books published by PrenticeHall. He is the author of numerous articles.

Roberta M. Siegel, is a computer consultant whose expertise includes both hardware and software applications, including network systems. Mrs. Siegel was formerly affiliated with Met Corporation and Lebensbaum and Russo, CPAs. She has written books published by Glenlake Publishing, the American Management Association, Southwestern, and Prentice-Hall. Mrs. Siegel's articles have appeared in the *CPA Journal, Computers in Accounting, Prentice-Hall Corporation Service, New Accountant, Personnel Management*, and *Retail Control*.

What This Book Will Do for You

The *International Handbook of Computer Networks* is a valuable reference for anyone directly or indirectly involved with networking, such as information systems and network managers along with support staff. It covers the state-of-the-art issues in network technologies in the business world so those involved with information systems are kept up-to-date. Future developments in network technology are also presented. The book spans the chasm between the network professional and management. Computer networks are an integral part of daily operations of companies. Network appplications and support are presented in detail. A failure of the network system is likely to have a critical impact on the organization. Potential vulnerabilities in a networked system could undermine operations and, therefore, must be minimized or eliminated. Fault tolerant networks and other methods for continued operation of the network are discussed. We provide ways in which problems may be avoided. Data integrity is crucial. Network security is a major concern which is fully addressed.

Network technology is changing rapidly. The purpose of this handbook is to help managers select, implement, and manage network technology. You will learn how to get more out of your network. Networks allow businesses to enhance productivity, improve quality and performance, and lower overall costs. The book will aid in either implementing a new network or upgrading an existing one with minimal operational difficulty.

This handbook covers all phases of network management. It clearly shows how to get the most out of networking in the organization. The book allows management to better communicate with network professionals. For example, we discuss how to support and manage the network department.

While discussing the recent advances in network technology, each chapter also focuses in on the needs of management. For example, when discussing Java Essentials, we emphasize personnel and human resource constraints in software development and support. We discuss network benefits, communication characteristics and applications, network configuration, types of suitable networks depending on the circumstances, client-server usage, network hardware and software, and data management. Tips are provided on getting the most from network vendors.

We describe the interconnecting of networks such as connecting smaller networks to create larger ones (e.g. Intranet, Extranet, WANs). Further, we describe how to best share information and files generated from the network throughout the organization.

A knowledge of the contents of this book will help in preventing problems before they arise. This includes careful monitoring of the system, preventive maintenance, benchmarking, and baselining. There is a presentation of troubleshooting techniques and approaches including what equipment to use and formulating the most useful strategies. We discuss how to keep the network operational and dependable even when problems arise such as server or hub failure.

The book includes coverage of the standard protocals used to access LAN physical media as well as those used to implement wide area networking. Bridging and switching relates to protocols and technologies applicable to connecting subnetworks. Routing protocols are used to route data through an internetwork. We discuss transmission and Internet protocols, control aspects, and applying protocols.

We discuss the wiring of networks, as well as taking into account possible alterations in floor plans. In making best use of a wiring network, consideration must be given to wiring standards, transmission, topology, access, and noting warning signs of potential problems. Included is a discussion of the use of 'wireless' technology applications including the pros and cons of such use, LAN standards, and considerations of infrared and radio transmissions.

Aspects of LAN switching must be considered by the manager to achieve maximum benefit of use. Of special interest in effectiveness terms are routing, switching modes, protocols, and results achieved. Switching may involve various layers.

The manager must consider WEB support and should be aware of Java and its utilization in the networking environment. Dealing with malicious applets, as well as security and control aspects are discussed.

We present the proper procedure of storing and managing data, including various kinds of backup, compressing and standardizing data, data migration, archiving, and recovery of information.

The reader learns how to best manage and control legacy applications systems that require special linkage to the network software. The best approach involves establishing a suitable profile to guide the system administration in upgrades to the network. Managing special backup procedures is discussed.

We provide the network administrator with the knowledge to control abuses and provide the optimum level of service to users with laptops and notebooks in the field who need access to the LAN for work.

Some of the other major areas discussed in the book are capacity design and management, upgrading the system, and operations management.

We anticipate future developments in networking such as satellite communications. This book is designed as a practical, 'how to' guide. We provide extensive examples to illustrate practical implementation and application of networks. The tools and techniques in this handbook can be adopted outright or modified to suit individual needs. Checklists, charts, graphs, diagrams, tables, exhibits, illustrations, and step-by-step instructions enhance the handbook's practical use. Answers to commonly asked questions are given.

This book addresses a wide range of computer network issues. It provides practical and thorough guidance. It helps managers improve network operations throughout their organization. This book is comprehensive and detailed covering all major areas of networking in a clear understandable manner. The managerial, technical, and strategic aspects of LANS and WANS, with practical advice on how to plan for, prepare, and manage the implementation of networks is provided.

A glossary of terms is provided. A specific area of interest may easily be found by referring to the index.

1 Introduction[1]

A network is a connection of computers and devices to share data, files, messages, applications, resources, software, and hardware. Networks have many business applications including collecting information and inquiry/update. Network architecture involves formulating the methods, operations, and functions associated with data formats and procedures to facilitate business applications. There is architecture switching and transmission equipment. The sharing of computers and peripherals is cost-effective and results in efficiencies. Most companies are geographically dispersed including multinationally. It is required that computers at each location transfer and exchange data daily and perhaps on a real-time basis. Networks exchange information in an expeditious manner as well as preserve and safeguard data.

A network stand-alone system is a dedicated network encompassing both remote and local data sources. An example is an arrangement in which branch offices of a company are interconnected to a headquarters computer. Another example is communications between departments and/or divisions of a company. The system includes provision for questioning/response processing of a database.

Network hardware allows for the interconnection of many smaller networks into larger networks including LANs and WANs.

Physical network design is designing how software and hardware components are combined to build a network to accomplish predetermined goals. The design should take into account growth opportunities. In a large network, network components include remote computer systems, remote terminals, host processors, and channels connecting these components. Network equipment and devices should be matched to maximize performance. When designing the network, the following should be taken into account:

- Reliability. Reliability should exist to assure data traffic is properly received. Network reliability consists of delivering error-free data as well as recovering from errors or lost data. The network's diagnostic system should be able to locate problems with components after identifying the affected network elements.
- Response time. Response time can be improved by lessening the delay between transmission and receipt. This is particularly crucial for interactive sessions between user applications. Throughput relates to transmitting the maximum amount of data per unit of time.
- Transmission of data along the least-cost path within the network. This is typically the shortest channel between devices and involves the use of the least number of intermediate components. Additionally, high-priority information can be transmitted over costly high-speed satellite channels, while low-priority information can be transmitted over relatively inexpensive telephone lines.

Backbone is a common way to connect networks as depicted in Figure 1.1.

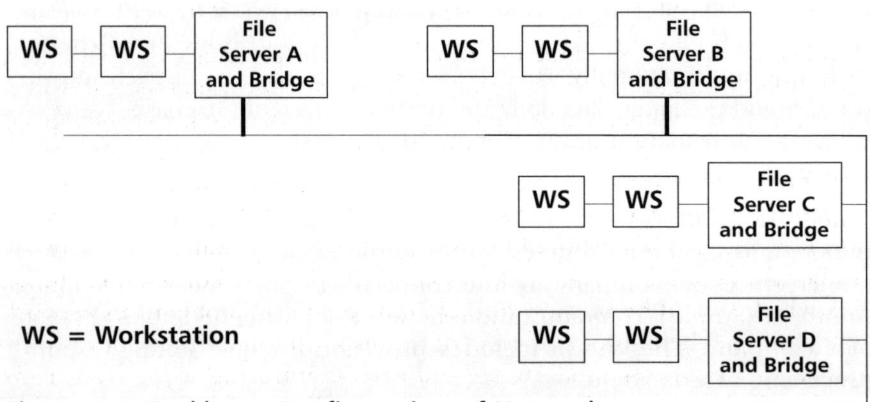

Figure 1.1: Backbone Configuration of Networks

Network binding refers to having a logical linkage of clients, protocols, and board drivers.

A network interface card (NIC) connects a computer to the network. The card must be compatible to the particular configuration. A deficient NIC results in a network slowdown. A transceiver is the component of a network interface sending and receiving signals.

Regional networks furnish a bridge between local access networks and various backbone networks.

System software directs computers. Application software performs functions and activities. Network simulation and modeling software appraises the network's configuration and undertakes what-if scenarios. Point products are software written to manage or rectify network management issues. Message control software is a program that controls the transmission of messages to and from remote terminals. A message data set is disk storage including messages awaiting communication to a specified terminal or host. To assure message security and control, one can feed received messages back to the initiating sender for comparison. A message queue is a listing of messages to be processed or about to be transmitted to a terminal.

Media refers to the medium over which the message is communicated. The media transmitting a message include telephone, radio waves, infrared, microwave, wireless, fiber optic, and satellite. Fiber optic cable is network cable in which transmission is by using light rather than electricity. The cable is typically used as the backbone in a large network especially over long distances. A bulk retrieval mechanism is the ability to retrieve various pieces of network data at a time from a particular agent. Nonprivileged mode refers to limited and restricted data access on the network system.

Bandwidth means how much data can be transferred across a wire at a particular time. As more information passes, the available bandwidth decreases. The type of media used impacts cost and speed. Network speed depends on the type of server, data to be transferred, connection speed, and number of users. Connectivity speed associated with the network depends on how much information is to be transferred (e.g., in gigabytes) and how often. There should be network tracking to watch over network traffic and identify bottlenecks. A bottleneck is the weakest or slowest network link which should be improved to prevent potential problems in network functioning. The bottleneck slows down response time. A network is only as fast as its slowest link! In pre-emptive monitoring a program runs continually to diagnostically test network traffic notifying the managing station when a failure is detected, and providing useful information about the occurrence. It is important to know file size to be transferred so as to properly ascertain hardware component specifications such as the size of the hard drive. You must know file size to be transferred so as to buy the right size hard drives and arrive at the network speed required. NOTE: The speed of a server's processor has little to do with the network's overall performance.

A packet refers to an information field having protocol data. A data packet is a data unit transmitted over a network among devices. The packet contains a header, addressing data, and the information itself. The final section of the data packet, the trailer, comprises an error-checking feature for control purposes. In a packet-switched network, data packets travel one at a time between the message source and message destination. A network analyzer appraises network traffic such as decoding data packets. Data bits are grouped and organized in a predetermined, structured manner. Virtual circuits are paths formed in connection packet switching networks.

The three methods of data flow between network computers are simplex, half-duplex, and full-duplex. With simplex transmission we move in one direction only such as the case with television or radio transmission. Because of the one-way approach of transmitting information, it is rare to have simplex transmission in computer networks. With half-duplex transmission, data flows in both directions. Unfortunately, though, the data cannot flow in both directions at the same time because once a query is transmitted from one device it must wait for a response to come back. Many network systems make use of half-duplex transmission. In the case of a full duplex system, data is transmitted in both directions simultaneously. There is not the intervening stop-and-wait feature of half-duplex. Full-duplex transmission is often used because of its fast response characteristic.

To direct data through the network to its ultimate destination, data switching equipment is used. Data is routed around busy or failed devices or channels. A network buffer is a storage unit to make up for a variance in data flow rate that is received vs. transmitted in the communications process.

Directory services organize the network according to a hierarchy of objects.

In a network, domain refers to the resources under the control of one or more host processors. Domains may be broken down into zones (subdomains). Primary domain controller refers to a primary server for services to network users and resources. In multi-tasking, multiple applications are run simultaneously.

A node is a network connection point or device. Workstations are often called network nodes. Workstations manage network traffic, run local programs, communicate with peripherals, and access the server. A node can act as the control area to transmit information to other sections of a network. In some instances, a node may undertake local processing activities. Trusted node refers to workstation-based security software.

Networking is connecting multiple sites for data transmission. A bridge refers to the connection between two similar networks. A subnetwork is part of a network through linkage by a bridge, router, or gateway. An internetwork is formed with routers linking smaller routers in the network. It is possible to route or bridge networks going at varying speeds.

Network hops refers to how many routers there are which join two nodes of the Internet. For example, assume host 1 is on network A, which routes to network B, and host 2 is on network C. This means that 1 is three hops from C. As the number of hops increase between two nodes, the more latency existing of transfer between nodes.

A network access device is a connection device to the Internet such as a modem or router which allows access to a client PC, server, or LAN. Interoperability assists in having incompatible networks perform together. In deciding on a particular network medium, consider the following:

- Speed.
- Physical layout of equipment.
- Geographic dispersion.
- Number of users who need to access different types of information and equipment. Who are these users and what do they need to access?
- How often updating is required.
- Type of network operating software to be used.
- Nature of business.
- Support for network.
- Number of workstations.
- Ability to expand network such as with additional workstations.
- Appearance.
- Cost.
- Compatibility.
- Processing requirements.
- Technical reliability.
- Types of applications.
- Application software to be used.
- Maintenance.
- Limitations placed on PCs.
- Ability of network to perform tasks without corrupting data moving through it.
- Requirement to access disparate equipment such as other networks and mainframes.

- Ability to store data.
- Connectability mechanism.
- Fire safety.
- Noise.
- Nature and size of files to be accessed.
- Frequency and time associated with user access by file.
- Internal and other considerations associated with backing up files.

A peer-to-peer network is when any computer on the network can act as a server. Peer-to-peer refers to two sides of a communication link using the same protocol interface to perform the network transmission. It is simple to implement this type of network. Peer-to-peer network operating systems are local area networks accomodating networking functions for workgroups, departments, and divisions. It is suitable for small businesses. In a peer-to-peer configuration, a central server does not exist to administer sharing. Instead each computer shares its resources. The configuration of individual workstations can either be used as a client or a server or a combination of the two. Peer-to-peer networking is appropriate if there is no desire to have the significant processing capabilities associated with a client/server framework. If a peer-to-peer network is to be expanded, then a dedicated file server should be added. Stand-alone computers can be turned into a peer-to-peer network by using cabling and a network interface card.

Peer-to-peer topology is usually used when there are less than fifty users. Figure 1.2 depicts the peer-to-peer configuration.

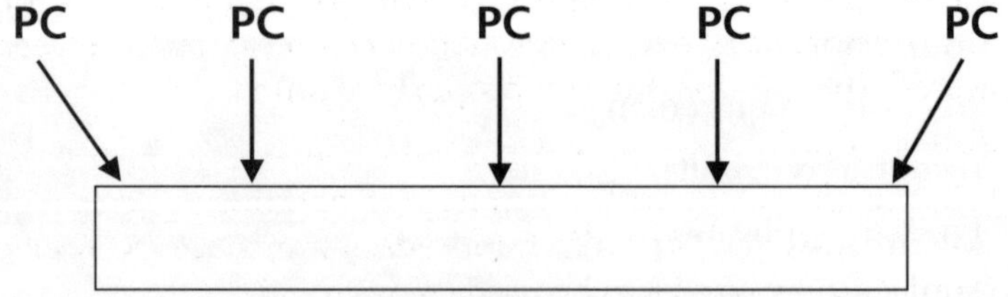

Figure 1.2: Peer-to-peer Configuration

The lack of centralized control results in greater network security risk. Password security is mostly all there is.

In conclusion, with a peer-to-peer network, there is a linkage of computers with no central control. Equality exists among the computers.

A hybrid network is a client/server network but also has peers sharing resources within the workgroup. A workgroup refers to either devices or users in a network grouped by location, position, or function. Workgroups of people could also be dispersed among multiple network segments. There may be many workgroups.

An extranet is a network using Internet technology to link a company to outside related parties such as customers and suppliers.

Many businesses have created virtual private networks (private WANs) via encrypted transmission over the Internet.

According to the International Standard Organization's OSI Model on the manner of communication and conformity of computer interaction on the network building block, there are the following layers in order: physical (communication hardware equipment), data link (data transmission and error correction), network (guidelines on handling different communication signals), transport (proper data transmission), session (user access of network), presentation (data translated and expressed in comprehendable form), and application (questioning and use of the database). The network control layer in the architectural setting is that layer of logic furnishing control among adjoining network nodes. A later layer depends on the earlier one(s).

Network Planning

Prior to initiating a network, a thorough system study should be undertaken. What is the funding for the network? What is the degree of security? What kind of activities should be accomplished? What is the traffic on the network? How many users will there be? After all crucial questions have been asked and answered, the network may be properly planned.

The network should be designed based on our expectation and the functionality of the network. For example, a nationwide retail store may need a data processing center with hundreds of workstations connecting to different retail locations to retrieve information daily.

In distributed processing, more than one computer system is used for processing. Each computer in the network could, if properly designed, process data locally. Thus, the data processing workload can be spread over many workstations. Communication cost can be reduced because most data does not have to be transferred to a central computer for processing. Distributed systems are more reliable than central processing systems because each workstation shares the workload.

Performance and Troubleshooting in the Network

The network administrator is responsible for establishing, maintaining, managing, controlling, and solving problems as they arise in the network. The manager must be alert for unanticipated occurrences and problems. Proper planning will result in quickly and effectively handling network difficulties before they get out of hand. The manager must watch out for possible hardware malfunctioning and conflicts.

Network baselining determines how actual network performance compares to expectations. Exceptions are identified and evaluated with a view toward future improvement. Network baselining tools monitor network performance over long time periods and report on deviations from baseline data. The manager should log important events affecting the network such as error incidence and suspicious logon attempts. A fault tolerant network is one that can recover from insignificant mistakes.

Network performance may be improved by altering the configuration such as by segmenting the network so throughput is increased. A segment is a string of computers. In segmenting, an additional network card is added to the server with workstations being transferred to the new card. Hence, we are adding another physical network. The segmentation of a network enables traffic on each part to be more efficient and quicker.

The symptoms of a hardware problem must be quickly noted. The manager should note that moving parts are less useful and efficient over time. Preventive maintenance or replacement at periodic intervals is called for. On the other hand, stationary parts typically will function during the equipment's life. Thus, the network manager should carefully scrutinize each component, identify signs of potential hardware failure, and take corrective steps to alleviate or prevent the problem from taking place.

Unfortunately, hard drives most often malfunction eventually. By the third year, failure rates may approach 40%. This is a major concern because the hard drive is where the information is stored. Hence, it is a must to back up the hard drive in either the form of a backup device, mirroring, or RAID. Further, backup devices should be placed in all clients and servers.

Although CD ROMs and floppies have greater reliability than hard drives, they consist of moving parts so failure is possible. Replacements should be available when failure occurs. Of course, all data should be backed up on at least one other CD ROM and floppy disk.

In looking at video aspects, keep in mind that the failure of a video card or monitor causes the other to have a problem because they are integrated.

Advantages

Advantages of networks are:

- Sharing of information and resources in an efficient manner. If one location has work in excess of what it can handle, the network can transfer the work to another computer in the network. This load sharing improves productivity via a more even utilization of a company's resources.
- Flexibility in the working environment. For example, employees can telecommute from home or when traveling on a bus to remote locations. No matter where the worker is, he or she can access key data quickly such as sales figures by customers and inventory stocking by product.
- Improved customer and client service.
- Better financial information about the business.
- Improved communication throughout the organization.
- Allows for data to be current because it is continually updated.
- Improved resource management.

Network Operating System

A network operating system (NOS) is at least two independent computers working together as one system. The NOS is software installed on clients and servers that deliver messages over the network and instruct the network on what to do. The NOS provides services for users and application programs. The NOS controls the activities of a network system, access to applications and peripherals, and resources. The NOS stores information as to those individuals who can access what data and in what manner and at what time. The NOS is software functioning on a network server to provide such client services as files, applications, and printing.

Connectivity enables applications to transparently communicate with other programs or processes irrespective of their locations. The NOS furnishes services including routing, distribution, and network management. Integration means that more than one NOS is running at the same time and interacting.

A NOS enables the following:

- Sharing of files and peripherals such as disks.
- Data integrity and security.
- Tools to add, alter, and remove users, computers, and peripherals from the network.
- Trouble shooting tools in monitoring the LAN.
- Interoperability between different NOSs and applications.
- Support functions such as database.
- Interoperability across other networks using telecommunications.

An example of a NOS is Windows NT. Lantastic is a popular peer-to-peer NOS.

Local Area Network (LAN)

A local area network (LAN) is a private communication network owned and operated by the user encompassing a limited territorial location linked such as one building, plant, or a few buildings situated near to each other. Computers are connected at the same site. Data switching equipment may be used. There may be multiple servers. The usual range is about one mile or less. A LAN is for a business segment, department, or division within a company. It is appropriate for a small business. There is usually substantial volume transfer. It links computers, terminals, storage units, printers, etc. A key objective of a LAN is to share applications and databases.

Internetworking refers to merging multiple LANs into one at a single location. In this case, each node becomes a cluster of stations (subnetworks). The LANs communicate with each other. An example is a company's internetwork of LANs located at ten floors of its headquarters building. Figure 1.3 depicts internetworking.

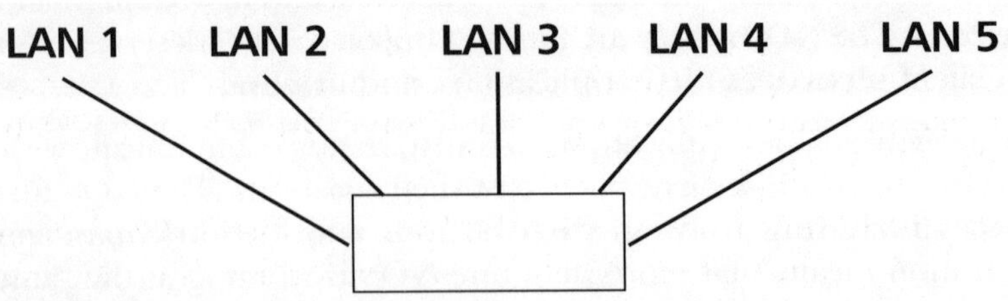

Figure 1.3: Internetworking

LAN configuration tends to be more structured. Because the channels in a LAN network are costly, the owners of a LAN are usually not concerned with the full utilization of channels. LANs are more structured and ordered than WANs. LANs are quick, flexible, compatible, reduce errors, lower processing cost, have ease in information flow, and maximize equipment utilization. LANs use ordinary telephone lines, fiber optics, coaxial cables, and other devices such as interfaces. While fiber optics are costly they do result in good performance and reliability. The performance of LANs is dependent on transmission bandwidth, protocols supported, and physical design. Bandwidth refers to the frequency range of a channel and takes into account transmission speed. However, as more devices are added to the network, the transmission speed will be decreased.

A LAN enables a company's computer-based equipment to communicate and share resources. There is a linking of workstations either with a PC-based server or a mainframe to share information providing high speed and effective client/server functionality. LANs provide greater reliability because each workstation is independent.

LAN resource management software enables access to shared network resources and services. LAN software architecture is comprised of network operating systems, network management software, application software, and security software. A network operating system is software enabling a user to manage the network's resources. The NOS runs on both the client operating system and the server computer. The activities performed by the NOS include network security such as controlling access and usage, printer management such as ranking printing jobs in importance, managing files such as storage and transferring software to clients, and administration such as changing and/or organizing client users as well as conducting maintenance functions like backing up data. A network interface card interfaces the computer and the network. The card serves as the communication protocol understood by the network architecture. As an example an ethernet card connects a PC to a bus network.

Network cost is tied to the degree of sophistication and size of the system. Some examples of popular LAN system software are Microsoft's Windows New Technology (NT), Novell's NetWare, and Unix. The Windows NT Server has such services as security, file sharing, print sharing, and limited-distributed computing.

Novell's NetWare is more prevalent in large and medium size businesses. However, it is more of a problem to obtain support and the cost of netware substantially exceeds that of a comparable Microsoft network.

Unix is the oldest network operating sytem. Unix connects all computers in a way that maximizes machine power by distributing data and resources across the network. Features of Unix include data sharing, e-mail, remote login, and file transfer. Unix can run on many different computer systems so there is user portability.

Linux is a free version of Unix. It has all the benefits of Unix and is a complete functional operating system. The features of Linux are file transfer protocol, interoperability, portability, multiuser capabilities, multitasking, virtual memory, speed, and utilities. Windows applications run many times quicker on Linux because of the operating system's advanced buffering capabilities. The Linux kernel is transferable to all kinds of hardware. Thus, Linux is ideal for "smart card" usage. A great number of companies and Internet Service Providers use Linux because of stability and reliability.

Two types of LANs are peer-to-peer and client/server. In a peer-to-peer arrangement, each workstation has equal priority. Data is exchanged by the workstations in the network. In the client/server structure, server computers furnish specified functions for client computers. Each server is designated a particular responsibility to perform.

The components of a LAN consist of a workstation, network operating system, and network interface card. A workstation may be used as a microcomputer, terminal, or server. There are different kinds of LAN servers based on the key activities to be performed in the client/server environment.

LANs can be custom designed to meet different application and site requirements. This can result in one or more network configurations (topologies). The most prevalent LAN topologies are mesh, star, bus, tree and ring.

A virtual local area network (VLAN) links computers by workgroups rather than by geographic area. A VLAN attempts to satisfy user needs. The workgroups can span many network segments. For example, financial managers and staff can be assigned to the same workgroup.

The benefits of interfacing networks are:

- If one LAN breaks down, other LANs continue to operate.
- More reliable and more expensive subnetworks can be used for critical functions.
- Lower total network costs.
- Flexibility in having individual subnetworks to satisfy specific needs.

The drawbacks of interfacing networks are:

- More complexity.
- Some network activities may not be able to go across network boundaries.

Benefits of a LAN include:

- Improved employee productivity.
- Lower cost.
- Accessibility of data by all employees on the network so files, software, and applications are shared throughout the company.
- Greater efficiency and effectiveness.
- Information tracking.
- Faster and better communication.

A company must select a proper network for its business and processing needs. In the next section, we discuss the various LAN topologies including hierarchical, bus, star, ring, and hybrid. Ethernet is also examined. Any given topology has advantages and disadvantages that must be addressed.

LAN Topologies

Topology is the physical layout and shape of the transmission media (e.g., wires) connecting the computers (workstations or nodes) in the network. The manner of the connections is the configuration. Topology also applies to the network's physical layout including where the computers are situated. We must consider the design or network protocol allowing data, video, and audio to travel over a network. The network topology affects its performance.

Hierarchical Topology

The hierarchical topology (vertical, tree structure) is a very common network. There is ease in the software controlling this network. The topology furnishes a concentration point for control and to resolve any errors. However, there is a possible bottleneck and reliability difficulties. There is a risk that network capabilities may be lost if failure occurs on a higher level. Figure 1.4 presents the hierarchical topology.

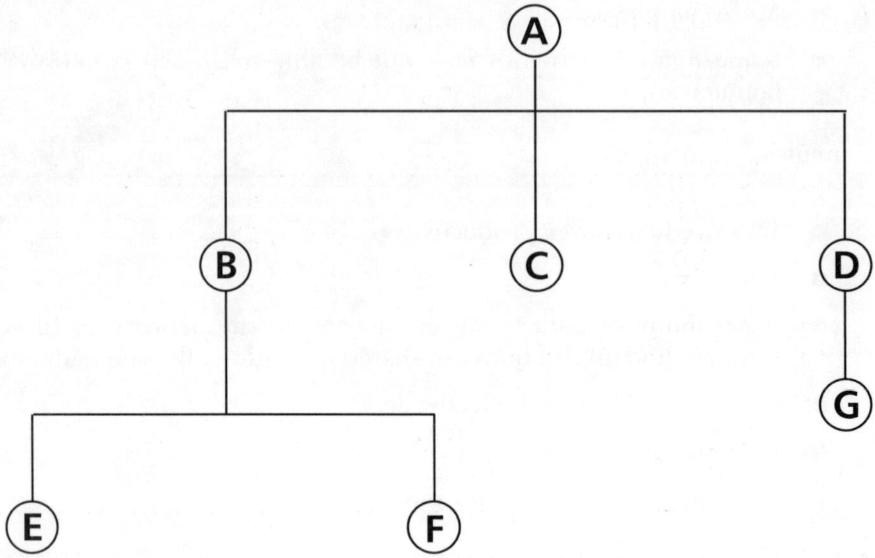

Figure 1.4: Hierarchical Configuration

Bus

A popular horizontal topology in LANs in the form of a linear channel with numerous connected workstations (nodes). There is one cable going from one end to the other end. There is simple traffic flow between devices. Further, data travelling on the bus by a workstation goes in either direction so other workstations connected to the bus receive the transmitted information. A central server is non-existent. Control functions are distributed to all stations on the LAN. Each workstation must detect faults. Each node transmits messages to all other devices. A node receiving a message that was not intended for it will discard the message. If the message was properly sent to the node it will be read. A message can only be transmitted by a single computer at a time so another computer must await the bus being clear if it wants to send its message. The manager must be assured that a signal is properly terminated because otherwise the network can act irregularly. A bus structure has ease in maintenance, and may be organized as a client/server or peer-to-peer network. The bus structure is based on broadcast topology because each message or data set transmitted on the bus goes to each node. Thus, a single station broadcasts to multiple stations. A bus configuration is passive because no strengthening of the signal occurs as it moves along.

Figures 1.5 and 1.6 present bus configurations.

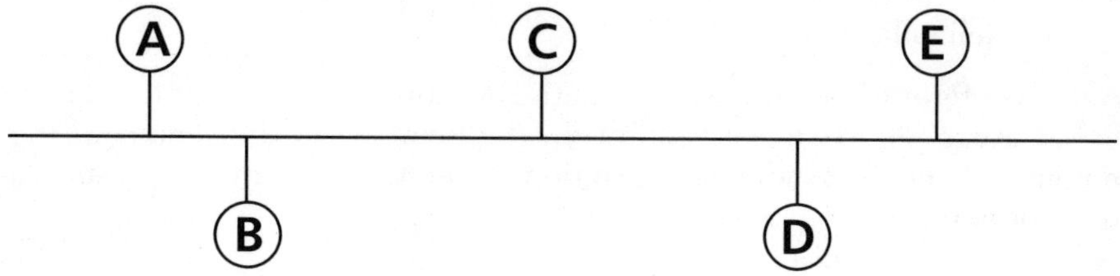

Figure 1.5: BUS Configuration

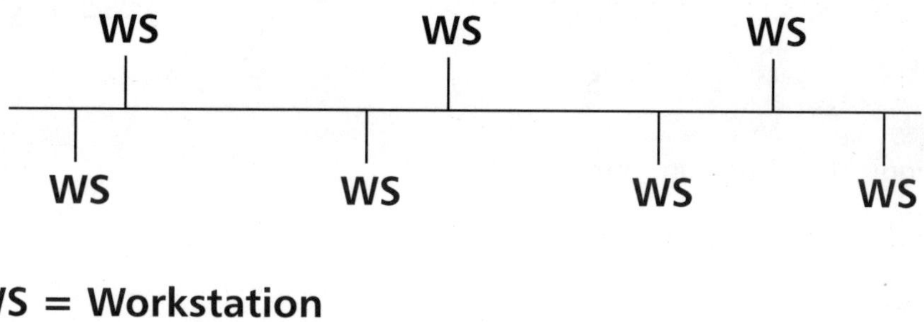

Figure 1.6: BUS Configuration

A problem with the bus configuration is that because all computers share a single channel, a failure in the communication channel or failed computer causes the loss of the network. A solution would be to use redundant channels. Another difficulty with bus configuration is by not having concentration points it will be more difficult to resolve problems. Hence, isolating faults to a specific component may be a problem. A bus network typically requires a minimum distance between taps to reduce noise. To identify a problem, each system element must be checked. Although the bus topology is recommended for shared databases, it is not advisable for single-message switching. It uses minimum topology to fill a geographic area, while at the same time having full connectivity. Another problem is slower operation when significant network traffic occurs.

The benefits of a bus topology are:

- Less costly because minimum cable for linkage is required.
- Simple to comprehend and apply.

- Extension through increasing cabling allows more computers to be added when needed.
- Reliability.

A tree configuration is an offshoot of the bus topology. Transmitted data is broadcast to all active workstations in the network on the shared wiring medium. There is no controlling station on the LAN. Figure 1.7 presents a tree configuration.

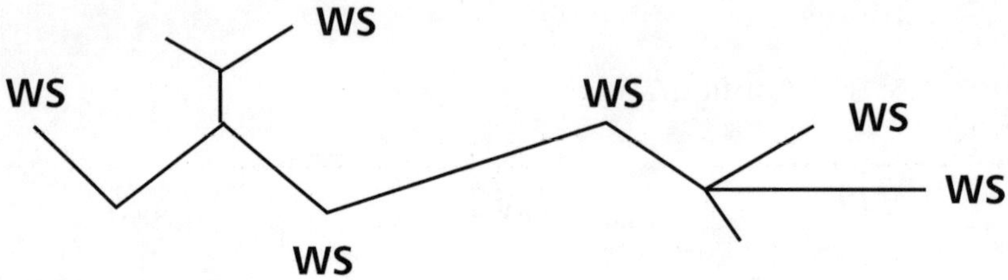

Figure 1.7: Tree Configuration

Star

Star configuration is popular and often used for data communication systems. The software used is not sophisticated. It is simple to track and control traffic because all traffic comes from the hub (center of the star). While the star topology has similarity to the hierarchical structure, it has more restricted distributed processing capabilities. The hub routes data traffic to other parts or elements. Faults are identified. It is possible to have bottlenecks at the hub which may detract from reliability. These difficulties may be minimized via redundant backup of the hub node. A star network is most suitable where we have to enter and process data at numerous locations with day-end distribution to different remote users. Data for general use is transmitted to the host computer for later processing. It is simple to find errors in the system because each communication has to go through the central controller to reach its destination. Even though maintenance is easily performed, the network ceases to operate when the central computer fails. To establish the system initially involves high cost because each node must be hooked up to the host computer. The cost of the mainframe is also substantial. It is easy to expand by running a wire from the terminal to the host computer.

Figures 1.8 and 1.9 present star topologies. Note in the diagrams that each workstation is attached to the hub (central switch) by point-to-point lines.

As the diagrams show, connections between workstations are set through the hub so that workstations can transfer data. The end points can be reached from the hub. The central hub can broadcast a message to all linked computers or just to the intended destination computer. A passive (active) hub does not (does) strengthen the signal to linked computers in the network. A hybrid hub is able to make different kinds of cable compatible in the network.

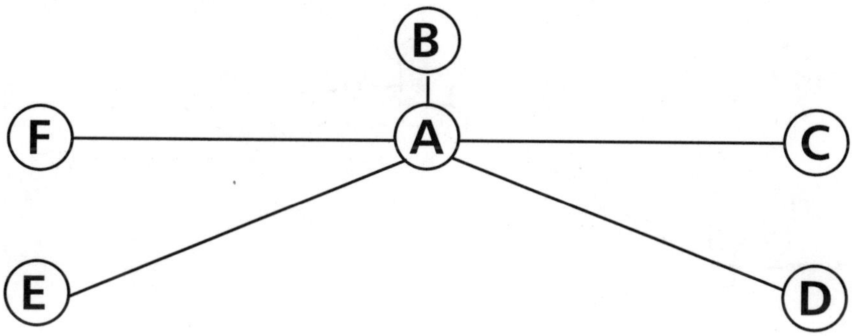

Figure 1.8: Star Topology

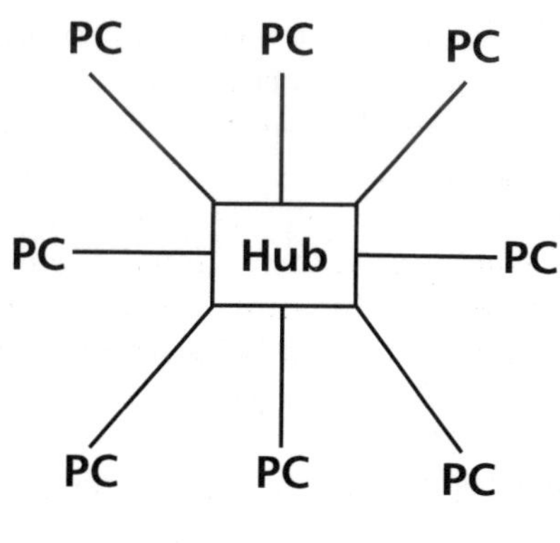

Figure 1.9: Star Configuration

A star network may be enlarged by having an additional star hub as depicted in Figure 1.10.

Figure 1.10: Hybrid Star Configuration

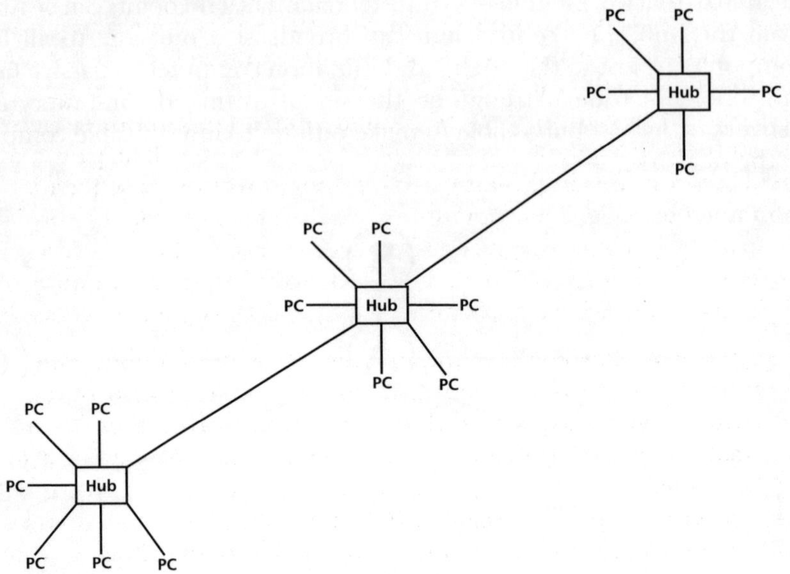

The benefits of a star configuration are:

- Intelligent hubs can track, manage, and control the network and identify problems needing correction. Further, an intelligent hub may be able to set a problem apart so the remainder of the network can continue to function.
- Different kinds of cable may be used with the proper hub.
- It is simple to adjust the configuration by adding additional computers to the hub. Further, a hub can be replaced with another one having more ports.

The drawbacks of a star topology are:

- A special device may be needed for switching and rebroadcast purposes.
- A malfunctioning of the focal hub can cause the entire network to go down.
- Higher expense of cabling.

Ring

Ring is a network structure in which connected computers are in a full circle. In other words, all workstations are connected in an unbroken or continuous loop. It is an active process. The final computer is linked to the first one so cessation does not occur. Each workstation regenerates (strengthens) the data signal as it retransmits the data packet that is circulating on the ring. Messages go in one direction in sequential order from node to node. The computer scans the message from the prior node for an address that it recognizes. A message is read if it has a correct address. Otherwise it is forwarded to the next node. It is uncommon in this configuration to have a bottleneck because of the organized structure.

A ring network has less cost and greater reliability if a minimum level of communication exists between microcomputers. This kind of network is best when several users at different locations have access to updated information on a continual basis. Here, more than one data transmission can take place at the same time. The ring network enables corporate managers to create and update shared databases. With a ring, there is more chance of errors relative to a star configuration since information is handled by numerous intervening stations. In consequence, the manager should suggest that data in a ring system do a full circle prior to being removed from the network.

A problem with the ring network is that one channnel ties all of the parts in a network. In the event the channel between two nodes fails, the whole network can be lost. This problem can be relieved by setting up a backup channel. Other means to rectify this problem are to install redundant cables, and use switches to automatically route traffic around the failed node.

The operation of the ring configuration requires that all connected nodes are operational. The failure of one node stops the ring network.

Figures 1.11 and 1.12 depict ring configurations.

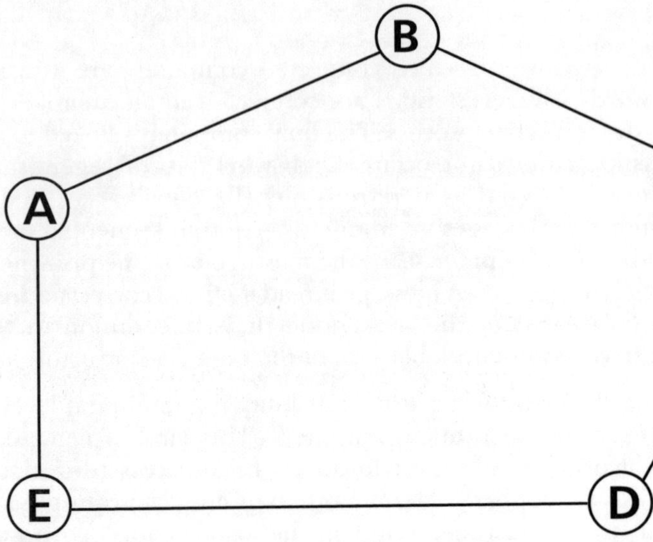

Figure 1.11: Ring Topology

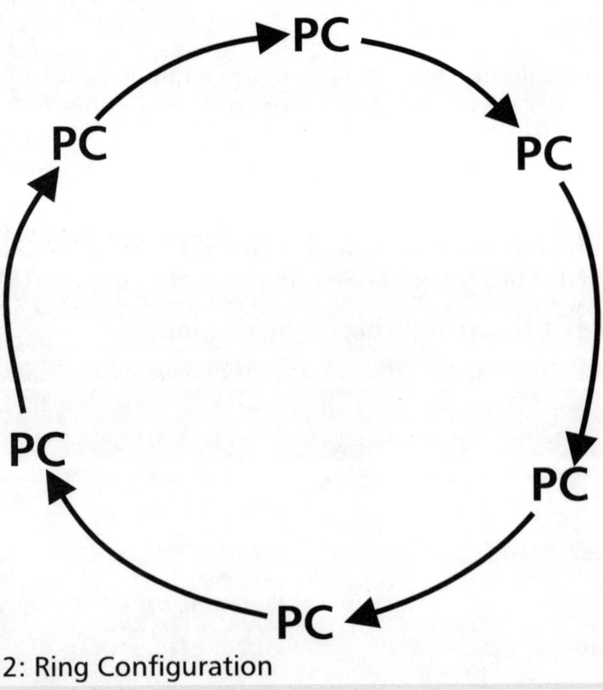

Figure 1.12: Ring Configuration

A ring structure is suitable when there are many clients, desire to maximize performance, and audio and video are needed.

Benefits to the ring configuration are:

- There will be no exploiting of one computer.
- Equitable sharing occurs because each computer in the system has similar access to the message.

Drawbacks to the ring configuration are:

- Identifying problems is not easy.
- A modification in the number of network computers is disruptive.
- If one computer goes down, the entire network is adversely impacted.

Some ring networks involve an orderly token (special data packet) passing from one computer to another one in the network while others do not. We will now discuss the token passing process.

The protocol usually used with a ring topology is token passing where access is controlled by possession of a circulating token. One token is only sent at a time so a message is kept at the local PC until there is a free token available to send that message to the desired destination. A computer can only send a data packet when it has the token. This avoids collisions. A token is a brief message (electronic signal) that goes around the ring until it reaches the desired computer. The token indicates which node is to send or receive the token. That computer will alter the message (token) by, say, adding information and sending the message on to the next intended computer. This movement continues in sequence until the message (token) goes to the original sender. When a PC receives the message it acknowledges it to the originator. Note that just one workstation can transfer data at a time. The next workstation must await its turn. In other words, after delivery, the message is taken off the token so the token is now free to take another message.

CAUTION: There is a degradation with an increase in nodes because the addition of a workstation detracts from the performance of the rest of the workstations.

A multistation access unit (MAU) is a central device in a token ring topology serving as the meeting point for all stations and resources in the ring.

In conclusion, the token goes around the network until a computer with information to send temporarily withdraws the token from the network, makes changes to it, and then reinstates it on the network with the data it must transmit.

NOTE: There are some fast networks that can accommodate several tokens at one time. For example, the Fiber Distribution Data Interface (FDDI) version of the Token Ring has two rings rather than one going through the network. There are primary and secondary rings with data moving in opposite directions.

Advantages of a token ring network are its simple installation, flexibility, support available, and reliability. A disadvantage is its higher cost.

MESH

In a mesh configuration, there are redundant connections among devices. Each device may be connected to the network. This redundancy allows information to be transmitted over diverse paths. Each workstation is connected to all the other stations on the LAN. As such, a lot of wiring is required.

The mesh topology is reliable but complex. The configuration of the network insulates it from bottlenecks and other failures. The many paths facilitate the routing of traffic around busy nodes or failed components.

As more devices are added, it is more of a problem to set up a mesh configuration due to the voluminous number of links required.

Figures 1.13 and 1.14 present mesh topologies.

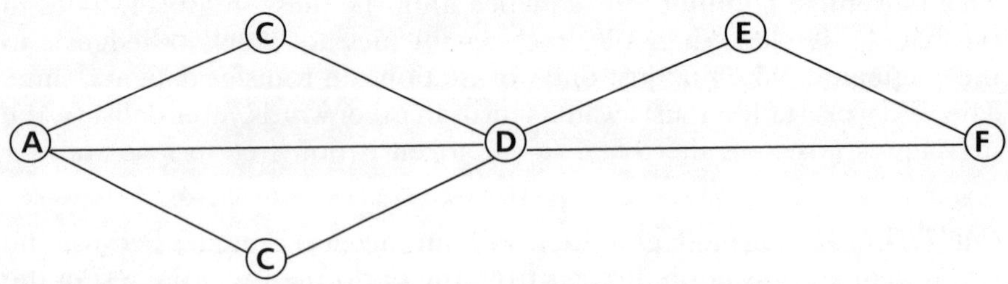

Figure 1.13: MESH Topology

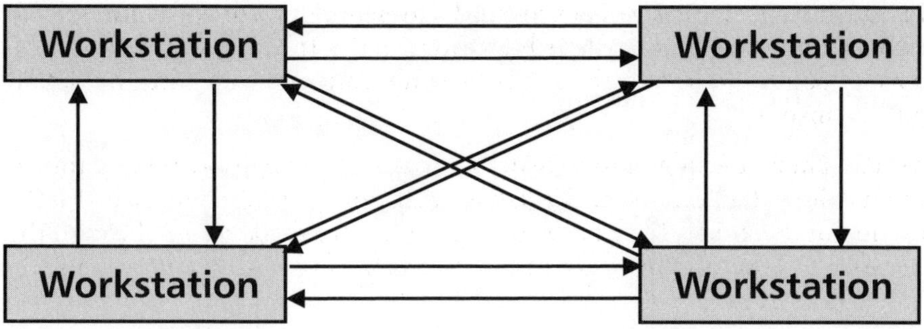

Figure 1.14: MESH Configuration

Benefits of mesh networks are fault tolerance and the ability to identify problems. Drawbacks are higher cost due to the increased redundant connections.

Hybrid

Network configurations may be combined so there may be a merger involving star, ring, and bus. For example, a bus network may connect with a ring topology which communicates with another star network. In determining which configuration to use, consider acceptable transmission time, amount of data to be sent, and the distance between points.

In a star bus arrangement, star hubs are connected to the bus. In the event of a computer malfunction, the hub identifies the problem and sets apart the affected computer. However, a malfunction with the hub will cause a termination of communication between linked computers. Further, the bus is fractured into two sections that are unable to communicate.

Figure 1.15 presents a star bus topology.

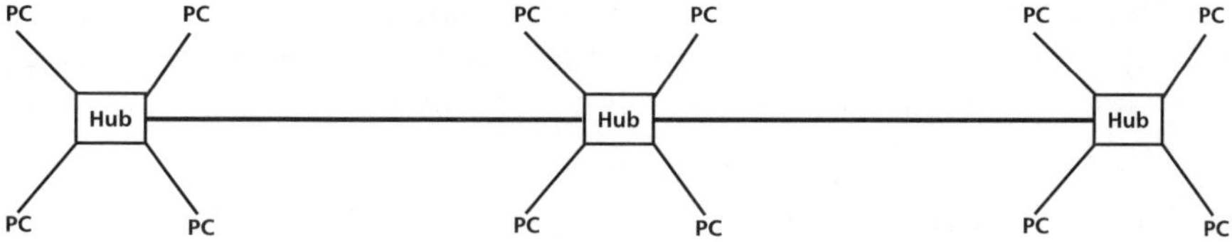

Figure 1.15: Star BUS Topology

In a star ring configuration, the cables are similar to a star topology but there is a ring in the middle hub.

Figure 1.16 depicts a star ring.

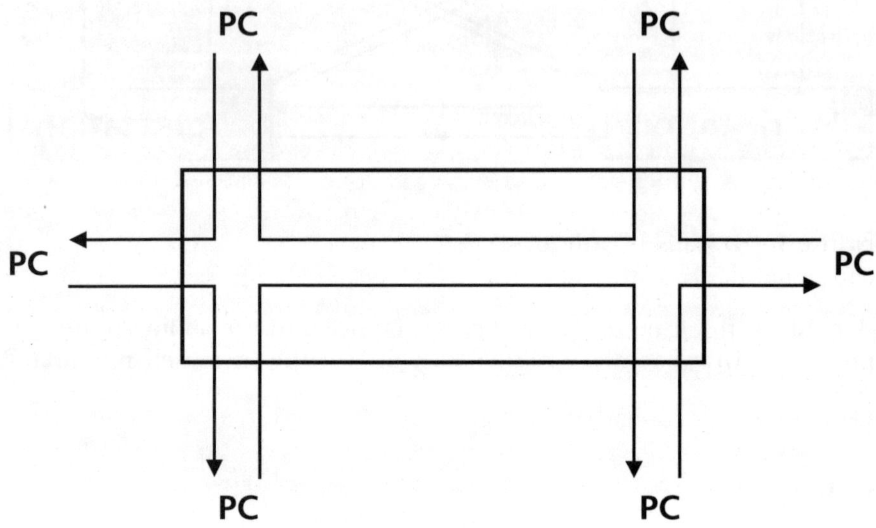

Figure 1.16: Star Ring

Ethernet

Ethernet is a standardized means to connect computers by cables in a network. The Ethernet is the set of infrastructure standards on which a network is constructed. It is the most prevalent network standard. Further, Ethernet LANs can be low cost.

Ethernet is a shared-media network architecture. In an Ethernet, there is communication among concentrators, terminals, workstations, and hosts. Ethernet functions with a star or linear-bus topology. Under Ethernet, there is competition among workstations to transmit data. As the number of workstations are added and become active, there is a deterioration in performance.

In an Ethernet LAN, a network hub is needed to connect all computers on the network. Each workstation must have an Ethernet card installed to provide network capability. The driver for an Ethernet should be installed in the computer. Either coaxial cable or twisted pair wire is required to connect workstations with the hub. An Ethernet in a form of twisted-pair configuration has the benefit of flexibility, isolation of faults, and ability to grow.

Each Ethernet-equipped computer (station) operates independently of all other stations on the network. A central controller does not exist. All stations attached to an Ethernet are connected to a shared signaling system (medium).

Ethernet networks work on the broadcast principle, sending data in units referred to as frames. Each frame contains data about a variety of items, including the sender's and the receiver's addresses. The sender broadcasts a frame that each receiver can see. At any particular time only one computer in the network is broadcasting and all others are receiving. Another computer may broadcast after the first is completed. While all machines on a network can see the broadcasting computer's frame, under ideal conditions only the computer whose address matches the receiver's address in the frame should be able to access the frame's contents.

Ethernet is a protocol standard in which all computers can communicate by sending their data at once. As a result, messages may interfere with each other and fail prior to being transmitted to their destination. As a safeguard, Ethernet LANs can use Carrier Sense Multiple Access/Collision Detection which after noting that a collision has taken place resends the message.

Wide Area Network (WAN)

A wide area network (also referred to as an enterprise network) involves multiple applications and locations. In a WAN, computers are connected over long distances. In other words, a WAN is when LANs are located in different places including remote sites. A WAN goes over local, regional, and global areas. The computers in a WAN may be anywhere from several miles to thousands of miles apart. A WAN utuilizes telephone lines, satellites, microwaves, or a mix of communication channels for signal transmission. However, if geographical separation is very significant, communication links will be less reliable and slower than in a LAN. WAN transmission is either local loop or broadband. Pay local refers to data, voice, and video transmission in a WAN. Typically, wide area networks are privately owned. However, a WAN structure tends to be more irregular than a LAN.

Because the company generally leases lines at signficant cost, an objective is to completely utilize the lines. To accomplish this, data is often routed for a geographical area through one channel. Channels are typically furnished by an interchange carrier (e.g., AT&T). These channels are typically slow and relatively error prone. Dial up services are the most popular connection in WANs because it is the lowest cost in connecting to the WAN.

WANs are important in the client/server environment because the applications typically apply to accessing data stored in separate locations.

A point-to-point network is when a site sends a message to all sites. In this case, it must send several separate messages. Most WANs are of this type.

A centralized WAN is one with a computer linking computers and dumb terminals to a central location.

Most network system software for WANs are custom-developed and typically require mainframe computers to operate.

Figure 1.17 depicts a WAN.

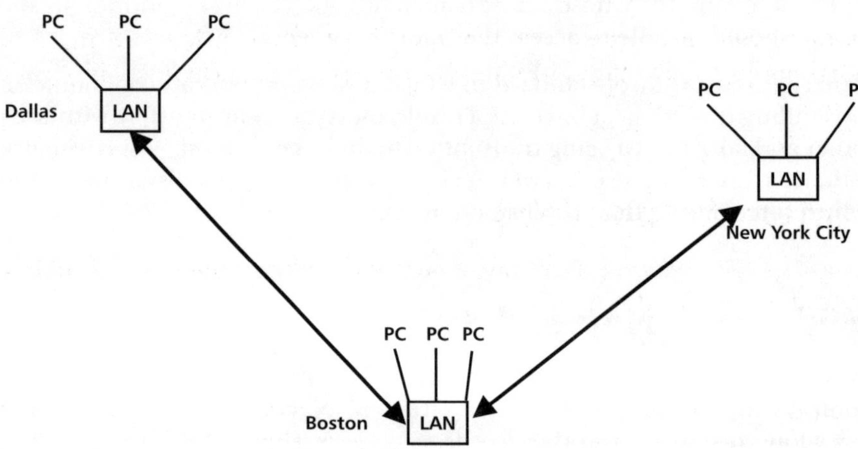

Figure 1.17: Wide Area Network

WANs use two basic principles in sharing data. These are the segmentation of data transmissions between devices into structured block or data packets that include control and address information besides the data itself. Multiplexing is used when these data packets are sent over a shared wide area connection along with other packetized data from other sources. Upon arrival these multiple source multiplexed data packets are demultiplexed and sent to their intended addresses.

Due to the large size of the multiplexed multi-sourced data packets, switching technologies need to be used in a WAN to direct the delivery of data to their addresses by passing the need for direct connections. The two kinds of switching technologies used in WANs are circuit switching and packet switching. With a circuit-switched network there is a switched dedicated circuit created to connect the two or more parties which eliminates the need for source and destination address information. A circuit-switched network is most appropriate when there is continuous data flows. Packet switching depends on a series of switches that relay data packets based on the routes the data packets must follow from their source to their destination. With a packet-switched network, data travels through the network one packet at a time through any available path within the network. Packet-switching is often used when there is batched data or in transaction environments. Frame relay involves transferring packets over a WAN based on packet switching.

A disadvantage of WANs relative to LANs is the difficulty in resource sharing because of limited bandwidth.

A metropolitan area network (MAN) is a WAN restricted to the area surrounding a city. In a MAN, LANs are typically attached within several miles.

Network Functions

Network functions include hubs, routers, bridges, gateways, switches, repeaters, and protocols.

Hubs

A hub (concentrator) is the central connection spot where the cables join. It is the central linkage point for network devices and where information from nodes pass. This is the interconnection point of stations and servers.

An intelligent hub can provide and manage information about how many data packets were transmitted and identify any problems in the transmission process. The intelligent hub comprehends the traffic moving through it. Intelligent hubs should be used because they are task-oriented, enable flexibility in the system, support many LANs and configurations, allow for LAN expansion, include bridges and routers in them, and furnish management and control capabilities. Benefits of this type of hub are fault tolerance and port switching. With intelligent hubs you can track how the network is performing and take corrective action against any difficulties that arise.

A bad hub will result in collisions and a slower network. CAUTION: Hubs if not properly situated in the network may result in bottlenecks.

Routers

Routers furnish data about resources in the LAN. There may be sequential or parallel routing. Routers may be software on a server, or physically-based internal or external devices. They are connection devices handling different protocols so that the best path between LANs is used. Routers can connect remote sites to form a WAN. Routers direct data flow. Routers at one site communicate with routers at other sites. It is possible to configure routers so that certain types (e.g., file transfer protocol) do not permit incoming or outgoing access. It is also possible to enable or disable certain routers to receive data just from specified network addresses.

Routers manage, filter, and control data packet transmission between networks in an efficient and effective manner. Routers decide on the particular information packet to send in an intelligent way among the numerous LAN segments. The best path is chosen from the sender to the destination address. For instance, if one segment of the network is busy, the router may send a message on a less busy route. Some routers are just computers having several network interface cards and specialized software conducting router activities. Routers can connect far networks together via modems. Routers create and maintain routing tables of destinations. Further, a routing information table presents where routers are situated and how far they are from the current router. The routing indicator is the address in the message header specifying the terminal which is to receive the transmitted message.

Many routers can filter out specified kinds of traffic you do not need. For example, information on a partiuclar server may be blocked out to users in a particular network. With certain software installed, routers can add security to the network in the form of packet filtering and firewalls. However, most routers do not provide an audit or security trail.

Routers can link multiple segments into a network. Routers can also link smaller networks into a larger one such as a company's WAN. Further, routers can link dissimilar networks such as Token Ring and Ethernet. Routers make communication easier.

Routers may support multiple protocols. They do not depend on a particular protocol or configuration. In other words, a router is able to receive a data packet in one protocol and then put that packet into another protocol before sending it on. Routers are typically used with protocols supporting a network layer such as TCP/IP. Multi-protocol routers support different mixes of network layer protocols.

A routine code is a mix of one or more digits to route a call to a prespecified area. A pass through a router is referred to as a hop. Data switching equipment routes data through the network. For example, such equipment can route data around failed or busy devices or channels. In dynamic routing, routers send special packets to update other routers.

The manager should answer the following questions about routers:

- Is your network better off with an internal, external, or software router and why?
- Is the router best for the particular protocol?

Bridges

A bridge can link heterogeneous network types. It is a device allowing at least two networks to transfer data among themselves. Bridges lessen any congestion. Thus, a bridge monitors, tracks, and controls traffic associated with data packets. A data packet may be directed to one location or to many locations. In the case of the latter, the bridge will broadcast the data packet to all intended locations at the same time. The configuration of the networks will not be an issue. Bridges also strengthen weak data packets. A defective packet or one without an appropriate address will be stopped. By connecting different types of local area networks, we derive an even bigger LAN. A bridge may also be used to segregate a network into smaller ones, and link segments of the same LAN. Bridges are particularly suitable in highly segregated networks.

Bridges can communicate with each other, and are protocol independent. Bridging is required if protocols do not support internetworking.

Figure 1.18 depicts a bridge.

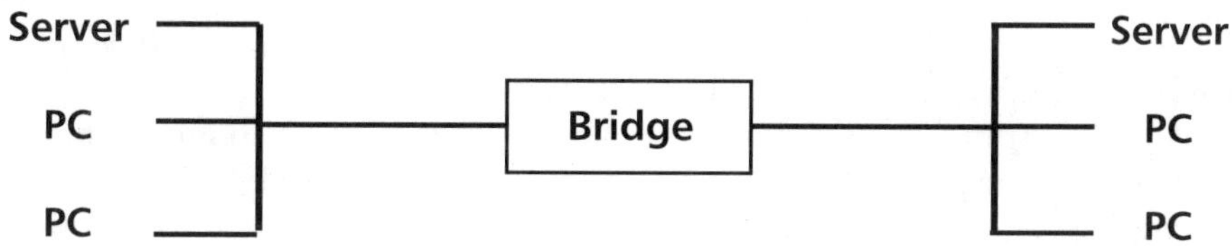

Figure 1.18: Bridge

Advantages of the backbone approach is increased velocity, filtering abilities, and efficiency of data packets. The disadvantage is the higher cost.

A brouter merges the operations of a router and bridge. It can link diverse network configurations and bridge them.

Gateway

A gateway is a device linking dissimilar networks so they can properly function with each other. The gateway is hardware and software required to join different network configurations for proper communication. Gateways are used when the backbone can just support one protocol so the other protocols are translated to it. A default gateway is the router in which data packets arrive at in the case a workstation is unable to tell which is the destination station. Gateways are usually used to connect Ethernet networks to mainframes.

Switching

Switching refers to a computer responsible for handling transmitted information to and from remote computers and terminals. The transmission is routed from sender to receiver. Switching manages the cabling needs of network connection. Each port functions independently. With switching there is a connection of workstations and servers among network segments along with the sharing of information packets. The data packets are sent expeditiously to the correct, intended node. This differs from a hub where such a packet passes all ports.

The advantages of switches are their ability to reconfigure the network to enhance its functioning and to isolate difficulties, and their disaster recovery capability, defect tolerance, and flexibility in topology.

Repeaters

Repeaters are used in a bus configuration to stretch out the distance on a cable run. Repeaters take the data packets transmitted by workstations and make them stronger so the packet can be transmitted over a longer distance without adversely affecting the contents of the data packet. Unless this strengthening process is done, weakness in the packet will occur making it susceptible to hindrance. If the signal is weak already before the repeater gets to it, the problem will be rectified before amplification takes place.

Repeaters are particularly suitable for very distant network stations.

A drawback of a repeater is its repeating all that is received at a port to other ports such as interference. Hence, the number of repeaters to be placed on the network should not be overdone.

Protocols

A protocol sets forth the communication dictates applying to the way network systems are classified by common wiring or geographical location. Network protocols are software implementations furnishing support for network data transmission. A protocol has standardized rules to achieve compatibility. Protocol packages may vary by type of network or for diverse servers on the same network. A protocol suite has numerous protocols that conduct specific activities. To have proper transmission among network computers there has to be a minimum of one protocol set. Protocols govern sequencing, timing, and format.

In selecting a protocol consider its efficiency in transferring data. An appraisal should be made of how protocols are performing. Any errors are noted with a view toward correction. Further, the packet approach allows for correcting errors on only one data packet instead of mandating a whole data file to be retransmitted. A protocol analyzer is hardware and software that searches for network problems and sets forth operational benchmarks.

Hardware protocols relate to the functioning and interaction of hardware devices. Software protocols enable different software to interact such as the communication between servers and clients. At the receiving end, the software takes the data off the wire and prepares it for the computer by taking off all the data added by the transmitting end. The protocols also enable computers to access network devices and services.

Multi-protocol routing software underlies a server network operating systems' ability to process multiple protocols at the same time. It enables interoperability to process and comprehend multiple network protocols as well as to translate between them. The server network operating system must have the capability to support multiple network protocols at the same time so as to maximize not only the different kinds of clients, but also the number and type of other servers with which a particular server can communicate. How many network interface cards can be supported?

Protocols have wide support and are implemented by many vendors, users, and standardized bodies. If computers use a proprietary protocol they can only communicate with proprietary equipment. However, network operating systems must support the LAN protocols.

Protocol conversion is a protocol translation so as to have adequate communication between two network nodes. A protocol translator device is used to convert the information transmission code of a computer or peripheral to that of another so we can have machinery or devices with varying data formats able to communicate with each other. An example may be when there is a transmission from a computer to a terminal.

Management protocols refer to the communication language within a network. Simple network management protocol (SNMP) carries out simple management of network components. It accumulates data about the network's health and conducts diagnostic tests on devices. Management protocols involve hardware arrangement, identifying errors and correcting them, software framework and data format. SNMP is the most commonly used network management protocol. The extended terminal access control system is a remote access management protocol supporting a three-tiered remote access management architecture.

A network file system is a protocol allowing user access to data on remote computers as if they were local.

Advanced program to program communications (APPC) is a series of operations that transaction programs can use to talk to each other. It enables user written programs to conduct transactions in a client/server network.

The Dynamic Data Exchange (DDE) is a protocol for data exchange between applications.

Lightweight directory access protocol (LDAP) is a popular protocol for directory access and administration. It is used to consolidate personnel data files and the management of devices, applications, and security.

Sequenced Packet Exchange/Internetwork Packet Exchange (SPX/IPX) is a protocol transmitting data packets to specified destinations (e.g., clients, servers) over the network. There is verification made that the data packet has been properly received by comparing the value of what was received to the predetermined value prior to transmission. A retransmission will occur if no response is forthcoming within a reasonable time period. Any failure in the connection is noted if a specified number of transmissions do not arrive at their destination.

Asynchronous Transfer Mode (ATM) is an efficient packet-switching protocol to mix different types of traffic such as data, voice, and video. ATM is a connection-oriented networking protocol used almost exclusively on WANs. When one user wants to talk to another, there is a call set up and a defined path between the two parties. ATM uses connection-oriented virtual circuits that enable message packets to follow each other sequentially through the same connection or physical circuit. Connection-oriented data packets follow the same routing or logical channel. ATM can support both real time and non-real time traffic at different speeds and quality levels. ATM provides a scalable bandwidth adapted to application requirements as well as a processor's performance.

Network Basic Input/Output System (NetBios) is a communications service used by client and server applications in LAN networks. NetBios furnishes applications with a programming interface to share services and information across various lower-layer network protocols. NetBios determines the server's network address, allows for the exchange of messages, and enables the exchange of data with a particular application as well as the broadcast of datagrams to a group.

Systems Network Architecture (SNA) network protocol identifies anything that may go wrong so as to formulate a proper response. Some kinds of anticipated errors are corrected automatically such as a modem failure. Other types of errors such as software difficulties are isolated, reported, evaluated, and corrected by the technicians.

External data presentation is the presentation layer protocol to format information consistently for client/server processing regardless of the operating system or platform.

The X protocol enables the display and management of graphical data. The X protocol distributes the processing of applications by enumerating a client/server relationship at the application level. The X client part of the application is separated from the X server which is the display part of the application. X clients usually run on a remote machine which has excess computing power and displays on an X server. The advantage is distributed processing and a meaningful client/server relationship. Thus, the X protocol defines a client/server relationship between an application and its display.

A post office protocol (POP) enables e-mail downloading from a server to a client. POP-3 is a client mail protocol used in offline functions.

Printer access protocol (PAP) enables linkage between workstations and file servers in addition to workstations and print servers. PAP has benefits of security, support for multiple protocols, and error control.

Multipurpose Internet Mail Extension (MIME) is a protocol providing document attachment to e-mail messages regardless of the network operating system, operating system, or source application program.

Tunneling protocols refer to private, secure channels between connected systems. A point-to-point tunneling protocol enables the formation of a virtual private network so users can access a secure server through a direct Internet connection.

A point-to-point protocol enables a dial-up connection to the Internet. Its operations include IP addressing, multiple protocol support, error monitoring and control, and safeguard features. Point-to-point protocol clients are remote clients linked to servers operating with different network operating systems.

A transport protocol is used to send messages among networks. This protocol sets the rules for network communications such as between client and server. A transport protocol may be connectionless so there is no virtual linkage between the sending and receiving stations.

A client/server protocol indicates the way in which clients request data and services from a server and the manner in which the server replies to that request.

Media protocols determine the kind of physical connection used on the network.

Transaction Control Protocol – Internet Protocol (TCP/IP) is a protocol suite of the Internet furnishing communication to various layers between different kinds of computers (clients and servers) that are widely distributed globally. TCP/IP is the most widely used protocol for networks. It allows for networks to be linked so that all computers can communicate. Transmission Control Protocol (TCP) segregates data packets into sections, gives them numbers, and transmits the packets in a random fashion. The recipient computer translates the data in a user comprehendable form. There is confirmation of receipt. Otherwise, the data packet is sent again. Internet Protocol (IP) is in charge of network addressing and routing. It is connectionless and searches for an address before delivering to a device situated on the network. In a connectionless IP protocol, there is transmission of data packets over various physical network paths on their way to a destination. There are no error-detection or flow-through techniques applied in a connectionless network. End-user devices such as communication software and modems are used to provide adequate flow-control and error-control. The dynamic host configuration protocol is used on a TCP/IP network to transmit client topology information. Telnet (TCP/IP) is storage in Internet servers of text-based data. The servers can be accessed by remote users. TCP/IP can route data around network failures.

TCP/IP is an industry standard suite of networking protocols. It allows different types of nodes in a dissimilar setting to communicate with each other in an understandable way. TCP/IP middleware makes networking choices transparent to application programmers.

The File Transfer Protocol (FTP) is a TCP/IP protocol and software enabling file transmission among different computer systems, generally connected via the Internet. FTP is an extension of Telnet specifications. An FTP proxy is a server uploading and downloading data from a server in the interest of a client. FTP applies to transferring files among internetwork nodes. FTP allows users to access remote hosts. FTP is a peer-to-peer protocol.

Shiva's proprietary protocol is a type of protocol with password encryption and call back features.

Apple Macintosh computers may be networked using either the ApplemTalk protocol or the EtherTalk protocol (Ethernet version for Macintoshes). The latter is used for more recent computers. Both configurations provide for data sharing and e-mail.

In conclusion, protocols relate to communication achieved through some kind of language or code in computer networks. Protocols are the collective languages. A network protocol is software consisting of the communication rules between computers.

[1] Todd B. Portello was coauthor of this chapter.

2

Capacity Design and Management

Introduction

If you have ever needed access to more specific information than what is available to you on your personal computer, you have probably needed the capabilities of a well-designed network.

The beauty of a network is the ability to share information quickly at a reduced cost to the user. The sharing of information is performed by almost all proprietors, whether it is an auto parts store to check the pricing of an oil filter, a large home center (i.e; Home Depot) to see what the availability is on a special sized drill bit, or a sailboat manufacturer on the status of a yacht purchased by a consumer.

All of these situations call for a well-designed network environment to exchange and share information. If these proprietors used single stand-alone computers they would have to get information physically sent to them, or brought by them, so that information could be exchanged and records or files could be updated.

Figure 2.1 shows a single stand-alone PC (Computer A) with no way to share the information it needs that resides on computer.

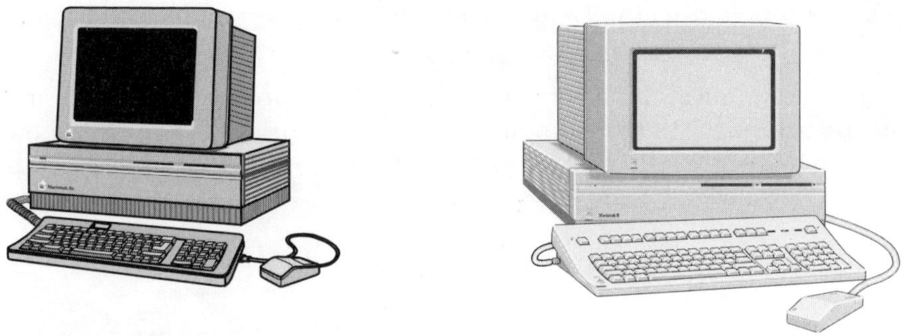

Figure 2.1: Stand-alone PC

The information needed on computer A that is stored on computer B would have to be exchanged by physical means (i.e. floppy disk/tape back-up, etc.).

With the enormous amount of shared information in the current workplace this is neither a logical nor a practical approach.

A more efficient means of sharing information between computers A & B would be a network arrangement such as the following:

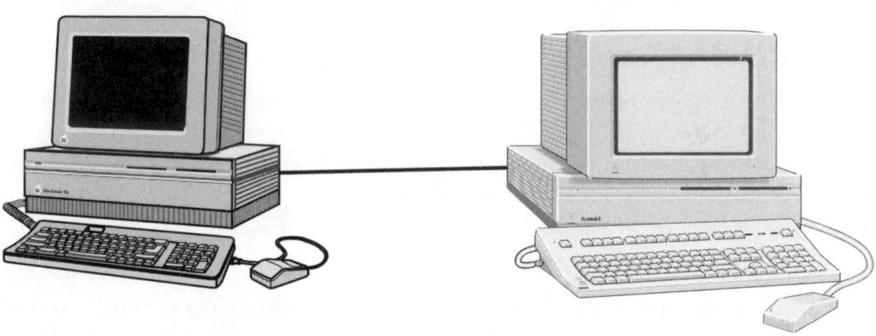

Figure 2.2: Simple Network

Figure 2.2 illustrates a single PC (Computer A) networked with a file server (Computer B).

Many small to medium sized organizations with the need to share large amounts of information are networked together by a LAN (Local Area Network) using a common method popular in these types of environments called bus topology.

The bus topology method allows a small number of computers to be joined together in one location.

A good example of this arrangement would be an accounting firm located in one building with five computers sharing one printer. Figure 2.3 shows All Hands Accounting, CPAs' computers linked together by a LAN with all departments sharing one printer.

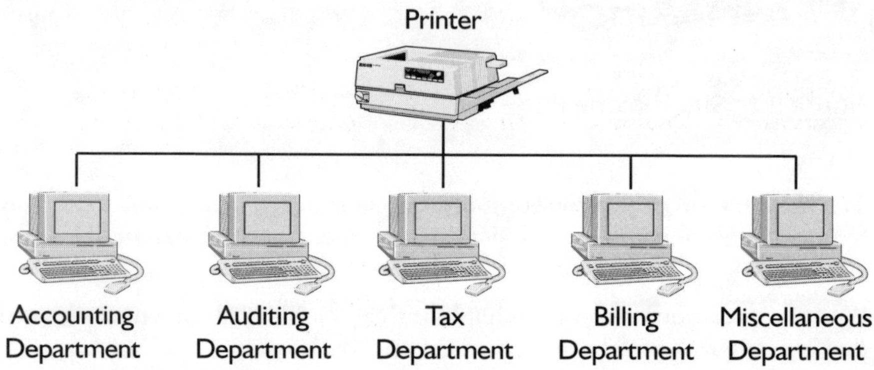

Figure 2.3: Network in an Accounting Firm

In All Hands Accounting, if CPAs had multiple locations in various states around the country, they would need a different arrangement to share information in these remote locations and to accomplish the same goals. Our accounting firm would need to put in place a well thought out and designed Wide Area Network (WAN). WANs connect LANs together by using telephone lines where available.

An example of a WAN used by a national accounting firm is illustrated in Figure 2.4. Please note that there can be many LANs linked together by a WAN. Also consider the possibility that this does not have to stop at the national level; it can be a global consideration as well. In All Hands Accounting, CPAs, due to a growth spurt, now have offices located in two major cities that are networked in the following manner.

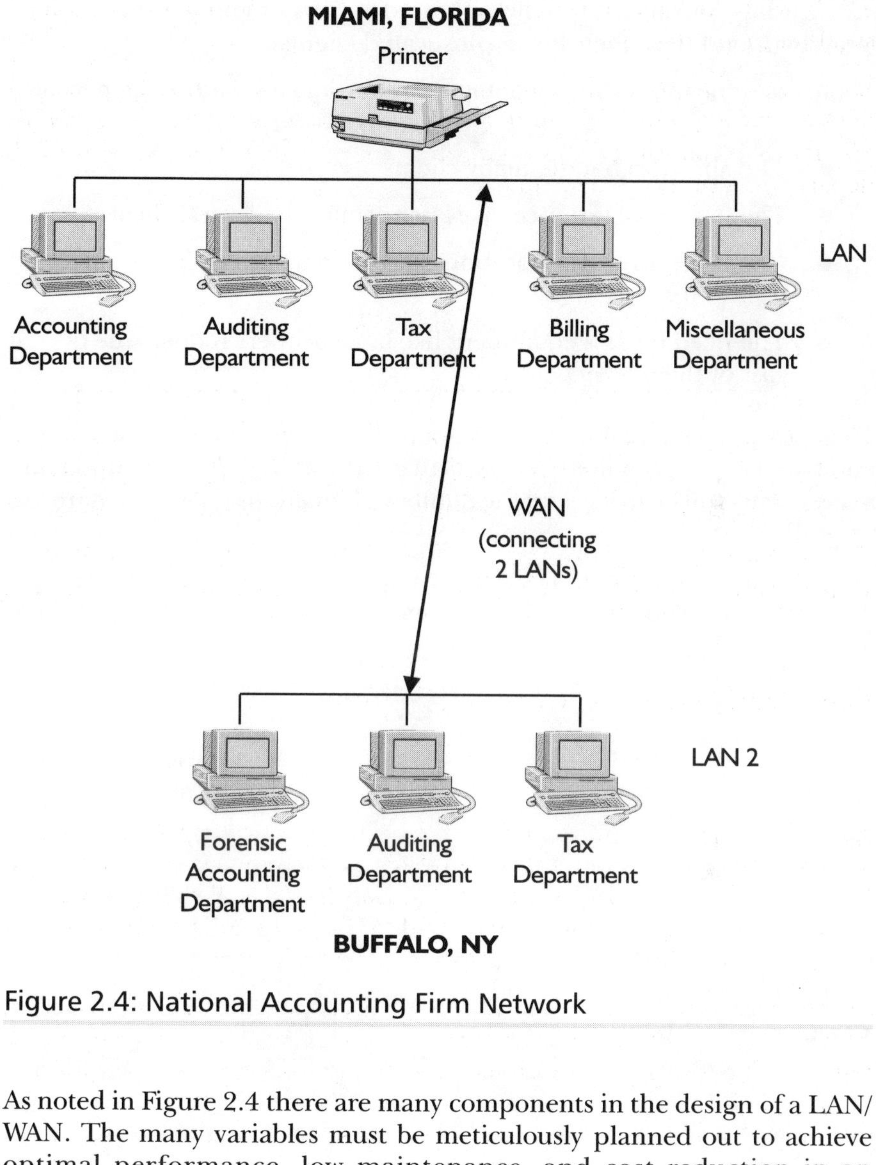

Figure 2.4: National Accounting Firm Network

As noted in Figure 2.4 there are many components in the design of a LAN/WAN. The many variables must be meticulously planned out to achieve optimal performance, low maintenance, and cost reduction in an environment that is subject to continuous change.

Servers

In general, a server is the source for resources that the clients on the network will be using. Servers are usually dedicated to file storage and are generally more high-end than their less sophisticated clients.

There are some important advantages in choosing a server-based network:

- The ability to handle many clients.
- The sharing of resources like laser printers by several client stations.
- Using a central file location (server) data is easily accessed and backed up.
- The need for less equipment like laser printers reduces the overall cost of the network.

These are just a few of the many advantages of a server-based network. The number of servers required in a particular network is dependent upon how many clients will be in use and the number of functions that are required to be performed.

A single server could perform all functions needed (smaller network) or a different server could be used to manage the database, printers, and files.

This network arrangement is illustrated in Figure 2.5.

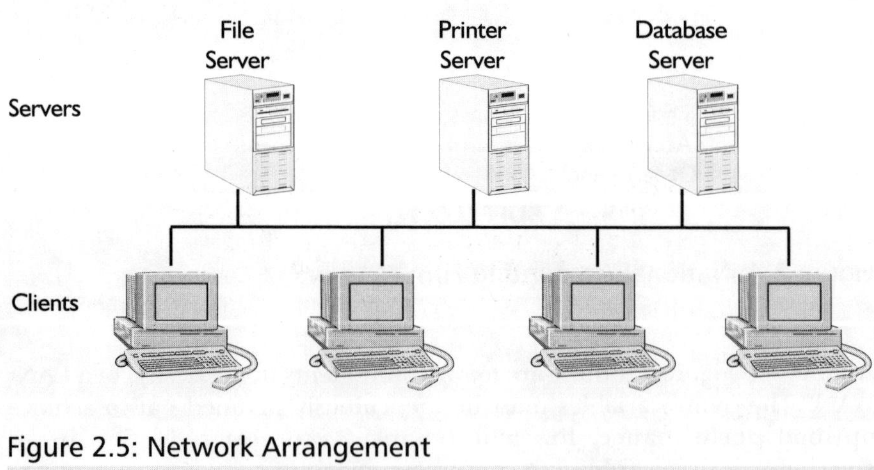

Figure 2.5: Network Arrangement

Figure 2.5 indicates a three-server system. Each server is dedicated to a specific function and, if cost is not a main consideration, this would be a preferred arrangement.

In reality, there would be many more clients in order to warrant the segregation of functions by the servers.

Systems Network Architecture

Local Area Networks (LANs) are structured and operable under several different network architectures such as token ring and ethernet. These network architectures are standardized by the IEEE (Institute of Electrical and Electronics Engineers).

Ethernet

To explain the Ethernet standard (802.3) is to visualize all peripherals and devices connected to the network as having the ability to send data packets to any other device or peripheral at any given time. In a large firm, this data packet exchange will be at a high volume and quite often at the same time as other data packets are being sent.

As illustrated by Figure 2.6, there are inevitable collisions with large amounts of data being exchanged throughout a network.

Figure 2.6: Inevitable Collisions

In essence, with the Ethernet architecture, and to avoid this problem when a collision occurs the CSMA/CD comes into play. CSMA/CD is a carrier-sensing multiple access with collision detection. This feature basically instructs all devices on the network to pay attention and see if other devices or peripherals are transmitting at that same time.

If other devices are transmitting, the workstation is told to wait and when the network is not sending data packets, the workstation will transmit at that time. Collisions of data packets can still happen after that if workstations send their data packets at the exact same time.

When data packets collide, they have a random wait time before they are sent again. When they are sent again and if they collide a second time the random wait time increases and keeps increasing until the data transmission is ultimately successful.

Figure 2.7 is a graphic illustration of the process of transmittal, data packet collision, random wait time, and retransmittal.

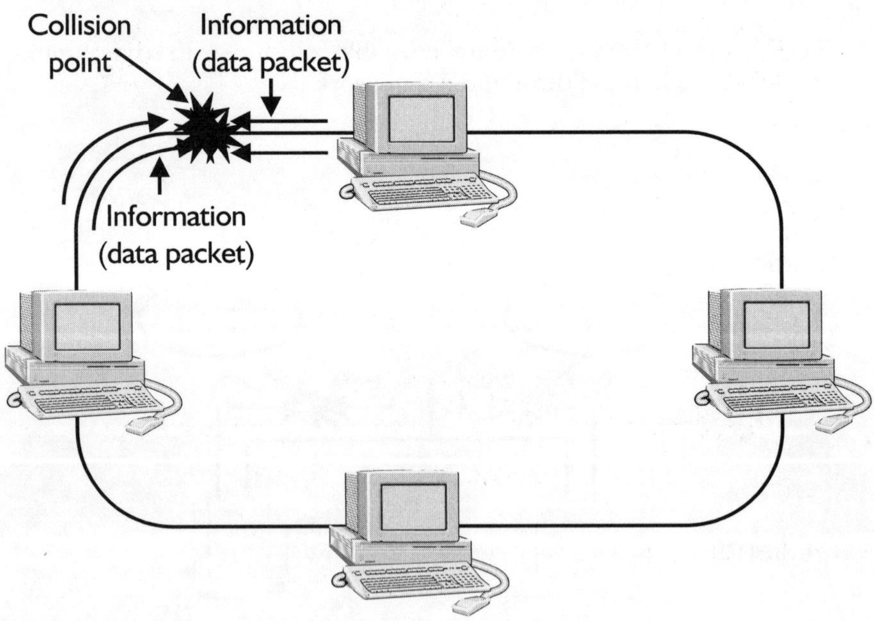

Figure 2.7: Data Packet Collision

In Figure 2.7, after the data packets collide, the client station in closest proximity to the collision will recognize there has been a collision and will emit a signal to all network devices notifying them there has been a data packet collision. The devices connected to the network will then wait a random time (fractions of a second) before sending the data packets again. If this keeps happening, the random wait time increases until there is a successful transmission.

Token Ring

The token ring standard (802.5) was designed to work with the token ring network design. A token is passed from device to device within a network. When a token containing information for a node on the network is passed around, that node will keep the information for that node contained on that token and pass along any remaining information not applicable to that node. If a workstation receives a token that has information on it for another workstation it passes it on until the proper workstation receives it. This process is illustrated by Figure 2.8.

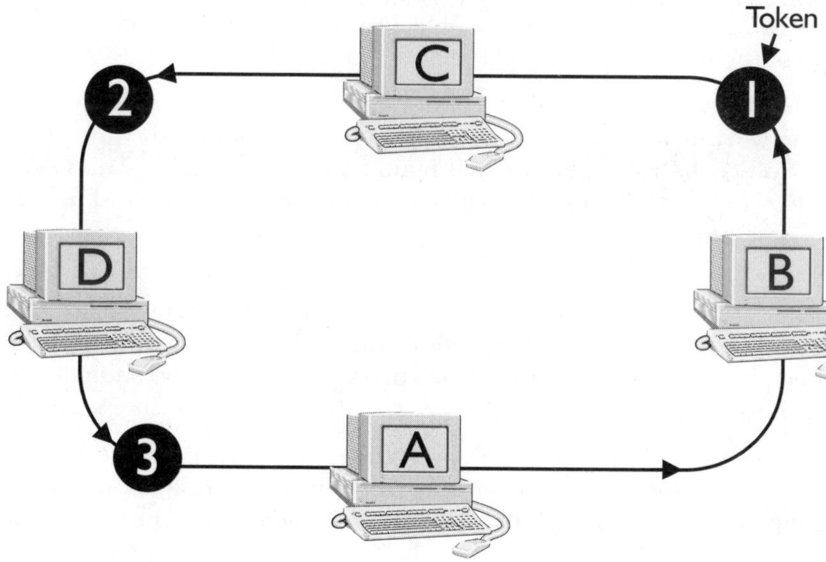

Figure 2.8: Token Ring

In Figure 2.8 above, the token between workstations B & C has data on it from workstation B for workstation D. When the token gets to workstation C it will get passed on to workstation D because there is no data on it for workstation C.

When workstation D receives the token it will copy the data for it and pass the token onto workstation A.

This process continues until all information finds its proper destination. These two types of architectures are just two of the more popular forms of design.

Network Design

Designing a network is a process of looking at the bigger picture and taking small steps to get there. A good network design will pay close attention to the physical layout of the network, the ultimate goal or purpose of the organization, and monetary or budgetary limitations or constraints.

The purpose, goal, or task to be accomplished should be considered first. The network designer needs to understand fully which direction the organization is headed and what goals need to be reached.

A clear-cut understanding of these objectives is the first step in the proper design of the network.

Consider the case of All Hands Accounting, CPAs. One of the objectives for their network is to transmit audit workpapers and findings from their satellite office in Buffalo, NY to their main office in Miami, FL. This needs to be accomplished because the partners that review the audit workpapers are all based in Miami.

The partners will need to review the work, make changes and recommendations, and retransmit the files through the network back to Buffalo. This exchange of information will go back and forth until the finished product is approved by the partners and given to the audit client. This is illustrated in Figure 2.9.

Another example may be an auto parts chain with a distribution center in another part of town. A customer may come in to "Getting it Done Auto Parts Inc." and request an oil filter for a 2001 Honda Civic. The proprietor checks the stock in his store and finds he sold the last one yesterday, so he gets on his workstation which is connected to a LAN and finds that the distribution center across town has fifty of them left. He requests a same day shipment and the customer has the proper oil filter in a few hours. This is illustrated in Figure 2.10.

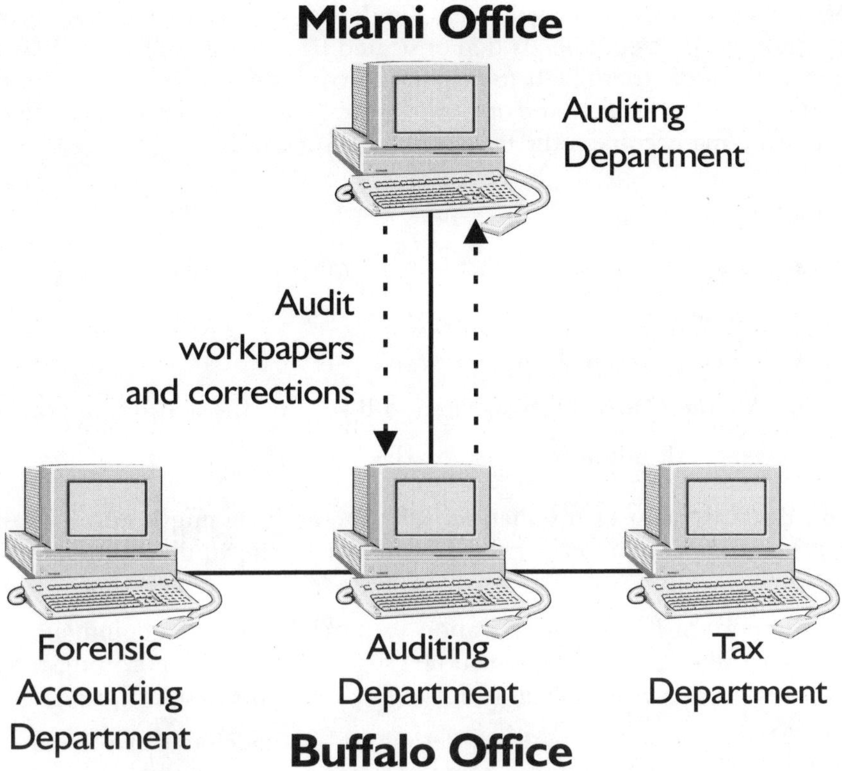

Figure 2.9: All Hands Accounting, CPAs

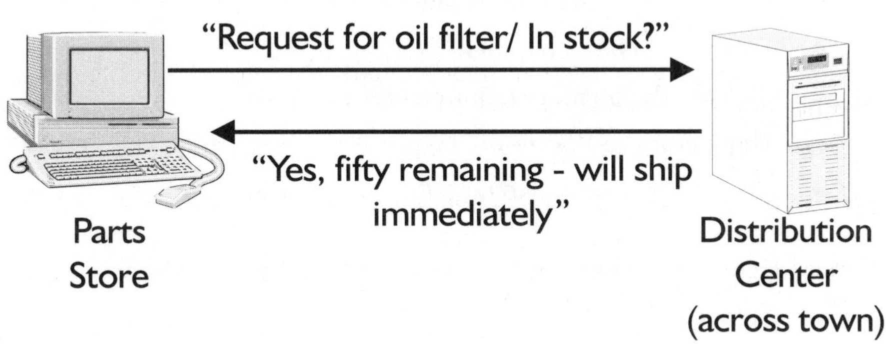

Figure 2.10: "Getting It Done Auto Parts Inc."

A wise network designer will keep the design simple with an ear to the ground for additional needs demonstrated by the organization. There is usually a budget that needs to be considered in the design phase. Noting the monetary constraints and operating within them will take some creativity. There are many costs in the design of a network such as:

- Network Software
- Network Hardware (Machines, Peripherals, Cabling, etc.)
- Training costs to familiarize users with the network
- Network Installation
- Maintenance costs associated with keeping the network operable
- Network Administrators' salaries

To project the cost of installation, software, and training is an easy task. Much harder is to project the long-term costs in the administration of the network.

It is important for the organization to purchase quality equipment and software without exceeding the budget so as to cut down on the long-range maintenance costs associated with keeping the network operable.

The physical aspect of network design and management involves many considerations and questions, some of which are mentioned below.

- What is the current physical layout?
- Can a network be installed in the office? (logistics)
- How many workstations will be needed?
- What size server is required?
- What are the memory requirements of the server and nodes?
- What network software will be used?
- Is the network software compatible with the application software in use?
- How many printers will be needed?
- Will it be one LAN or a few? How will they be connected?
- What will the network operating system be?
- What network architecture is to be used?

- Is there a central place to keep the server, and can cabling be run throughout the present location? Can cables be snaked behind walls, in ceilings, etc.?
- Is there reliable technical support for the equipment/software purchased and can that support be relied upon at all times?
- Does the network administrator have the proper administrative and technical skills? Does he or she have the best interests and vision of the organization?

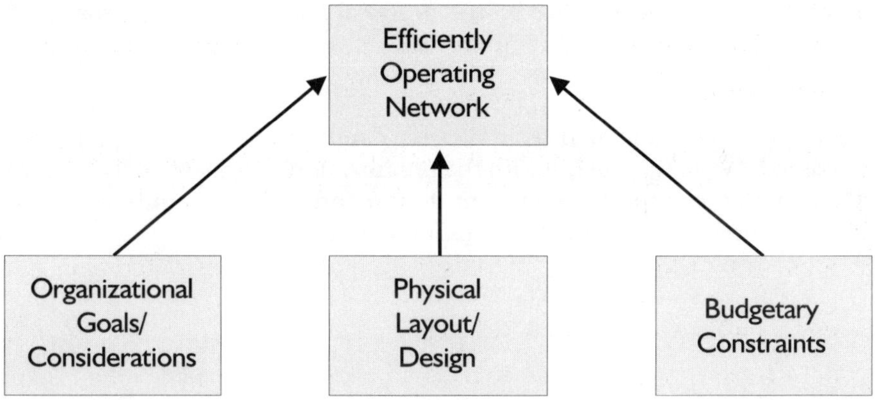

Figure 2.11: Efficiently Operating Network

Figure 2.11 is a graphical depiction of the actual makeup of an efficiently operating network. As noted by the graphic, a network is not just a series of cables, computers, and software.

Server Needs and Memory Requirements

Other points to consider when designing an efficient network are the number of servers required to handle the present network traffic. A good margin for expansion in accordance with the business plan and growth outlook for the organization should be factored in when selecting the size and number of servers to be used. Memory requirements for servers and nodes also need to be addressed when designing the network. Memory will be dependent upon what network software is being used and also on the application software in place that will accomplish the daily tasks of the organization.

Network Administrators

A good network administrator is the heart of an efficiently running network. A network administrator is responsible for the daily functioning and maintenance of the organization's network. The network administrator should be chosen based on education, parallel visions with upper level management as it relates to the organization's direction, and technical expertise.

A network administrator should be able to upgrade the necessary areas of the network with the newest technology while operating within the organization's budgetary constraints. A network administrator needs to be a problem solver with an eye on the practical and simplistic approach.

Conclusion

Design and management of a network is a daunting task. The design process needs to be well thought out with the organization's ultimate goals in mind. There are many aspects to efficient design and management that need to be considered before even one computer is purchased.

3
TCP / IP Basics

Use of TCP/IP

TCP/IP or Transmission Control Protocol / Internet Protocol in essence is a set or group of protocols that link computers to the Internet. Generally, it is arranged as a stack of protocols in layers.

TCP/IP makes it possible for Deflecto Appraisals Inc. in Figure 3.1 to connect to the Web and perform various other functions.

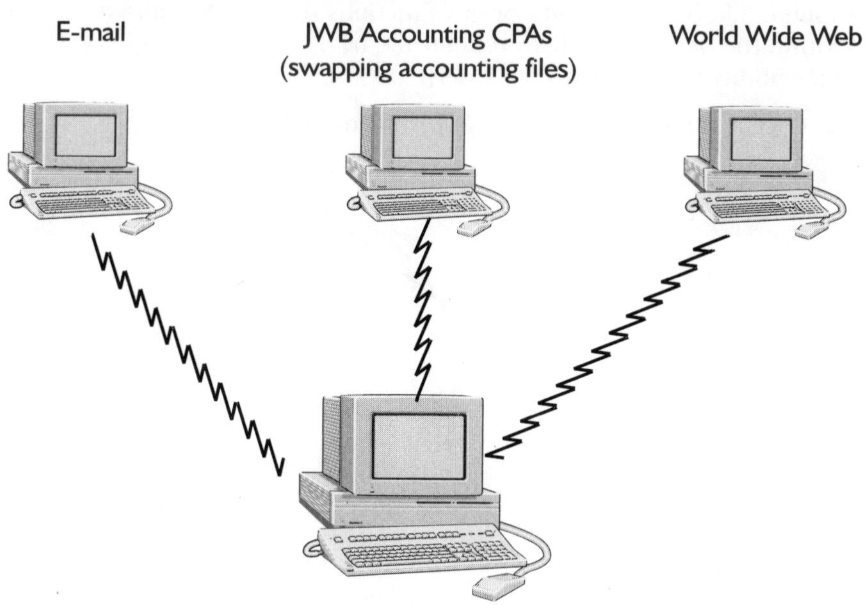

Figure 3.1: Deflecto Appraisals Inc

TCP/IP was brought about by the DARPA network (Department of Defense Advanced Research Projects Agency) for advanced Federal projects. It is currently used in a commercial sense to communicate with the Internet.

TCP/IP is not the only game in town, there are other protocols out there, but TCP/IP is generally regarded as the most common in a networked environment. Many different businesses are using the Internet to exchange information and TCP/IP is the protocol that is getting it done.

Because the Internet is really in its infancy stages by all comparisons, TCP/IP will be used for years to come. It will be important for current users to continue to gain access and share information as well as for new users to find more in-depth and creative uses for the Internet.

TCP/IP is a set of rules or protocols that tell computers the way in which your data and information should move, how files are shared with other remote computers, and for communicating with others on the Internet.

TCP/IP is the most popular set of protocols used in networked systems today. There are others such as Novell's SPX/IPX or IBM's SNA but none used more widely than TCP/IP. An advantage of TCP/IP is that it runs on any computer or network. It does not matter what the components of that network are, it can use cables or radio signals, Ethernet architecture or token-ring.

Generally, TCP/IP is found in the operating system that comes with your computer. There are four important functions that TCP/IP allows you to do within the networked environment: file transfer, remote login, E-mail, and the ability to connect to the Internet.

This concept is illustrated graphically in Figure 3.2.

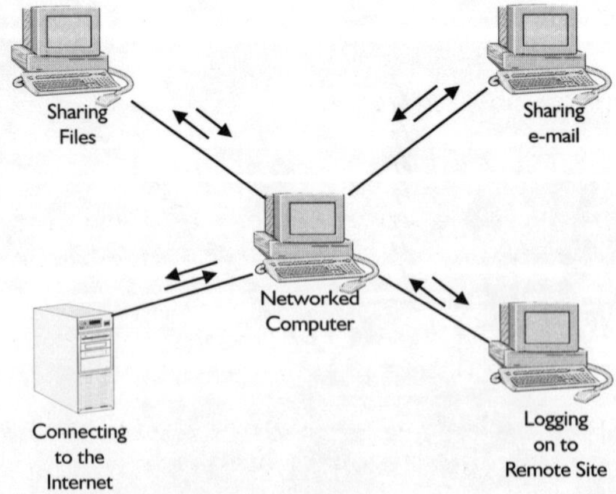

Figure 3.2: Network Environment

File Transfer

The rules for the transferring of files are guided by File Transfer Protocol (FTP). This protocol was born out of the need to share files between remote computers no matter what size the file was or how far away the systems were from each other.

The need for a method of sharing files that is transparent to the user was obvious. When you upload (send files) or download (receive files) you do it between your local computer (the one you are using) and a remote computer (a computer in a different location). A tremendous advantage of TCP/IP and the fact that is so universally used is that no matter what operating systems the two file sharing computers are using, the file sharing will work.

When file sharing between remote computers, two connections must be established. The control connection is initiated to activate the commands in the file transfer protocol. The data connection is initiated to allow the transferring of data.

Figure 3.3 shows these two connections being established.

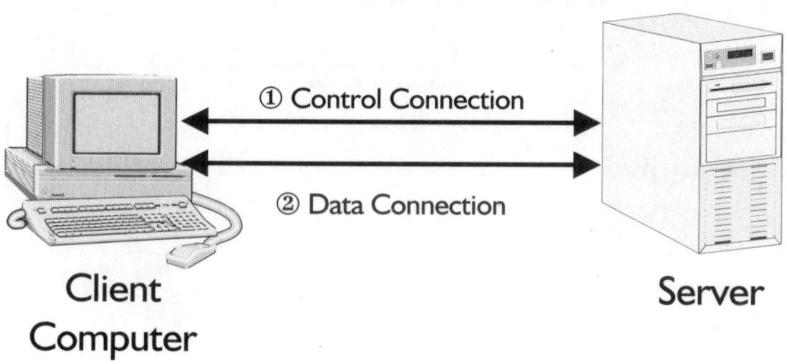

Figure 3.3: File Sharing Connections

The control connection sets up the process by which the file transfer protocol rules and guidelines are put into effect. These FTP rules dictate the way files are transferred. The control connection is required to remain established throughout the entire file transfer process. If the control connection breaks down, the ensuing data connection will be lost.

When the control connection has been established in a normal file transfer session, and the file transfer protocols have been exchanged, the remote computer reads a pre-established port and the local computer starts the data transfer.

When the control connection is established and the data is being transferred, it is exchanged in a variety of ways – three to be exact.

In general, when the data transmission process is complete, the session is terminated with an end of file (EOF). Both computers involved will recognize this occurrence and shut down the data connection.

The data file transfer process can be a short one as in the case of an MS Word file (a meeting agenda), or it can be a long one as in the case of an application program.

The first means of data transmission is block mode. In block mode, the data is sent as a group of data blocks. This method breaks out the data into headers that define things like the length of the block, starting and ending points , and the integrity of the data. Block mode is graphically depicted in Figure 3.4.

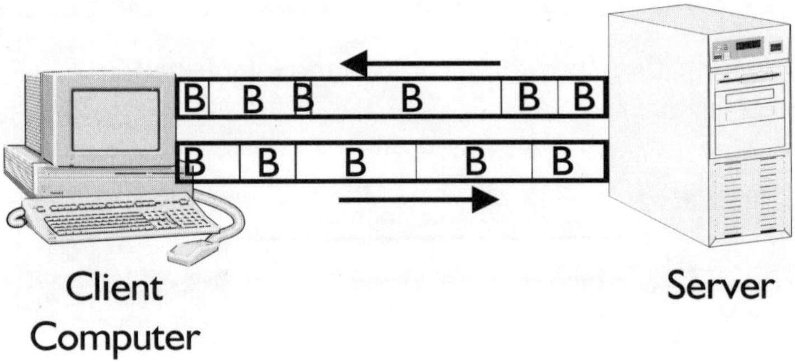

Figure 3.4: Block Mode

Notice in Figure 3.4 that the Bs represent a series of blocks of data. As noted by the illustration, the blocks are of varying sizes. As mentioned before, the job of the headers is to determine where each block starts and ends and to verify the integrity of the individual blocks.

Stream mode is another form of data transmission possible within the FTP guidelines. Before explaining stream mode we need to define the term octet. An octet is a group of eight bits of data which in turn will comprise a singular character. It is sometimes used interchangeably with the term byte.

In stream mode, data is sent in a stream of bytes or octets. The beginnings and endings of the data are determined by predefined control codes.

The stream mode is illustrated in Figure 3.5.

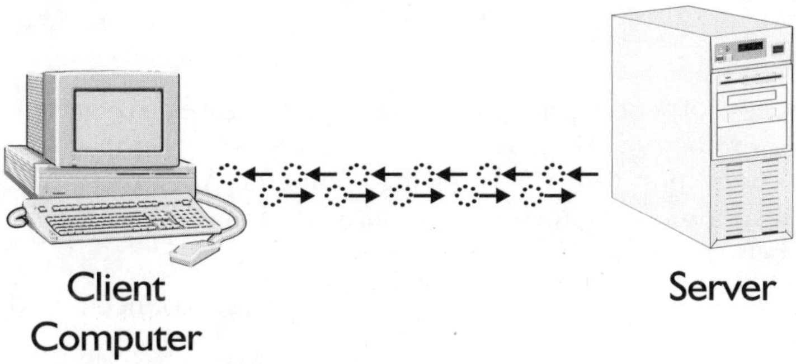

Figure 3.5: Stream Mode

An octet in Figure 3.5 is shown by the eight dotted circles being exchanged between the client and server. Sometimes, millions of octets make up a single file that is being transferred. An appreciation of the speed and accuracy with which this is accomplished should be noted at this point.

The last mode of data transmission is compressed mode. Some data cannot be compressed, but other data can. When data can be compressed it is done by defining escape sequences and replicating data into a smaller package.

Compressed mode is illustrated in Figure 3.6.

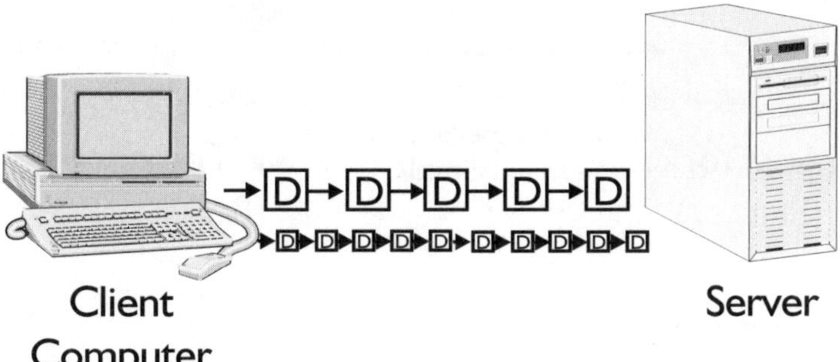

Figure 3.6: Compressed Mode

As can be seen in Figure 3.6, data can be sent in a normal fashion (slow, inaccurate) or it can be sent in a compressed mode with the data transfer speed and accuracy of data increasing.

E-mail

The process of sending and receiving e-mail is governed by Simple Mail Transfer Protocol (SMTP). This works by the sending computer setting up a TCP connection with the receiving computer. After this connection is established a set of protocols comes into effect and the transfer of e-mail occurs between the sender and receiver.

In contrast to FTP the SMTP establishes only one connection where the mail is exchanged, not two as in FTP.

An illustration of the e-mail transfer process is shown in Figure 3.7.

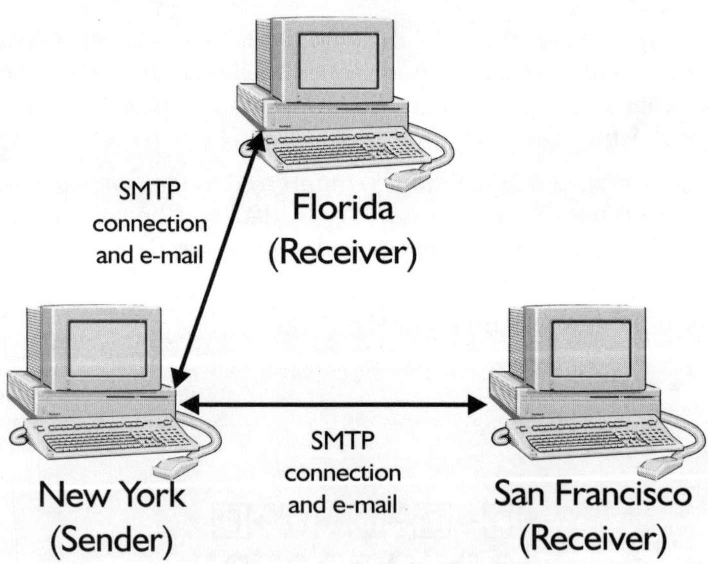

Figure 3.7: E-mail Transfer Process

E-mail can be sent across the street or across the world. This concept is extremely important and necessary for business reasons as well as personal reasons. Sometimes e-mail cannot be delivered directly to a certain site, in that case there are systems established that are intended to forward that e-mail to its proper destination.

Note the system established in Figure 3.8 to forward an undeliverable e-mail to its proper destination.

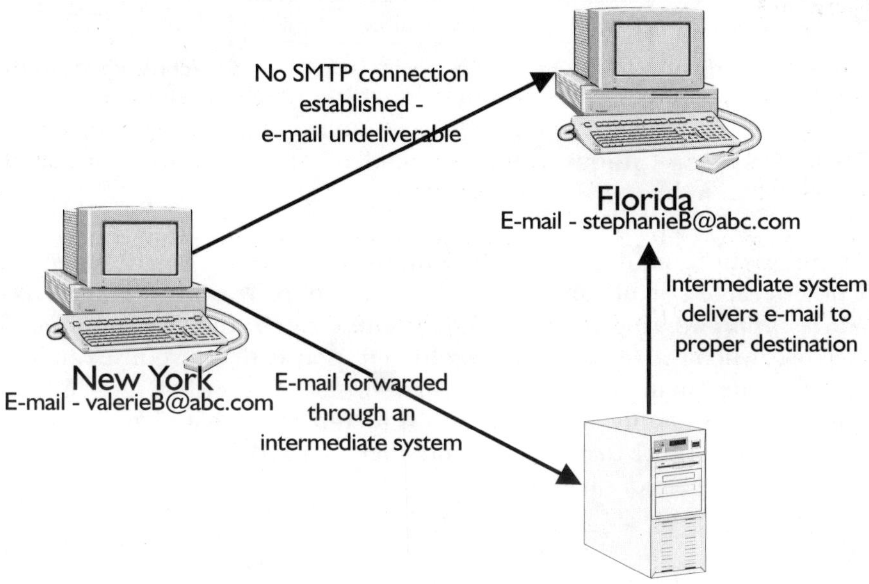

Figure 3.8: E-mail Delivery

These intermediate systems that forward mail to their rightful destinations are called mail servers. These mail servers don't always forward the e-mail right away; often they will store the mail and forward it at a later time.

The actual process of sending an e-mail is accomplished in three stages indicated in Figure 3.9.

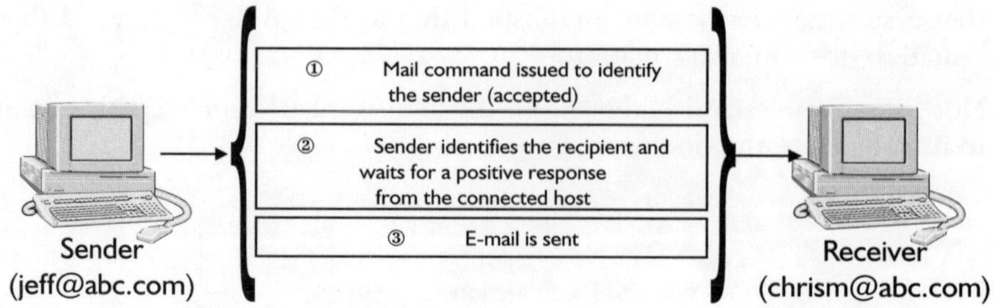

Figure 3.9: Three Stages of E-mail

At times the e-mail being sent cannot be delivered to the receiver. Undeliverable e-mail can happen for a variety of reasons (i.e. incorrect address, non-working intermediate system, etc.). In the sending process step two, when the sender is waiting for a response, that response can be a negative one. An important ingredient of this process is when the sender is identified in step one, a reverse path is identified so that error messages can be rerouted back to the sender and when mail is delivered, the recipient knows the address of the sender.

Remote Access

Remote access to networks is an invaluable tool today, given the mobile nature of our civilization. Many business people need access to networks and information after hours and while traveling. This transfer of information is generally accomplished over telephone lines.

The user at a remote location dials into the server and establishes a connection. The server in this case is called an access terminal. This server has a modem set-up so that it can communicate with remote login users through the serial lines.

Serial lines are provided in many different sites for travelers due to the high usage of mobile computing platforms.

Figure 3.10 is a graphical depiction of a remote user accessing a server to retrieve information on a LAN.

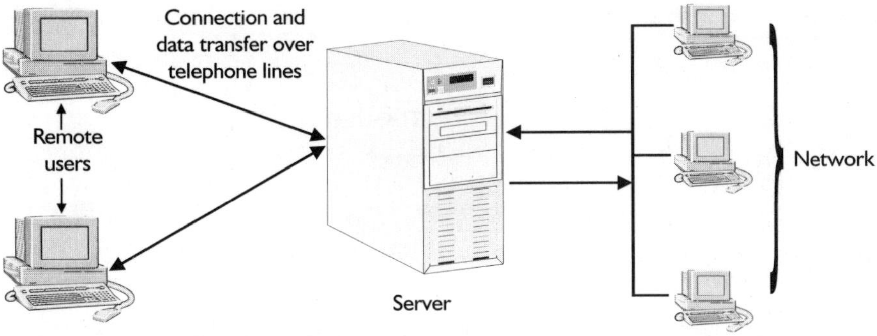

Figure 3.10: Remote Users

IP Addressing

When you connect to the Internet and receive e-mail from someone across the world, how does that e-mail make its way directly to you?

The only way this could occur is to have a specific address to connect to. IP addressing is a numeric way to assign a unique address to each and every computer on the Internet. A computer can be on one network or be connected to a number of networks – so in turn it could have one IP address or a few.

Every computer that is connected to the Internet has an IP address. Usually it is an assigned numerical address or sometimes it can be a chosen name by the user. The way an IP address is set up is by name and by number.

A typical IP name would be something like:

Cba. Sju.edu

This type of address has three parts to its name. The first part – Cba is the name of the computer. The second part of the name – Sju is the name of the organization, in this example it is St. John's University. The third and last part of the name – edu is the top level internet domain which in this case is an educational institution.

When you want to connect to another address on the Internet you usually type in an address name and it is converted into a numeric address by TCP/IP. This process is necessary due to the tremendous number of IP addresses in use today. In IPv4 the address is written in four fields broken up by dots. This arrangement supports many more addresses than an alphabetic only arrangement.

This conversion from alpha to a numeric address is illustrated in Figure 3.11.

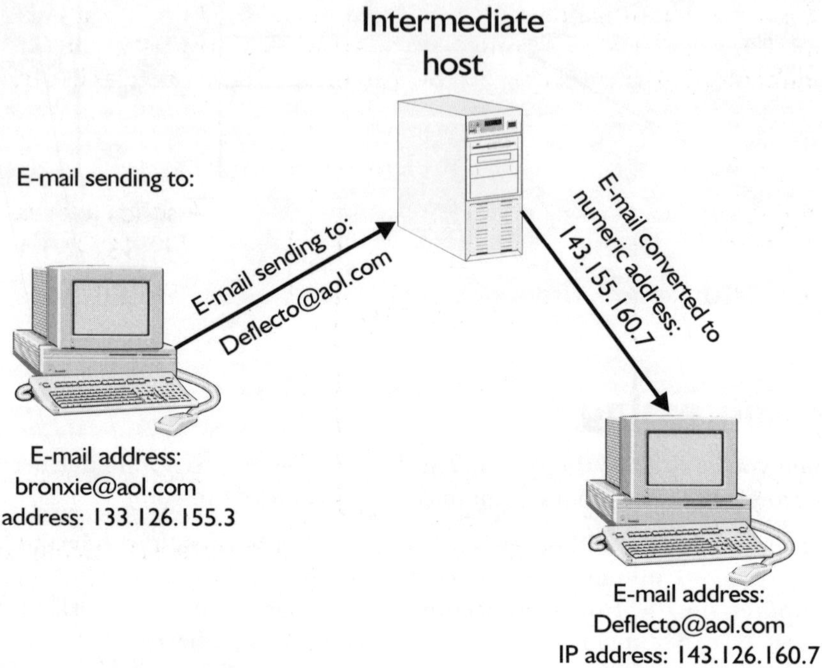

Figure 3.11: Conversion of Alpha to Numeric

As noted previously, two of the same IP address cannot exist, if they do e-mail would never arrive at the correct destination or may wind up in the wrong place.

Address Resolution Protocol (ARP)

ARP is a common networking protocol that maps the IP address to the MAC address when dealing with the various networking architectures such as token-ring or Ethernet. The MAC layer (Data Link Layer) of the OSI model controls the movement of information between systems on its way to the data's ultimate destination.

The way this works is that computers A & B want to communicate with each other. A is sending data to B, so A will establish that computers A & B are connected to the same network. A will then send the address out on the network (Ethernet) and all machines on the network will receive and examine the destination address. All computers that don't recognize the destination address will disregard the transmission, computer B will recognize the destination address as its own and establish a transmission link between the two computers. After this link is established, communications between the two can commence. This process is illustrated in Figure 3.12.

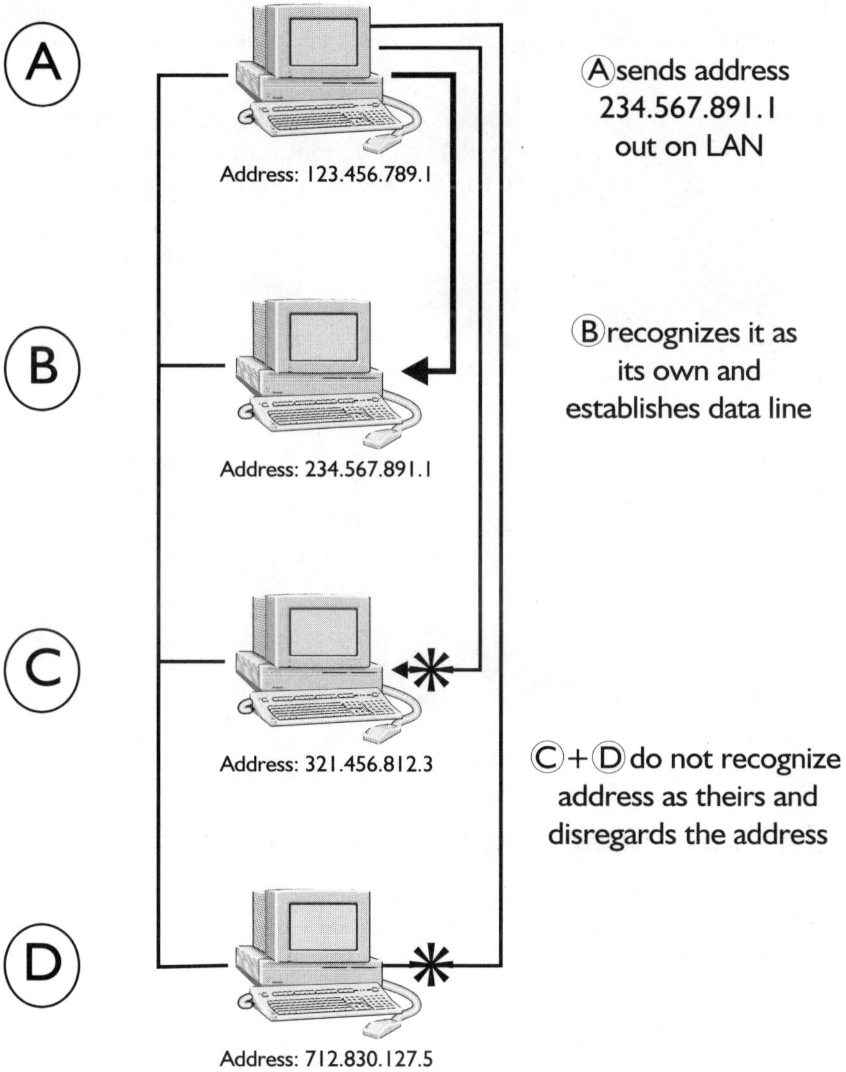

Figure 3.12: Communication of Computers

Routing Fundamentals

Routing is a way for data to reach its final destination no matter what happens to various parts of the network system along the way. There are many ways for your data to make it to its destination.

A router directs data in different ways so that it bypasses trouble spots and gets there accurately and efficiently. Routing directs data traffic between various networks. Routers keep track of the networks that are working and all other intermediate systems in between that can direct data to its destination.

Routers keep this information on routing tables that list the relay systems and available networks where data can be sent unimpeded. An illustration of a file being routed through a network to its correct destination is illustrated in Figure 3.13.

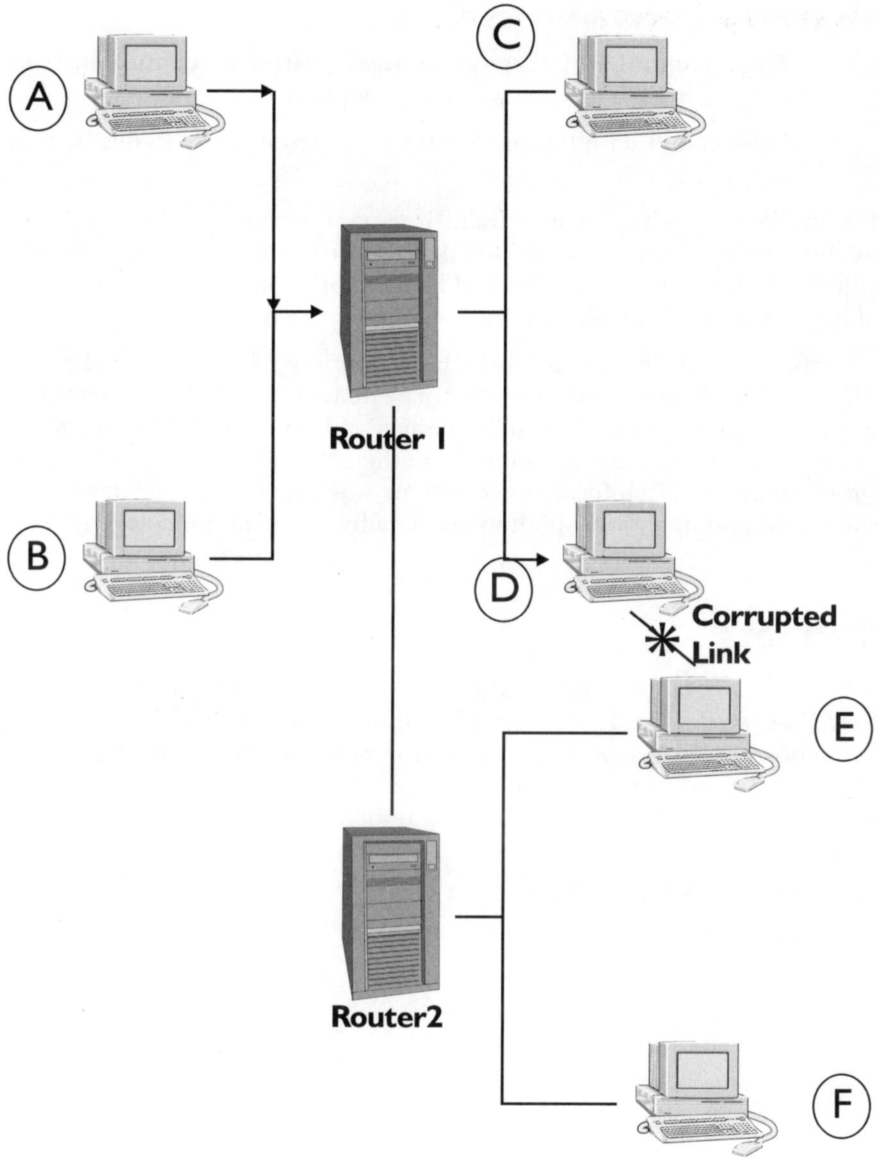

Figure 3.13: Routing

In Figure 3.13 above, computer A wants to set up a link to computer F and establish an e-mail connection. When a router is employed there are a few choices on the path that the data will take on its path to computer F from computer A.

The connection could go two ways:

1. From computer A through router 1, through computer D to computer E, to its destination at computer F.
2. From computer A through router 1, onto router 2, to its destination at computer F.

By employing a system that contains routers, the router table established in this case will not choose option 1 due to the corrupted link between computers D and E. This corrupted link will be listed in the routing table along with other available options for sending the data.

The routers used in this case will choose option 2 because it is the one option of the two that is working. As can be imagined, routers are extremely important given the number of machines and possible data routes to be taken to get data from one point to another. Routing is based on a set of protocols that collect information on the network in question and determine the fastest and/or shortest path to get data from one point to another.

Subnets

Often within a large company there may be many LANs in use. Subnets break networks down into smaller, more efficient and manageable subnetworks. Each subnet has its own network address made up of the main network address and a section of the host. Generally, a subnet is easier to control and maintain due to the reduction in size and diminished network traffic.

A subnet is illustrated in Figure 3.14.

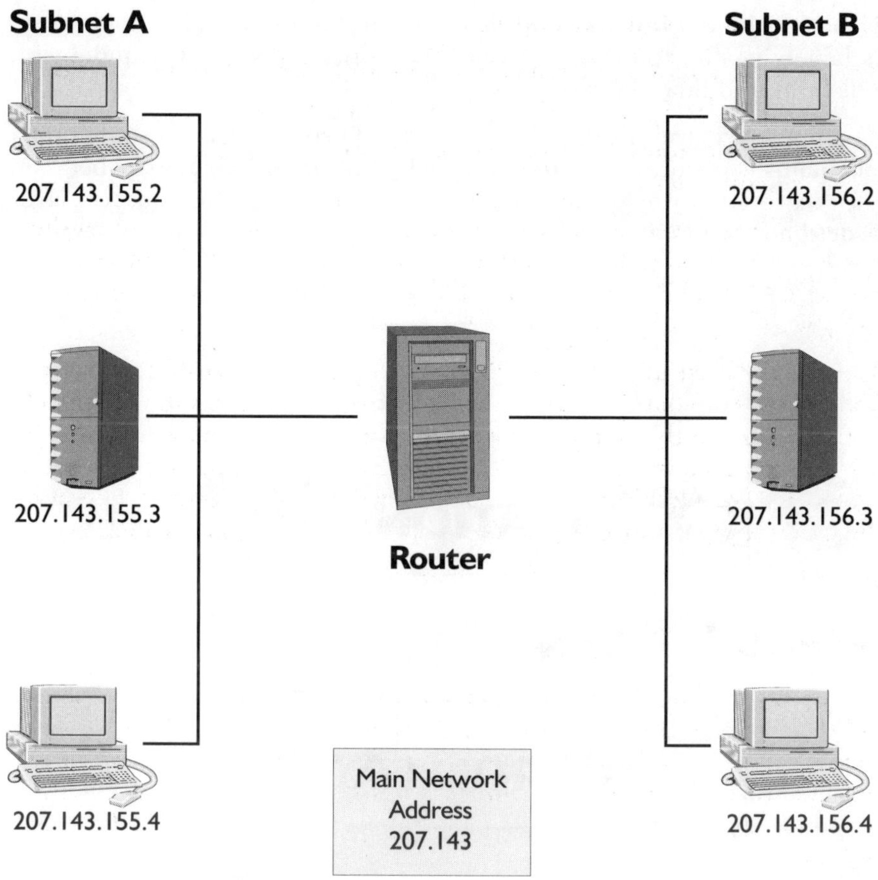

Figure 3.14: Subnet

These two subnets operate independently of each other, yet are part of the same network address. Subnets are practical in some networking situations, but not in all. Subnets are an alternative to establishing additional networks and incurring additional costs.

Transport Layer Protocols

The transport layer of the internet protocol suite is somewhat similar to the transport layer in the OSI model. The transport layer in the OSI model is the fourth layer from the bottom in a seven-layer stack. This architecture is largely geared towards providing the upper layers of the model with reliability and flow of data.

The transport layer also exhibits a degree of error correction so that data maintains its accuracy and reliability. The transport layer resides between the Network layer below, and the Session layer above. The transport layer is designed to provide a reliable bridge between the two adjacent layers.

A graphic depiction of the OSI model is shown in Figure 3.15.

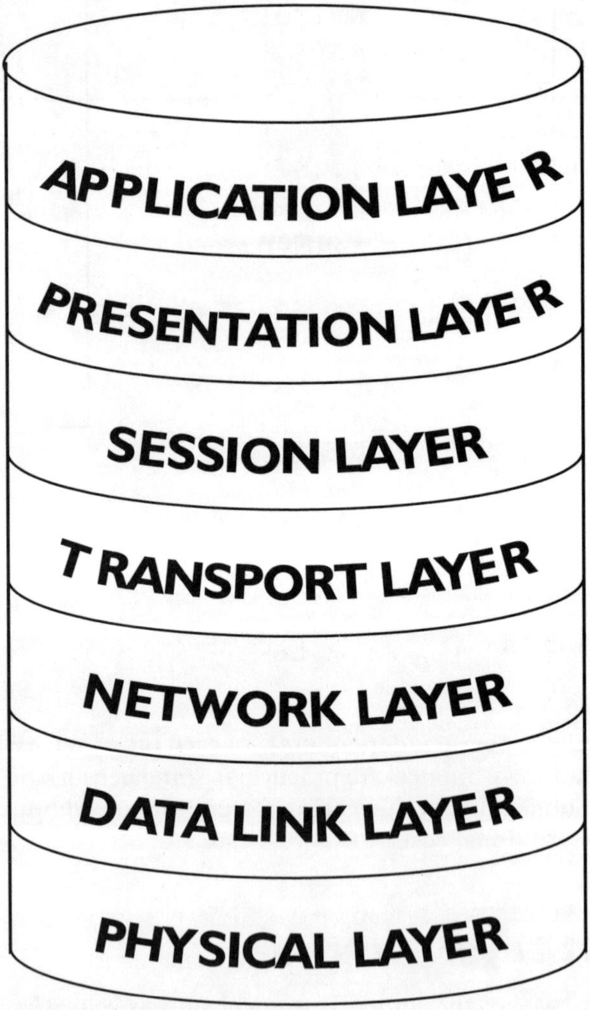

Figure 3.15: The OSI Stack

In contrast, the transport layer of the internet protocol suite is one of only four layers in the internet protocol suite stack. The transport layer in this architecture is combined with some features of the session layer from the OSI model and performs basically the same functions as the OSI model. The difference with the transport layer in this architecture is that it provides reliability and flow of data to a predetermined top layer port – the application layer.

Generally, the set of rules that are followed in this model and layer are the Transmission Control Protocols (TCP). This is a reliability-based concept with smooth flow of non-corrupted data. The thrust of this layer is transmission of unimpeded data through reliable connections.

An illustration of the Internet Protocol Suite is shown in Figure 3.16. It is important to note that many of the layers in the internet protocol suite perform multiple functions thus reducing the stack from seven layers to four layers.

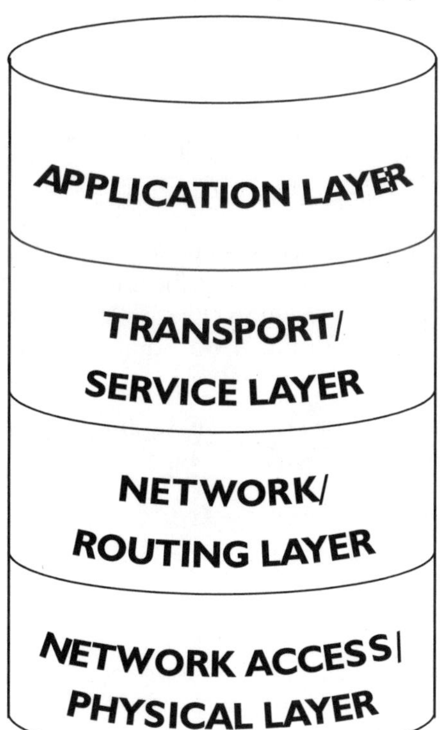

Figure 3.16: Internet Protocol Suite Stack

Also note that the transport layer lies between the application layer and the networking layer and like the OSI model provides a reliable transmission flow between the two.

Application Layer Protocols

The application layer in the internet protocol suite is communications-based and provides a means with which to establish a communications link. The File Transfer Protocol (FTP) mentioned previously is an application layer protocol. The FTP is a means by which files are sent, copied or extracted from other machines.

Another application layer protocol is Simple Network Management Protocol (SNMP). SNMP is a network management protocol that monitors your network to make sure that all peripherals connected to your LAN are functioning properly and the network is operating at an optimal level. Another one of many application protocols is Telnet which gives the user the ability to log onto a remote computer in a transparent fashion. This application protocol is invaluable in this age of mobile computing and the travel oriented business environment.

Security

Almost every organization is linked in some way to the Internet. They may just receive e-mail or they may do all their business on the internet as well as receive orders for goods and services. Every organization no matter how small it is, has some sensitive information that needs to stay within the organization.

Note the comparison of the two companies – one large and one small in Figure 3.17.

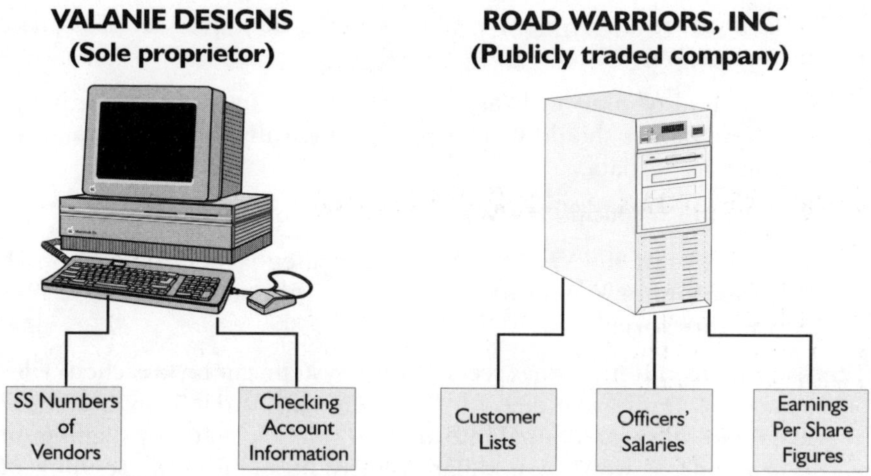

Figure 3.17: Comparison of Companies

As noted in Figure 3.17, Valanie Designs has some security issues. Although they may not seem like much, security is relative to the organization and therefore quite important to this sole proprietor. Take for example the checking account information. If someone had this information they might be able to access the account online and misappropriate funds to their account. This is an important security issue for Valanie Designs.

On the other hand, Road Warriors Inc. is a publicly traded company with offices throughout the U.S. Road Warriors Inc. has a high volume of activity between intercompany e-mails, financial information, and sales orders being placed on-line. For Road Warriors Inc. the security of their system is of major concern. For this public company security topics may include, but are not limited to, safeguarding customer lists of databases – if other companies were to gain access to this information they could steal customers. Another sensitive area is salaries of officers and employees. The salary issue may be more of an internal security concern rather than for outsiders. Finally, earnings per share information is a well-guarded secret until the figures are released every quarter. The implications of leaking EPS information could cost millions to an organization. Some things to address when implementing a security strategy are:

1. Procedures regarding dos and don'ts on the network must be put in place. (Where can you go on the LAN?)

2. Computers and servers must be secured at all times. Consider using a security service.
3. A system of passwords and domain names must be put in place. These names should be changed periodically so prior employees can't access data.
4. Software applications – What programs can you use?
5. Files within the system need to be safeguarded. The network administrator should be the only one with access to certain files and directories.

There are many different ways security on a system can be breached. Files or sensitive R & D information could be stolen. Password information could be leaked allowing confidential information to be accessed by logging on as the original password. A popular security problem is the sending of viruses that corrupt files and bring networks to a standstill.

Thankfully there are a series of protocols established by the Internet Engineering Task Force (IETF) called IP security protocols. There are specific security resources for each level of the TCP/IP stack. The lower levels of the stack contain the physical and data link layers. The most common way of securing data at this level is to encrypt the data.

Encryption is a process where data is converted into code so the data cannot be read. The person who has clearance to read the data has a key to descramble the data so they can read it. The next two levels of the TCP/IP stack are the Internet and transport layers. At these levels the password and computer addresses are encrypted to thwart violators from gaining access to this commonly used information. A good thing to do is *never* reveal your password. The upper levels of the TCP/IP stack contain the application, presentation and session layers. Many of the programs used in these upper layers provide their own password security. IP Security is paramount at every level of network operation and continues to change as different avenues of security breaking are discovered.

Network Performance

Networks are a fabulous tool used to accomplish many tasks. A network needs to operate at optimal efficiency in order to provide the services required. Optimal network performance starts from the purchase of the first computer and is an ongoing project that needs to be continually monitored. From intercompany file transfer to operating on a WAN, all aspects require monitoring by the systems and network administrator.

Network administrators use a combination of physical checks and software applications to measure IP performance. A popular theory in network performance is to keep the system simple, while leaving room for upgrades. System performance and maintenance is a daily task regardless of the size of the network. A good network administrator is an invaluable part of the network environment.

BOOKS USED

1. *Networking Complete*, Second Edition, by Sybex $19.99
2. *TCP/IP Explained*, Digital Press by Philip Miller $39.95
3. *TCP/IP for Dummies*, 4th Edition, IDG Books by Candace Leiden Marshall Wilensky $24.99

4
Wiring Networks

This chapter contains information concerning the wiring of networks, as well as taking into account possible alterations in floor plans. In making best use of a wiring network, consideration must be given to wiring standards, nature of transmission, topology, access, and noting warning signs of potential problems. Included is a discussion of the use of "wireless" technology applications and use, the pros and cons of such use, LAN standards, and considerations of infrared and radio transmissions. Topics to be discussed include:

- Transmission Media
- Wiring Standards
- Topology
- The Need for Cabling
- Wiring Management Software and Diagnostic Equipment
- Anticipating Changes in Floor Plans
- Wireless Networks

Designing the Physical Network

Connecting computers, devices, data, and applications through networks presents endless opportunities for the administrator. In theory, you should be able to connect to, use, view, or update any device or bit of information anywhere on the network. But remember, your own success will vary depending on how well your network was designed and installed.

Good network design will allow your network to grow as large as you need it to. Bad design, especially if followed by slipshod installation, will plague both you and your users forever. Your system may work, but it won't work well, and when the inevitable need for upgrades surface, you'll be faced with hit-or-miss results. Some additions may still be easy but others simply won't work.

We cannot overemphasize the need for comprehensive planning and design. Of course, few people have the budgets to put top-of-the-line components at every twist and turn of the cable, and even if they do, rapidly changing technology can make those components quickly obsolete. Nonetheless, we stress that you familiarize yourself with the consequences your decisions create in selecting network components.

In this chapter, we provide detailed descriptions of the various network topologies to consider when planning a network strategy, as well as comprehensive suggestions for selecting actual cable types from today's market. To round out the chapter, we close with a look at dial-up capabilities and other connection options for wide areas.

The Need for Cabling

Everything electric emits electronic signals. These signals can conflict with each other if allowed to coexist on the same plane. In other words, two objects cannot occupy the same space at the same time. A perfect example is when you travel between two cities and get bits and pieces of two different radio stations on the same channel. The problem is only solved when your car moves completely out of range of one station and into the range of the other. Thus, it is important for the network cable to have the resilience to thwart such interference. After all, we don't want radio and TV signals interrupting and possibly corrupting data traveling over our network cable. By the same token, we don't want our data signals leaking from the cables only to be intercepted by those devious enough to capture our valuable information for their own use.

Another very important consideration is installing enough cable in the beginning. There's nothing worse than beginning the installation of additional equipment and discovering you don't have enough spare cable to support it. In the long run, doubling the amount of cable you feel you need at the start is very cost-efficient. Even if you don't foresee your network having much immediate growth, you can count on significant expansion over time. Remember that it is not the cable itself but the labor that is usually the most expensive part of the cable installation. Installing ample cable to begin with will save you money and time.

The type of cable you choose will hinge on your preferences for security, data reliability, and topology.

Cable Topology

As you may have guessed by now, there is a lot of organization that goes into networking computers. For cabling (wiring) purposes, there are three main schemes to organize how the cable connects to the nodes: linear-bus, star-shaped, and ring topology.

Unlike the phone system in your home or business, network wiring must be installed in a certain pattern in order to quickly diagnose malfunctioning cables and to compensate for the signaling pattern of certain network communication.

Linear-Bus Topology

In a linear-bus arrangement, a single cable called the bus or trunk, is installed – usually along a cable path in the ceiling or wall – and terminated at each node on the network. Linear-bus topology may also be referred to as "daisy chain." A good analogy for how this topology works is to think of electrical extension cords. Whether stringing lights for the holidays or linking those orange extension cords when edging the yard, you are using a linear-bus topology. The plug-in points are the terminators.

T-connectors are T-shaped metal tubes that serve as connection points for each network node. They work in conjunction with the terminators, which are placed at the beginning and end of the cable (the first and the last PC).

As you might imagine, linear-bus topology is low-cost and simple to design. Assuming you can keep track of the primary cable, it is easy to set up additional PCs by simply tapping into and expanding the trunk cable (like adding more holiday lights to the rooftop). Since each PC is attached directly to the cable, there is no need to purchase additional hardware such as hubs or wire centers. Another advantage is that you will use less cable than with star-shaped topology, discussed next, because the length of cable only has to connect to the next device on the network.

Unfortunately, any break in the cable can cause the entire network, or a good part of it, to fail. A failure can be caused by one end or the other losing proper termination, which can be caused by a pinched cable, loose T-connector, or malfunctioning terminator. As with a string of holiday lights, these failures can be difficult to diagnose.

Convenient and inexpensive, a linear-bus topology is appropriate for very small workgroups of between three and ten members.

Star-Shaped Topology

Utilizing a star-shaped topology, an individual wire is run from each PC to a central location. Here, all the wires are connected to a hub device, which completes the electronic connection. You can think of a hub in a star-shaped topology as a Greyhound bus stop. It provides a central location for all of the signals to meet and then branch out to their individual destinations.

Star-shaped wiring uses more cable than linear-bus wiring, but has several important benefits. Since each PC's cable is unique to that machine, if there is a cable fault, only the PC attached to the broken cable is likely to be affected. For example, if you start with a star-shaped network cabled with common unshielded twisted pair (UTP) wiring for Ethernet (covered later in this chapter), and then later decide to replace the Ethernet with Token Ring, you won't have to rewire your offices.

Star-shaped topology is easier to manage and administer than linear-bus. Once it is set up, it can be easily documented. Most hubs that connect star-shaped networks contain some form of status lights that assist you with locating and diagnosing faults.

The disadvantage of a star-shaped configuration is cost. More cable is needed, as well as a hub to connect all the wires together. However, if greater control over cable network faults is what you're looking for, it may be well worth considering this type of network topology.

Even though Ethernet is a bus technology, the most common installation method uses a star-shaped topology over unshielded twisted pair.

Token Ring Topology

Ring topology networks, where the cable is laid in a circle, are nearly extinct. Although you may hear the word ring used as a synonym for networking, it is mostly used to refer to a media standard such as Token Ring. While token ring networks do actually use a ring topology in the signaling scheme (each signal has to pass by each node sequentially in the ring), they are typically configured with a specialized star-shaped topology.

Cable Types

Once you've decided on the topology, you're ready to select the wiring to be used. There are several types of wiring that network electrical impulses will travel over. Each type has physical differences, length requirements, preferred topologies and signaling mechanisms, and a range of pricing structures. The following section will provide both the technical information referred to above as well as descriptions of the most commonly used in peer-to-peer and client-server networks.

Coaxial Cable

Coaxial cable infrastructure accounts for five percent of the computer networks in the United States. Data is communicated across the network on a single strand of copper wire that is protected from electronic eavesdropping and interference by either more copper wire or foil similar in appearance to that used in your kitchen, surrounding the original strand in a weaved pattern. The topology of choice for coaxial cable is a linear-bus, and it is consequently used mostly for peer-to-peer networks.

There are two types of coaxial cable: thinnet, which looks like the kind used for cable TV, and thicknet. Their main differences include size, price, ease of installation, resiliency to electromagnetic interference, and effective signaling distances.

Thicknet Coaxial Cable

Thicknet is commonly referred to as frozen yellow garden hose because of its color and stiff texture; nonetheless, it has the same basic physical make up as thinnet, only with a much thicker protective shielding. Thicknet's inherent installation limitations – it is difficult to put in walls, around corners, and so on – have made it the least popular choice in recent years. Its primary use today is as backbone wiring between buildings, floors of buildings, or areas with high levels of electromagnetic radiation, such as factories and warehouses. Thicknet's ability to carry a signal approximately 541 yards without a repeater makes it especially appealing for use in the factory or warehouse setting. It typically connects to the desktop through an attachment unit interface (AUI) connector, which looks similar to a serial port. From the desktop, the AUI cable hooks into the thicknet with what are called vampire connectors, and these can be very touchy if not attached within a few inches of the designated marks along the cable. It is for these reasons thinnet is the more popular of the two.

Thinnet Coaxial Cable

From the outside, thinnet (also called cheapernet because it's less expensive than thicknet) looks exactly like the cable used for cable TV but do not let appearances deceive you; they're completely different. The correct type of thinnet cable is labeled RG-58 for Ethernet and RG-62 for ARCnet, both of which are described later in this chapter.

Another similarity thinnet has with cable TV coaxial is the connecting ends. The ends of the cable have what are called BNC connectors. Affixing these connectors to the cable requires no more than a wire stripper and a pair of pliers. Another difference between cable TV wiring and that used for networking is that cable TV wiring can't be connected to network devices.

Think of T-connectors as though they are pipefittings used in plumbing. Data flowing into one connection point continues and flows out of the other two connection points in equal amounts. These connectors aren't valves, but rather three-way converging points and must occur at every signaling base. In other words, on the back of every network interface card there must be a T-connector in order to branch the connection. Otherwise the signal passing inside the wire will continue on without an opportunity for the node to transmit its data.

Thinnet can support data streams up to 200 yards without a repeater. Granted, this is considerably shorter than the 541 yards of thicknet, but with its more pliable exterior, thinnet is easier to install. Naturally, the type of T-connector is also important. There are many inexpensive imitations available, so make sure they match military specification UG274.

Shielded Twisted Pair

Next to coaxial, shielded twisted pair (STP) is the most electromagnetically resistant copper-based cable used in LANs. Like coaxial, the main data conducting wires are protected with woven copper or metallic foil. In addition to that, the wires are twisted around each other, thus providing the most protection from interfering signals (coming or going) and giving it its name. One pair of wires is used for receiving data while the other pair is used for sending data. STP can be used in both peer-to-peer and client-server networks. However, because of its higher cost of purchase and installation, it is generally only used in large client-server based networks with star-shaped topology and a Token Ring signaling scheme.

As part of the increased shielding properties, STP requires specialized connectors to the various nodes (PCs, printers, and so forth). Rather than the cable being attached to a single connector that then attaches to the

NIC (as on coaxial cable), STP cable requires a D-shell connector to the NIC and an IBM data connector to the medium attachment unit (MAU) whose end then goes into the wall towards the hub.

To prevent data loss from cable deterioration or human error (cutting the cable), as well as crashing the entire network, STP is installed independently between each workstation, server, and wiring hub. In other words, rather than having one long, extended coaxial cable, STP is split between each connection point. The extra cable between the nodes and its relative thickness means it quickly fills up the wiring conduits in the walls. All in all, both the purchase and the installation of STP are more expensive than coaxial.

Unshielded Twisted Pair

For low cost, an abundance of qualified installation experts, and an installed base (number of users) of 90 percent within the United States, unshielded twisted pair (UTP) is a great choice for cable. However, UTP is not without complexities. For example, unlike a coaxial cable that runs directly from PC to server, UTP must have an intermediary in the form of a hub. This can be expensive depending on how many connections it supports. Furthermore, the installer's "rule of thumb" suggests limiting UTP to 330 feet between the network node and a hub, almost half the usable distance of Ethernet coaxial (550 feet).

UTP looks deceptively like a regular telephone line, although UTP cable is slightly thicker. Also, UTP uses RJ-45 connectors whereas telephone wires require RJ-11 connectors. Telephone-grade twisted-pair cable is composed of two wires twisted together at six turns per inch to provide electrical interference shielding and consistent impedance (electrical resistance). Because existing buildings usually contain plenty of this wire, it is often used as an inexpensive, easy way to link computers.

However, using telephone wire for your network, especially when it is already in place, can lead to several major problems. First, unshielded twisted-pair cable is sensitive to electromagnetic interference. In addition, low-quality twisted-pair cables have a varying number of twists per inch, which can distort the expected electrical resistance. Though this cable is usually quite adequate for most telephone communications, network data transmission demands much smaller tolerances.

Also it is important to note that telephone wire is not always run in straight lines. Cable that appears to run a relatively short distance between two offices might actually run through half the building to get from one to the other. A misjudgment could cause you to exceed the maximum cable length specifications.

The technical terms for the differences between UTP types are the category levels. Category 3 is a typical phone line and Categories 3 to 5 are network-grade. The differences are in both the speed and amount of data they can conceivably carry. Category 3 has been tested at 16 megahertz (MHz) for attenuation (which describes the rate at which the signal becomes weaker the further it travels) with data capabilities up to 16 megabits per second (Mbps). Category 4 has been tested at 20MHz but has a low installed base, because most people usually move from Category 3 straight to Category 5 when upgrading. Category 5 cable is capable of bandwidth speeds up to 100 MHz and can handle up to 100Mbps, which is important when we discuss Fast Ethernet later in this chapter.

While not as shielded as coaxial or STP, all UTP wiring is doubly protected from encroachment by insulation over the copper wires as well as twisting every pair of the wires together. The plastic sheath provides absolutely no electromagnetic shielding, although the twists in the wire provide more protection than if it were untwisted.

In short, unshielded twisted-pair cable is inexpensive, easy to install, and provides a variety of uses. But be careful: The money that you save may be offset by additional costs later if the network doesn't function properly. To better determine if UTP is worth using, hire an experienced wiring or cable installation contractor to test the signal quality and actual cable lengths of the wiring in your building. This will add to the cost of your network, but it is the best way to avoid cable-quality problems later on.

Fiber-Optic Cable

Fiber-optic cable, the most expensive, uses pulses of laser light to transmit data over glass cables. It is not subject to electrical noise and can handle exceedingly high data rates. While the price of fiber-optic cable has decreased over the years, it still requires skill and expertise to ensure proper connections. The ends of the fiber must be polished with precision to correctly transmit data.

The core of fiber optic cable is made from silica glass or plastic. The latter is often called plastic optical fiber (POF), and is less expensive and easier to terminate, although it does have bandwidth and distance limitations.

There are two types of transmission methods that use fiber-optic cable-multimode and single mode. Multimode transmitters are typically used with larger cables such as 62.5/125 fiber. This number indicates a cable with a core diameter of 62.5 microns, and a cladding (layer surrounding the core) of 125 microns. Multimode is used for shorter distances, and requires a less powerful laser light. In fact, most multimode transmitters actually use light-emitting diodes (LED) instead of lasers. Single mode transmitters enable high bandwidth at great distances using a smaller core, typically 8.3/125 fiber, and a powerful laser transmitter. Fiber-optic cable usually carries several warnings relating to the dangerous nature of some laser transmitters. Read and follow warning labels carefully.

Fiber-optic cable has a significant advantage over all metallic cable options. Fiber-optic cable is more reliable because it is not susceptible to packet loss through electromagnetic or radio interference. This feature makes fiber-optic cable appropriate for those who need a high level of security and for electromagnetically noisy environments. Because the light pulses that carry the data are restricted to the cable itself, it is virtually impossible to surreptitiously tap into the medium without getting down into the actual core of the cable. In electromagnetically noisy environments, like those around large cranes and massive mechanical equipment, fiber-optic cable is often used to bypass the interference caused by engines and generators. In hazardous, explosive environments, fiber-optic cabling is suitable since it does not conduct or transmit electricity.

Fiber-optic cable is also very thin and flexible, making it easier to move than the heavier copper cables. Because signal losses while traversing the medium are less than those associated with metallic conductors, fiber-optic cable can support network segments up to 3.5 kilometers (approximately 2.2 miles) apart. Where long runs or links between buildings are required, fiber-optic cable is often the only viable choice.

While it is likely that you or someone on your staff is competent at cable installation for any of the wire-based types, we strongly recommend that you allow only qualified professionals to install fiber-optic cables. Not only does the technology demand close tolerances for tapping and termination, it also requires sophisticated, expensive test equipment to test the quality of the fiber-optic lines installed.

In addition to the preceding examples, there are options available for LANs that are more secure and have higher throughput (the amount of data that passes through). Due to their high costs, however, they're usually reserved for MANs or WANs, where ends are separated by miles rather than feet.

Dial-Up and Dedicated Remote Connections

Generally, any network confined to distances measured in feet comprises a LAN, a few miles between major connections could be considered a MAN, and any network that involves ten or more miles can be thought of as a WAN. For WANs there are several wiring options.

Analog Connections

Simple cable setups use standard analog modems. The modem connected to my PC dials your modem, and through specialized software I can use all of the resources on your PC. This type of connection has many uses in a network, including giving remote users access to it and transferring information between databases or e-mail systems, just to name a few. Modem speeds have peaked at 56Kbps, with 53Kbps being the maximum speed allowed under current FCC regulations. This type of connection is great for establishing quick links and transferring small files between the machines or using remote control to troubleshoot. Analog modems are not good, however, for maintaining static connections, transferring large files (larger than 2MB), or, if your office only has a single phone line, allowing for simultaneous data and voice connections. Analog modems have these limitations because in addition to transferring data they must also convert it.

While it is true that most phone lines in large metropolitan areas are digital (they transmit an electronic signal in binary code either as on/off pulses or as high/low voltages), they are only digital between the dialing central offices of the phone company. From the central offices to homes and businesses the lines are analog, so that voice can be transferred. Thus, the signal pulses that the computers communicate must be converted from digital to analog (which is the static noise you hear) so they can be converted on the receiving end from analog back to digital. For higher bandwidth, more reliable connections, faster data transfers, and a more simplistic, cleaner dial-up data exchange, one of the digital solutions listed below is a better choice.

Integrated Services Digital Network (ISDN)

ISDN can transmit data up to speeds of 128 kilobits per second (Kbps), maintain static connections, and allow for multiple connections (data and voice). It does this by combining a total of three digital lines.

Two of the lines are for data and transmit at 64Kbps each, for a total of 128Kbps, while the third line has a speed of 16Kbps and is used for telephone communications.

Unfortunately, due to its newness, ISDN is not available everywhere. However, most major U.S. cities offer ISDN services. Call your local phone company for details on availability and pricing.

Digital Subscriber Line (DSL)

Digital subscriber lines use standard copper telephone wiring to bring high bandwidth to offices. DSL comes in a variety of "flavors," differentiated by their speed and transmission types. The generic term, used when referring to multiple DSL technologies, is xDSL.

Similar to ISDN, DSL can carry both voice and data signals over the same cable. Unlike ISDN, DSL maintains a constant data connection, with no need for the end user to dial a telephone number to connect to the Internet. Once an Internet service provider is set up, the user will have instantaneous access to the vast resources available online.

Voice calls, whether inbound or outbound, come through on the line without interfering with the data connection. This allows, for example, users to send a fax and receive a phone call, all while maintaining a constant Internet connection.

Two main types of DSL exist today - splitter-based and splitterless. Splitter-based DSL requires a splitter installed at the physical end location. The splitter allows data to flow simultaneously with other analog communications, such as voice, for instance. Splitterless DSL uses similar technology, but places a splitter at the central office, thereby reducing the installation and hardware costs of individual splitters at each office.

Cable Modem

Cable modems provide high-speed connectivity over a standard Cable TV (CATV) coaxial cable. The cable terminates at a cable modem in your location, which acts more like an NIC than a modem. Much like DSL, a constant connection is maintained between your PC and your Internet service provider.

Today, Cable TV lines pass by 96 percent of American homes, but so far businesses have much less chance of access. So while this connectivity option will probably be available to part-time home workers, it may not be available for a remote office.

The current dominating standard for cable modem operation is DOCSIS (Data Over Cable Service Interface Specification). DOCSIS standardized the protocols and modulation schemes for carrying bi-directional data over coaxial cable. This standardization encouraged the rapid production and acceptance of cable modems.

Cable modem bandwidth is shared between your location, and everyone else on the same local loop. So while the total bandwidth of your local coaxial cable may reach 27Mbps downstream, only a fraction of that speed is available at any time. The more subscribers on your local loop, the slower the data will travel. Upstream rates may range from 128Kbps to 2.5Mbps depending on your subscription type and rate.

Additionally, since the line is shared with others in the area, privacy issues arise. Most cable modems do not address privacy issues, so make sure to research the cable modem that is functional with your cable provider. A firewall, implemented by hardware or software, may be your best bet in thwarting those who seek to compromise your PC or network. Chapter 5 discusses firewalls in more detail.

Next we'll cover the three types of cable modems: external, internal, and set-top box.

With an external cable modem, the coaxial cable is connected to the cable modem, and an Ethernet port (typically RJ45) allows for connection to the PC or LAN. If the cable modem is connected directly to a PC, an NIC will be required. Alternatively, the cable modem can be connected to a hub to share the high-speed connection on the network.

An internal cable modem provides the most direct method of connectivity. The CATV cable connects directly into the cable modem installed into a single PC.

A set-top box connects directly to a television and usually has e-mail, Web browsing, and intelligent television control built in. An interactive set-top box will allow the novice user to easily get on the Web, without even knowing how to use a computer, because it acts as a stand-alone Internet interface.

T1 and T3

If larger pipes of information over long distances are what you need for transferring data, then a TI or T3 may be what you're looking for. Unlike analog modems or ISDN, T-carriers provide a constant connection, so no dial up is required. TI is a 1.544Mbps channel that can handle 24 voice or data channels at 64Kbps. The standard TI frame is 193 bits long, which holds twenty-four 8-bit voice samples and one synchronization bit; 8,000 frames are transmitted per second. T3 is a 44.736Mbps channel that can handle 672 voice or data channels at 64Kbps. T3 requires fiber-optic cable.

Basically, TI and T3 are direct connections from one end of the country or state to another. Typically this technology is used in the medical community for the remote diagnosing of patients in hard to reach or poverty stricken areas, where families cannot afford to make the trip to a large city. It is also used as a high bandwidth solution for Internet connectivity for large businesses that can afford the exorbitant fees and require the constant connectivity to maintain competitiveness in business.

SMDS

One more option worth considering is Switched Multimegabit Data Service (SMDS). SMDS is a high-speed, switched data communications service offered by the local telephone companies for interconnecting LANs in different geographic locations. It was introduced in 1992 and became generally available nationwide by 1995.

Connection to an SMDS service can be made from a variety of devices, including bridges, routers, and CSU/DSUs (the digital equivalent of a modem), as well as via frame relay and ATM networks. SMDS can employ various networking technologies. Most implementations use rates up to 45Mbps.

Data is framed for transmission using the SMDS Interface Protocol (SIP), which packages data as Level 3 Protocol Data Units (L3-PDU). The L3-PDU contains source and destination addresses and a data field that holds up to 9,188 bytes.

That covers the gamut of cabled options for networks. However, there is still one other option to connect PCs together with a network: wireless.

Wireless Networking

Probably the simplest form of a wireless LAN is the one that is created between a laptop and a desktop (and even a printer) using infrared technology. Basically, there is a red panel on each device that looks and acts exactly like the remote control that goes with your TV set. Some of the low-powered systems behave just like your remote control unit: the infrared ports must be in plain view of each other and within a certain minimum distance in order to transmit and receive data properly. However, high-quality, high-powered systems do not have the same line-of-sight restrictions, and will transmit around objects and over much greater distances.

The next simplest form of wireless LAN technology is that developed by such companies as Proxim, Lucent, and 3Com. These products use radio frequencies to connect PCs. Unless your LAN is on a peer-to-peer network, however, most of your wireless LANs are only one part of your network as a whole, which includes cabled nodes. In other words, some of the PCs will have wired connections to the network while others will be tethered with radio waves. Transmission rates over a wireless LAN can go up to 50Mbps.

The wireless standard for LANs is defined by IEEE 802.11. Purchasing a wireless solution that is 802.11-compliant ensures product selection from many vendors.

Some vendors, however, also provide products utilizing proprietary protocols. While these products may have higher transmission rates, they are normally not interoperable across vendor lines. Radio frequency LANs are generally much slower than cabled LANs but it is the capability of mobility that makes them attractive. For example, industries with disparate users or structures that don't support standard wiring are prime candidates for wireless networks. In medical institutions all over the United States, doctors and nurses use hand-held wireless devices to record patient data or laptops with wireless adapter cards (PCMCIA cards) to transmit data back and forth to the servers or printers because running cable in patients' rooms is not feasible. Another good example of where wireless technology is used is in the manufacturing industry. With huge warehouses of products, networked computers are a necessity to keep track of inventory, and wireless networked computers allow merchants to transmit data to the network while staying mobile.

Sometimes, though, mobility is not the issue. Earlier we discussed the bandwidths of TI lines for usage with long distances and large bandwidths. Another option, and probably the ultimate in wireless networking communication is the use of satellites. Satellite communication is primarily used for WANs.

Communication satellites orbit the earth at the equator so that they are motionless relative to earth-based receivers (that way the terrestrial antennas or dishes don't have to constantly restructure themselves to point at different areas of the sky). In comparison to TI lines, satellite connections are more feasible for several city connections, but only for certain types of data. Because it takes a signal 0.27 seconds to make the trip from the satellite to the receiver, only data transfers like e-mail or overnight database changes are feasible, although someone trying to type in real time using the 0.27 second delay wouldn't really lose much speed on their words per minute. Although the wireless world is a new one for PCs, through continual advantages they are bound to become more prolific.

Wireless Links for Internetworking

Before we talk about the various types of phone lines and transmission options, we must also consider connecting without wires. Depending on your physical environment, you may find the most cost-effective connection is the one you make through thin air.

Two primary wireless technologies are available for internetworking: microwave and spread spectrum. Either can make a cost-effective option, especially when the only other choice might be to lay your network cable and conduit by digging up a parking lot. Day-to-day operating costs of wireless technologies are low too. Because no one owns the air, you don't pay fees for airtime, as long as you control both sending and receiving units. Of course, that also means if your wireless link goes down, you can't call the telephone company. You'll have no one but yourself to fix it.

A microwave signal can travel several hundreds of miles, as long as the transmitters have line of sight with each other. They must be pointed directly at each other without obstructions from other buildings or trees. Speeds of 2Mbps to 34Mbps are available from this technology.

Regulatory issues may affect your ability to use microwave systems. Licenses from national authorities are always required, and some transmission schemes may be restricted altogether. Construction activity may also affect your ability to use microwave technology. If someone decides to build a high rise and disrupts your transmission, you may have to find an alternative technology, or rent roof space on the newly constructed obstruction.

Spread spectrum is a wireless, radio technology originally developed by the military for sécure and interference-immune communications. Spread spectrum has been available to commercial users since 1985. There are actually two types of spread spectrum in use today, both developed for the military: frequency hopping and code division multiple access (CDMA). Frequency hopping is an analog technique, whereas CDMA is digital.

Frequency hopping is based on the fact that transmissions are hard to intercept. Using short transmissions on many frequencies spreads the message out, making it hard to intercept the whole message. CDMA is far more useful, but much more complex.

Spread spectrum technology also promises some help to interconnect certain nearby LANs. For example, the Proxim RangeLAN2 family of high-speed, wireless data bridges connects LANs in buildings up to three miles apart. This technology is faster than long distance phone lines, but slower than LAN speeds. Most spread spectrum LANs operate at about 2Mbps or less, compared to a standard Ethernet, which is 10Mbps. However, Proxim's 1.6Mbps radio data rate delivers a throughput that rivals most leased line connections, such as 56/64Kbps, Frame Relay, and TI.

RangeLAN2 supports a transparent bridge architecture that works with any protocol or network operating system that supports Ethernet, including Novell NetWare, TCP/IP Microsoft Windows NT, AppleTalk, and others.

Cabling And Interfaces

At the most basic level of any network communications lies the medium by which data is transmitted. For the purposes of data transmission, the term "media" can include both cabling and wireless technologies. Although physical cables are the most commonly used media for network connectivity, wireless technologies are becoming increasingly popular for their ability to link wide area networks (WANs). The type of media you use in your network is a key consideration. Media vary in several ways, including data transmission speed, ease of installation, and expense. When planning your network, consider these and other factors carefully.

Cable Types

There are several different cable types that are used in modern networks. Size, cost, data transfer rates, minimum/maximum lengths, and ease of installation vary for all of them. Various networking situations and requirements may require distinctly different cable types. This section describes the uses and limitations of each type of cable.

Twisted-Pair Cable

Twisted-pair cable is a media type used in many network topologies, including Ethernet, ARCNet, and IBM Token Ring. Twisted-pair cabling comes in two types: shielded and unshielded.

Probably the most commonly used type of networking cable in America, twisted-pair cabling was originally developed for use in telephone lines. A familiar example would be the cabling used to connect the telephone to the wall jack-CAT1, two-pair, unshielded twisted-pair (UTP) cabling, also known as "silver satin" cable. This is one type of twisted-pair cabling that typically consists of two pairs of insulated copper wires, twisted around each other, then enclosed within a plastic sheath. The twisting of the wires around each other provides a degree of protection from crosstalk (we'll cover this shortly) and other types of outside interference.

Unshielded Twisted-Pair Cabling

The Electronics Industries Association and the Telecommunications Industries Association (EIAJTIA) have created standards (EIAJTIA 568 Commercial Building Wiring Standard) that define UTP cable categories. Five types of UTP are available. They are referred to, in ascending quality order, as CAT1 through CAT5. The higher category cables usually contain more wire pairs, and these wires contain a higher number of twists per foot.

The CAT1 telephone cabling that we have already mentioned, although adequate for voice communications, does not support digital data transfer, and, therefore, should not be used in this capacity. CAT2 cabling is an older type of UTP that is rarely used. It supports data transfer rates of up to 4 Mbps. CAT3 cable, with a data transfer rate of up to 10 Mbps, is the realistic minimum grade of UTP required for today's data networks. In fact, CAT3 cable is the lowest category of UTP that meets the IEEE 802.3 (explained in the next chapter) standards for a 1013aseT Ethernet network.

CAT4 cable is an intermediate UTP cable specification that supports data transfer rates of up to 16 Mbps. For new network installations, UTP CAT2 through 4 have been largely abandoned in favor of the newer CAT5 UTP, which can handle data transfer rates up to 100 Mbps.

Unshielded twisted-pair cabling is connected from each host computer's NIC to the network patch panel, which is then connected to a network hub, using RJ-45 connectors at each connection point. An RJ-45 connector is an eight-wire (four-pair) media connector. It is slightly larger than (but similar in appearance to) the RJ-11 connector used to attach a phone line from a wall jack to your telephone or modem.

Note: It is possible, though unusual to use an RJ-11 connector data networking because RJ-11 is usually used to attach lower grades of cable.

A good example of this configuration is the Ethernet 1OBaseT standard network, which is characterized by UTP (CAT3 through 5) cable that uses RJ-45 connectors. Shielded twisted-pair (STP) cable can also be used, but it is less common. This cable type supports a data transmission rate of 10 through 100 Mbps and can transmit data up to 100 meters without a repeater. 1OBascT networks are a popular all-around choice because they are supported on most platforms, use inexpensive media, and are easier to troubleshoot than other network types. It is also possible in many cases to run UTP through already-existing telephone line conduits, thus adding to its ease of installation.

Although it is inexpensive and easier to install, UTP cabling is not without its drawbacks: namely, interference from outside electromagnetic sources, and signal crossing between adjacent wires, called crosstalk. To some degree, the wire's design – the twisting of one wire around the other – cancels out much of the natural signal overflow and interference that exist from one wire to the other. Although electromagnetic interference and crosstalk can occur on other media types, UTP is particularly susceptible because it lacks the shielding present in other cable types. This lack of shielding also makes UTP particularly vulnerable to wiretapping – a point you should examine carefully if data security is a priority.

UTP is also subject to a greater degree of attenuation, or lessening of signal strength over distance, than other cable types. This means that cable segments using UTP have the most stringent distance limitations – no segment may exceed 100 meters in length. Cable length maximums exist because signals weaken over distance, as they are partly absorbed by the media on which they travel. This attenuation causes the signals to become unreadable after the specified distances, unless a repeater (a device that cleans up and retransmits the signal) is used.

Shielded Twisted-Pair Cabling

Shielded twisted-pair (STP) cabling has traditionally been used in several network types, including AppleTalk and Token Ring. STP, which is of similar internal construction, is subject to the same 100-meter restriction as UTP. In addition, shielded twisted-pair cable usually contains, at its core, four or more pairs of twisted copper wires. STP differs from UTP in that it contains shielding – an electrically grounded woven copper mesh or aluminum foil that surrounds the cable's internal wires. This shielding separates them from the cable's outer sheath and provides resistance to external electromagnetic interference (EMI). Some types of STP also use shielding internally around each wire pair to keep the pair separated from the others, which further reduces crosstalk. Additionally, STP is considered to be more secure than UTP, because its shielding makes it somewhat less vulnerable to wiretapping.

STP cabling data transmission rates and distance restrictions are identical to those of UTP. Although STP provides more protection from EMI than UTP, it is not often used in newer network installations because it is more difficult to install and maintain. One reason for this is that the shielding makes the cable less flexible; another consideration is that STP usually requires electrical grounding.

Coaxial Cable

Coaxial cable was the first cable type used to connect computers to a network, and it helped form the basis of the original Ethernet standard. This cable type consists of a copper conductive center wire that is thicker than the wires found in twisted-pair cable, thus enabling higher data transmission rates over longer distances. The center conductor is covered by a layer of plastic foam insulating material, which, in turn, is surrounded by a second conductor, usually woven copper mesh or aluminum foil. This outer conductor is not used to transfer data, but it provides an electrical ground and shields the center conductor from internal and external interference.

Although not used as much as UTP in newer network installations, coaxial cabling is still common in much of the already-installed computer network base. It is also the type of cabling used for cable television hookups. Coaxial cabling can transmit data at 10 Mbps, for maximum distances of 185 to 500 meters. Coaxial cable manufacturers have created specifications that separate coaxial cables into categories (see Table 4.1) depending on characteristics such as impedance (current resistance, measured in ohms) and cable thickness. Coaxial cabling schemes usually require terminators, the impedance of which must be properly matched to the cable type. The two main coaxial cable types used in local area networks are Thin Ethernet (also known as "Thinnet") and Thick Ethernet (also known as "Thicknet"). In coaxial configurations, Thinnet and Thicknet are often combined within the same network, with Thicknet cabling used for the backbones and Thinnet used for the branch segments.

Table 4.1: Coaxial Cable types

Type	*Name*
RG-8 and RG-11	Thicknet (50 ohms)
RG-58 Family:	Thinnet (50 ohms)
RG-58/U	Thinnet, solid copper center conductor
RG-58A/U	Thinnet, wire-strand center conductor
RG-58C/U	Thinnet, military grade
RG-59	Broadband/Cable television (75 ohms)
RG-59/U	Broadband/Cable television (50 ohms)
RG-62	ARCNet (93 ohms)

Thinnet

Thinnet cable, also referred to as RG-58 cable, is the most commonly used coaxial media in computer networks. Second in popularity only to UTP, it is the most flexible of the coaxial cable types, about one-quarter inch in diameter. Thinnet cable can be used to connect each computer directly to the others on the LAN, using British Naval Connector (BNC) T-connectors and 50-ohms terminators. Because Thinnet configurations require no special equipment or external transceivers and can be used without hubs, Thinnet cabling schemes are an easy and relatively inexpensive way to set up a small network quickly.

According to the IEEE specification for 1OBase2 Ethernet networks, the BNC T-connectors and BNC barrel connectors are used to attach RG-58 AX or RG-58 C/U cable segments to each other. They also link the network cable to the transceiver on each computer's NIC. The BNC barrel connector is similar to the T-connector except that the barrel does not have the bottom part of the "T". Terminators, which are resistors that prevent signal echo, are required at both ends of each segment. This configuration supports data transmission speeds of up to 10 Mbps, with maximum cable lengths of 185 meters between repeaters.

Thinnet's greater transmission distance and shielding (which provides better security than twisted-pair cabling), make it a good choice in cases where these qualities are critical. However, because it is less flexible, it is somewhat more difficult to work with. It is also not the best choice in a situation where the network cabling must be installed in existing telephone wiring conduits-if this is an issue, UTP is preferable.

Thicknet

Thicknet, which was used for the original Ethernet specification, is a thicker and more expensive cable than Thinnet. It is similar in construction to Thinnet, but is a great deal less flexible. Thicknet cabling is used as the basis for a standard Ethernet (1013ase5) network. The IEEE Thick Ethernet specification for 1OBase5 networks uses either RG-8 or RG-11 cable (approximately one-half inch in diameter) as a linear bus. The difference is that it uses attachment unit interface (AUI) external transceivers connected to each NIC by means of a "vampire," or tap that pierces the cable's sheath to access the wire. Each AUI is connected to a compatible AUI (called a DB 15) connector on its computer's NIC. Thicknet cabling has a thick center conductor core, which allows it to transmit reliably at a distance of up to 500 meters per cable segment – a significant distance advantage over Thinnet. For this reason, it is often used to create backbones that link Thinnet networks. Thicknet media can transmit data at a rate of up to 10 Mbps.

Although the thickness of Thicknet cabling allows for longer transmission distances, greater security, and resistance to interference, its rigidity and bulkiness make it very difficult to install. It is also quite expensive. For these reasons, Thicknet networks are rare today. Thicknet is not a good choice for a new network installation if you can use another solution.

ARCNet

ARCNet token-passing networks generally use RG-62 A/U coaxial cable. RG-62 cabling is not used for Ethernet networks. ARCNet cable is similar to cable television cable and at one time was popular in networks. Today, support for ARCNet networks is minimal.

Plenum Cabling

The plenum is the crawl space in a building that lies between the ceiling and the roof. The special grades of fire-resistant cables used in this area are called plenum cables. This area is commonly used for telephone and network wiring. Fire codes require that cabling installed in the plenum area be fire resistant and have casing material that will not give off hazardous fumes if it does burn.

Fiber Optic Cable

At the cutting edge of network cabling technology, fiber optic cable provides superior data transmission speed over longer distances. It is also immune to interference and eavesdropping. Fiber optic cable consists of a glass or plastic center conductor, surrounded by another layer of glass or plastic cladding, and a protective outer jacket. Data is transmitted across the cable by a laser or light-emitting diode (LED) transmitter that sends oneway light pulses through the center glass fiber. The glass cladding helps to keep the light focused into the inner core. The signal is received at the other end by a photodiode receiver that converts the fight pulses to an electrical signal that the receiving computer can use.

Data transmission speeds for fiber optic networks range from 100 Mbps to 2 Gbps (Gigabits per second), and data can be sent reliably up to a distance of 2 kilometers without a repeater. A fiber optic cable can support video and voice, as well as data transmission. Because the light pulses are completely closed within the outer sheath, fiber media are virtually impervious to outside interference or eavesdropping. These qualities make fiber optic cabling an attractive option for networks that must be very secure or require extremely fast transmission over long distances.

Because light pulses can only travel in one direction, fiber optic cabling systems must contain an incoming cable and an outgoing cable for each segment that will be transmitting and receiving data. Fiber cabling is also rigid and difficult to install, making it the most expensive of all network media types. You should probably consider alternate cabling types first, when they are adequate for the situation. Fiber media require special connectors and highly skilled installers, factors that further contribute to the high implementation expense. One way to minimize the expense is to limit use of fiber cabling to network backbones, or in areas where EMI, flammability, or other environmental issues are of concern. You must carefully analyze cost factors versus requirements before deciding on fiber optic cable for a network installation.

Wireless Technologies

In addition to traditional physical media, wireless data transmission methods can provide a convenient, and sometimes necessary, alternative to network cabling connections. Wireless technologies vary in signal type, frequency (higher frequencies mean higher transmission rates), and transmission distance. Interference and cost are significant considerations. Because of the increasing numbers of WANs, and the need for mobile computing solutions, wireless network technologies comprise an ever-growing segment of the network population. The three main types of wireless data transmission technology are radio, microwave, and infrared.

Radio Waves

Radio technologies transmit data via radio frequencies and have practically nonexistent distance limitations. They are used to link LANs over great geographical distances. Radio transmission is generally expensive, subject to government regulation, and is quite susceptible to electronic and atmospheric interference. They are highly susceptible to eavesdropping, and thus require encryption or other transmission modifications to attain a reasonable level of security.

Microwaves

Microwave transmission uses higher frequencies for both short distance and global transmissions; its main limitation is that the transmitter and receiver must be within the line of sight of each other. Microwave transmission is commonly used to connect LANs in separate buildings, where using physical media is impossible or impractical. A good example of this would be two adjacent skyscrapers, where using cables would be impossible. Microwave is also used in global transmissions, which use geosynchronous satellites and ground-based dishes to adhere to the line-of-sight requirement. Microwave transmission can be extremely expensive, but microwaves are less susceptible to interference and eavesdropping than radio waves and provide higher bandwidth.

Infrared Transmissions

Infrared technologies, which operate at very high frequencies approaching those of visible light, can be used to establish close-range point-to-point or broadcast transmissions. They typically use LEDs to transmit infrared waves to the receiver. Because they can be physically blocked, and can experience interference from bright light, infrared transmissions are limited to short distance, line-of-sight applications. Infrared transmission is commonly used within stores or office buildings, or sometimes to link two buildings. Another popular use of infrared is for wireless data transfer in portable computers. Infrared technologies range from inexpensive to very expensive.

Cabling Considerations

When planning a network or adding to an existing network, you must consider several points concerning the cabling: cost, distance, transfer rate, ease of installation, number of nodes supported, and resistance to interference. We have discussed several types of cabling in this chapter, so let's compare the cable types using each of these considerations (see Table 4.2).

When designing or adding to a network, you should take each of the factors in Table 4.2 into consideration. Also, you should know how many computers you will have on the network and on each segment.

Table 4.2: Cable type comparisons

Type	*Speed*	*Distance*	*Installation*	*Interference*	*Cost*
10BaseT	10 Mbps	100M	Easy	Highly susceptible	Least expensive
100BaseT	100 Mbps	100M	Easy	Highly susceptible	More expensive than 10 Base T
STP	16 Mbps to 155 Mbps	100M	Moderately easy	Somewhat resistant	More expensive than Thinnet Or UTP

Type	*Speed*	*Distance*	*Installation*	*Interference*	*Cost*
10Base2	10 Mbps	185M	Medium difficulty	Somewhat resistant	Inexpensive
10Base5	10 Mbps	500M	More difficult than Thinnet	More resistant than most cable	More expensive than most cable
Fiber Optic	100 Mbps to 2Gbps	2 K	Most difficult	Not susceptible to electronic interference	Most expensive type of cable

Understanding NDIS And ODI

Network software architectures contain several distinct layers. Based loosely on the seven layers of the OSI Reference Model of network protocols, from the bottom layer, which deals with the hardware that makes up the physical connection, to the top layer, which is comprised of the network applications. At one time, the code for all functions of the OSI model was entirely rewritten for each new combination of NIC, protocol, and redirector. This provided for an overly large number of combinations, required a great deal of reworking the code, and was inflexible; each NIC driver could only be bound to a single protocol stack, and vice versa.

To solve this problem, a more modular approach for each layer's function was required. This was addressed by creating device interfaces-blocks of program code that act as a standard communication interface between the functional layers. These device interfaces act as interpreters: one layer does not have to understand the way an adjacent layer performs a certain task for the two layers to communicate and cooperate in getting a task done.

The main benefit of interfaces is that, once written, they provide the same services to, and understand the communications of, drivers and protocols written by any vendor that has adhered to the interface specification. Vendor driver developers do not have to understand the application programming interfaces (APIs) of other vendors in order to develop compatible software. This means that proprietary layers of code can be removed and replaced to allow for changes and upgrades. The OSI Reference Model also allows for multiple pieces of proprietary software to reside together at the same functional layer, thus enabling the binding of multiple NIC drivers to multiple network protocols, using the same intermediary device interface.

Microsoft, in combination with IBM, developed the Network Device Interface Specification (NDIS) as its implementation of the device interface concept, specifically to reside between the NIC (also called MAC or Media Access Control) driver and the protocol stack in the Data Link layer of the OSI Reference Model. NDIS allows for the binding of multiple NDIS-compliant network cards to one protocol stack or multiple protocols to a single NIC, or binds multiple protocols to multiple NICs.

Novell and Apple developed Open Datalink Interface (ODI), Novell's implementation of NDIS. It is used in Novell's NetWare networks, and provides functionality comparable to NDIS. It allows Novell's IPX/SPX protocol (and Microsoft's IPX implementation, NWLINK) to be bound to multiple NIC drivers. It also provides support for NetBIOS names.

Table 4.3: Coaxial Cable types

Network Type	*Nodes Per Segment*	*Nodes Per Network*
10BaseT	2	1,024
10BaseF*	2	1,024
100BaseT	2	1,024
10Base2	30	900** (5 segments, 3 populated)
10Base5	100	1,024 (5 segments, 3 populated)

* The 10BaseF specification is similar to 10BaseT except fiber optic cabling is used. The fiber cable can be run at much higher speeds, but if the computers only transfer at 10 Mbps, then the specification is 10BaseF.

** Although the theoretical maximum for a single 10Base2 network is 1,024 nodes, practical limitations on devices per segment result in an effective maximum of 900, or 30 devices for each of 60 segments that may have nodes attached.

Two Types of Networks – Peer-to-peer and Server-based

Peer-to-peer Networks

Computer networks arose out of the need to share data and capabilities between computers. To fill this need two basic types of networks have evolved. The first network type known as a peer-to-peer network is a simple, cost effective way to share resources. Pee-to-peer networks consist of workstations connected together through a network cable. This simple network design does not utilize a dedicated network server found in the server-based networks. Instead, in a peer-to-peer network all of the computers have equal functionality and act as both the client and the server. Each machine dedicates some data that will be shared to other machines on the network.

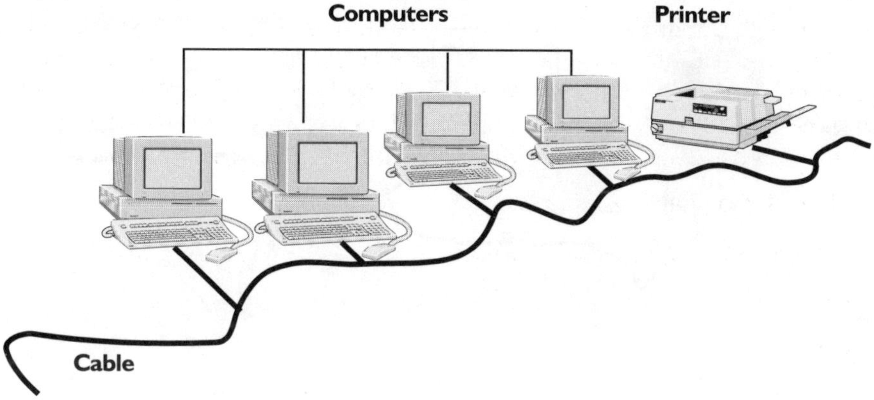

Figure 4.1: Peer-to-peer network

Peer-to-peer networks are best used in an environment with less than ten workstations where the sharing of data and resources is not common. Centralized control on these networks is not an option because of the lack of a server. This means that network security is extremely limited. The operating systems used in peer-to-peer networks allow very basic security levels. Security consists mainly of a password. That is usually all that is required to access data that has been dedicated to be shared across the network. Microsoft Windows 2000, and Microsoft Windows NT and Microsoft Windows XP are most commonly used Operating Systems (OS) in this simple network design. Only a network interface card (NIC) and cabling are required to turn your stand alone computers into a peer-to-peer network.

Server-Based Networks

Today's offices require a more powerful, more secure solution to their network computing needs. These offices utilize large databases of information that simply cannot be stored and accessed on a simple workstation. The solution is the server-based network which will be discussed in great detail. These networks offer greater opportunity for growth via resource sharing and solve many of the security issues companys need to address. By centrally administering and controlling resources, server based networks can be much more efficient.

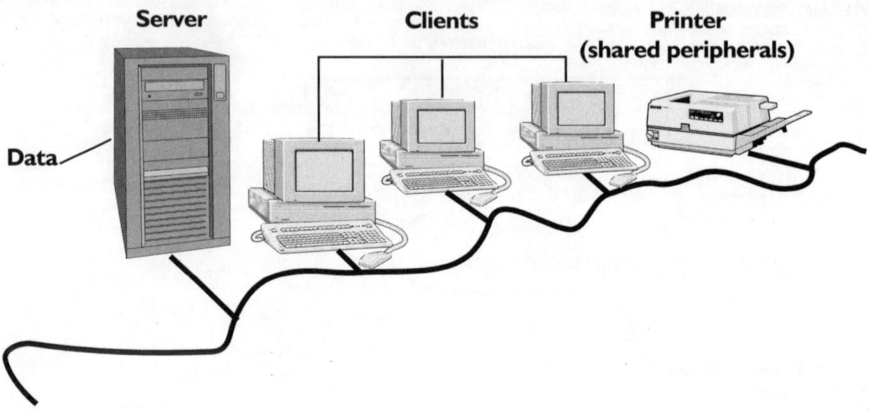

Figure 4.2: Server-based Networks

At the heart of these networks is the server. The server is a machine dedicated to manage the flow of data across the network. This computer can be in charge of several tasks including file and printer sharing, application sharing, and communication sharing. File and print sharing servers manage user access to data stored across the network. Client machines can access files located on the server for manipulation or printing by downloading the data from the server and storing the file on the client's local hard drive. Application servers allow client machines the same abilities as a file and print server except the files are never downloaded from the server. Instead client machines have the ability to access and manipulate files stored only on the server. This is effective for networks with large databases of information that need to be shared between many client machines. Communication servers allow routing of email messages and access to the Internet. This functionality is extremely important in today's wired world. The communication server establishes a simple gateway for all incoming and outgoing data.

Configuring your Server-Based LAN: Preliminary Considerations

Before you go out and purchase the most expensive, most powerful hardware available it is essential to examine what you are trying to accomplish. Asking yourself several simple questions will save you thousands of dollars and greatly increase network performance. Listed below are several preliminary questions that need to be addressed:

1. What types of applications are needed over the network?
2. Who will need access to the network data?
3. How large are the files being accessed?
4. How often will users access the data and will they need to manipulate it in any manner?
5. How often is it necessary to back up the data?

Application Servers

The first question above gives you an insight as to the software your LAN requires. The sharing of applications is the entire purpose of a LAN. The problem is that most managers do not understand software well enough to decide which application is necessary to get the job done. LANs could require the sharing of a large database of information. In this instance you need to consider using database servers such as Microsoft Exchange Server. Other LANs are established to share information regarding project management functions. Other LANs require only sharing of files and the printers. No additional software would be required for this LAN other than the server software. The key is to take a look at what other people are using to run similar applications. You can then use their experiences to help you decide what combination of software and hardware is necessary.

Network Security

Question two needs to be addressed as soon as possible. By determining who can have access to what information, you will be able to protect your network from security breaches. Most networks will allow local users access to information on the server and other client workstations. However, there are security issues that must be considered between client machines in your

LAN. Windows Server 2003 is a popular back-end server for small office LANs. The Windows Server 2003 comes equipped with security passwords and permissions to prevent local users from gaining access to specific files and network settings. When installing Windows Server 2003 onto a computer there exist a wide variety of security options. The most basic security level is the login name and password prompt that appears every time the machine is turned on. Without this information an unauthorized user will not be able to enter the operating system. Windows Server 2003 allows much higher levels of security once you enter the operating system.

Firewalls

Remote clients must be considered when setting up your LAN. The majority of LANs will allow some access to outside users. It is essential to limit the access to only those individuals you specify. Also you need to make clear to your network administrator the files and applications available to those remote users. The best tool to prevent an illegal network break-in is a "firewall". A firewall is a mechanism that acts as the traffic cop between your network and the outside world. Firewalls accomplish two main functions: permitting access to your LAN and preventing access to your LAN. The moment you connect your network to the outside world you are opening it up to viruses, hackers, and unauthorized entries. A firewall may not be able to prevent all break-ins but it does give you the ability to control what information is available to the outside world. Many corporations use a firewall system as a place to store public information about products, services, and files available for download. In this fashion, the firewall is acting as a proxy server for your network by allowing outsiders access to only specified information. Firewalls can be established to permit only e-mail messages entry into the LAN. Less stringent firewalls can be set up to allow local users to communicate freely with the outside world while at the same time preventing anyone from the outside world entering into the LAN. This may limit the functionality of your network. Therefore, you can use your firewall to block entry by certain users that are known to cause problems. For instance, if there is an outside user that is consistently breaking into your network, the firewall will be set up to block that specific user's Internet Protocol (IP) address from entering into your network.

Firewalls will not completely solve network security issues. Firewalls cannot protect against security breaches that do not go through the firewall. A firewall is only a small portion of a corporation's total security system. Network security breaches can easily occur over the telephone and fax machine. Individuals inside your network may be the cause of your security breaches. They can easily circumvent the firewall by storing important information on a floppy disk, zip disk, or compact disk. Once the information is recorded onto one of these types of media it can easily be transported out of the building through a coat pocket or briefcase. A firewall must be part of the organizational security design. Organizations that spend thousands of dollars on firewalls but allow people to enter and leave the office with unchecked pieces of media have wasted precious resources.

Can a firewall prevent data destroying viruses? Unfortunately there is some confusion as to the ability of a firewall to protect against viruses. Most viruses today are spread via e-mail. These e-mail messages contain an attachment that contains the virus. When the user opens the attachment, the virus is free to run its destructive course. According to Matt Curtin and Marcus Ranum, authors of *Firewalls FAQ*, "Firewalls can't protect very well against things like viruses. There are too many ways of encoding binary files for transfer over networks, and too many different architectures and viruses to try to search for them all". (http://www.cis.ohio-state.edu/hypertext/faq/usenet/firewalls-faq/faq.htm) Even so, there are several companies that advertise firewall products that can scan for and detect viruses entering the network through email attachments. DataTel Communications Technologies advertise their Interscan product line as being able to stop the spread of viruses into the network. Interscan performs the following function:

1. Internet e-mail is routed through the network to the antivirus gateway, where e-mail attachments are detached and scanned for viruses.
2. If an infected attachment is detected, the antivirus firewall routes the infected attachment to an antivirus engine, where it is disinfected. The cleaned e-mail is then passed through to the mail server, and an alert is sent to the network administrator, the recipient, and the sender.

Most corporations are deeply concerned by the threat of viruses. The best solution is a combination of firewall protection along with installing virus scanning software at every single computer. McAfee and Symantec are two companies that produce widely recognized virus scanning software. Either product will be useful as long as users continually download the latest updates from the Internet. These updates can be found at either http://www.mcafee.com or http://www.symantec.com depending upon which piece of software you are using. Without updating for the latest virus checks the software is completely useless against current viruses.

Network Speed

The amount of data being transferred and the frequency of transfers determine the connectivity speed required for your network efficiency. The amount of data refers to size either in kilobytes (KBs), megabytes (MBs), or gigabytes (GBs). Word documents, Excel spreadsheets, and graphic images typically range from less than 100 kilobytes to several megabytes in size. Larger databases are measured in terms of gigabytes. It is essential to determine the size of the files being transferred over your network. This information is the key for determining the specifications of hardware components such as hard drive size and speed. Today most hard drives range in size from 20 gigabytes to 120 gigabytes. If you understand the size of the files being transferred you can purchase the proper size hard drives. The size of the data being transferred will help determine the necessary network speed. The majority of networks run at either 10 or 100 MB/sec. Networks that move a few hundred gigabytes of data need to spend the extra money because it will be much more efficient and cost effective in the long run.

Configuring Network Server Operating Systems

There are three commonly recognized types of LAN server operating system software: Windows Server 2003, Novell Netware, and Unix. Windows Server 2003 is a product of Microsoft Corporation. It is promoted by Microsoft as a "multipurpose server operating system that integrates a variety of network services businesses need today." (http://www.microsoft.com) These services include file sharing, print sharing, security features, and limited distributed computing features. Microsoft's following is very strong. There are thousands of people holding Microsoft's Certificate in System Engineering (MCSE). Finding support for Microsoft's Windows Server 2003 will not be a problem. The cost depends on the size and complexity of your network.

Novell's Netware is another option for LAN server software. Netware is more commonly found in medium and large businesses. Netware offers similar functionality as Microsoft's Windows Server 2003 and is said to be more stable. Support for Netware is more difficult to find and could cost three times as much as supporting a comparable Microsoft network. Novell has been losing ground to Microsoft because of its higher maintenance expense. A reason could be the graphical user interface employed by Netware is not as straightforward as your Windows OS. To find out more information on Netware visit http://www.novell.com/showcase/netware.html.

Unix is the oldest network operating system. It was developed in 1969 by Bell Laboratories to deliver a simple operating system written in a high level programming language that allowed for the re-use of code. In the eyes of the creators, Unix would serve as a means to connect all computers in a fashion that maximized the collective power of the machines by distributing information and resources across the network. Unix networking support includes remote login, file transfer, electronic mail, and data sharing. Unix's biggest advantage lies in its ability to "run on so many different computer systems ranging from small desktops to the largest computers in the world". (http://www.hsrl.rutgers.edu/ug/unix_history.html) Once a user has learned UNIX, the skills can be used on many different systems. This ability for a user to work on many different makes of computer systems without re-training is called "user portability". Many users of other operating systems have converted to using UNIX because they felt that UNIX would be the "last" operating system they would have to learn." No other operating system is as easily transferable to different hardware architectures.

There are many different varieties of Unix. Sun base their Solaris OS on the Unix kernel. Hewlett Packard install their proprietary version of Unix on large network servers. The licensing fees for both HP and Sun's Unix are astounding. The cost to install and maintain each of these network server packages could cost over $10,000 annually.

Linux

Linux is a free version of Unix. Linux was originally created by and named after Linus B. Tovald at the University of Helsinki in Finland. Inspired by his interest in Minix, a small Unix operating system, Linus hacked his first kernel in hopes of creating "a better Minix than Minix". Linux aims towards POSIX compliance, a set of standards that show what a Unix should be. It contains all of the features that you would expect from a Unix and a fully functional operating system. These features include true multi-tasking, utilizing virtual memory, and multi-user capabilities allowing hundreds of remote users to simultaneously access a single Linux workstation. In addition, Linux contains the world's fastest TCP/IP drivers, making Linux an excellent choice for any server. Benchmarks have shown Linux outperforms traditional servers based on Windows NT, Novell and other versions of Unix running on similar hardware. This feature can be attributed to the fact that Linux is a child of the Internet. Linux developers utilized the World Wide Web as a means of communication. Speed, therefore, became a necessity. Dr. Mike Holst, a Professor of Mathematics at the University of

California at San Diego, views networking and Symmetric Multi-Processing (SMP) support as significantly more developed in Linux than with the Windows NT Server. He tested his result by testing performance of both operating systems on his quad processor server.

Linux is the choice for thousands of Internet Service Providers (ISP), university computer laboratories, financial institutions, and software developers. These companies need reliability and stability to run their network and Linux is becoming the LAN software of choice. Usenet News, E-mail, and File Transfer Protocol (FTP) are supplied with Linux by default. It also comes with an X-Windows interface conforming to the X\Open standard. X-Windows is a Graphical User Interface (GUI) that looks similar to the familiar Microsoft desktop. Redhat's Linux version 6.0 has gone as far to emulate windows as anyone else. Redhat has created a user interface complete with start button (found in the lower left hand corner), pop-up menus, and even their own version of windows where applications run.

Linux possesses all of the features of Unix including interoperability, small size, and portability. Interoperability allows Windows programs to run under Linux through X-Windows with the help of an emulator called "Wine". Windows applications could run up to ten times faster on Linux due to the OS's advanced buffering capabilities. The Linux kernel is extremely small and can be easily transferred to all types of hardware. This means that Linux can be carried around on a floppy disk that could be installed on any machine. This is exactly what Jason Valdry, a computer support consultant at the University of California, Irvine does. He keeps a portable Linux OS on a floppy drive in his pocket. Imagine the opportunity Linux presents for "Smart Card" technology. A device the size of a credit card could be running a computer application, connecting to the Internet, and databasing all your personal information at the same time.

Linux is a free Unix clone. The Linux kernel is protected under the General Public License (GPL) which states that the source code must be freely distributed and everyone is allowed to make copies for their own use or for redistribution. Thousands of FTP servers on the Web are available to download the Linux kernel. Linux is available through commercial vendors such as Redhat (http://www.redhat.com), Caldera (http://www.caldera.com), and Slackware (http://www.slackware.com). The cost is less than $50.00 for an installation disk and manual. This low cost makes Linux all the more impressive. Sun and HP offer almost identical Unix OSs that cost a minimum of $10,000 annually. This saving is especially important for "financially challenged" companies. Support for Linux is available free over the Internet. Linux user groups, news groups, and chat rooms provide an area where Linux users can post their questions to the world. The quality and reliability of the responses is the Achilles heal of Linux according to Windows supporters. However, the aggregate wealth of knowledge found on the web dwarfs any amount of source code released by Microsoft.

Linux is supported on Personal Computers (PCs) powered by the Intel Pentium and Digital Equipment Alpha processors. Although Linux promotes itself as being supported on a wide variety of architecture, hardware compatibility in your PC is still an issue. Most computer manufacturers offer little or no support for the fledgling operating system. News groups warn people to avoid large computer manufacturers such as Dell and Gateway because of "poor treatment of the Unix community." More knowledgeable companies, such as PSSC Labs (http://www.pssclabs.com) have embraced Linux as a viable alternative to Windows because of its adaptability and powerful networking potential. PSSC Labs has manufactured the Linux based Aeneas Supercomputer for Dr. Herbert Hamber of the University of California at Irvine. The Aeneas Supercomputer is considered the fastest computer in Orange County with a maximum speed of 20.1 Gigaflops. Today's fastest PCs can reach only 0.55 Gigaflops. A good source for compatible hardware and available software can be found at Redhat's Web site, http://www.redhat.com.

Server Hardware Requirements

At this point you have chosen which network OS server software to power your network. Now you can configure your hardware to meet your needs. We will assume that you have selected either Windows Server 2003 or a Linux Server. Although the functionality of these two competing products are very similar, you need to pay particular attention to the different hardware requirements.

Most people consider the network server the central point of the network. The network server houses the network operating system, stores data from all points on the network, distributes information to clients on the network, and backs up relevant data on a piece of media. If you take a look at any large computer manufacturer's Web page they have a specific line of computers dedicated as servers. In reality a server is simply a computer installed with network server software. Managers should again examine their speed, power, and back-up needs before purchasing any server hardware. Many people believe the more powerful the server the faster the network. This is simply not the case. Network speed is limited by many factors other than the server including connection speed and the amount of data being transferred. Let's lift open the case on the typical network server. PSSC Labs (http://www.pssclabs.com) advertise their POWERSERVE as the "ultimate tool for network management". Listed below is the complete configuration for POWERSERVE.

PSSC Labs POWERSERVE includes:

Case: Full Tower with 250 Watt UL Power Supply

Motherboard: Dual Processor with Intel 440BX chipset

CPU: (2) Intel Pentium III 500MHz with 512k pb Cache

Memory: 512MB SDRAM PC 100

Floppy: 3.5" 1.44 drive

Hard Drive: (2) 18GB Ultra Wide SCSI 7200 rpm, 8.5ms access

HD Controller: Adaptec 2940U2W

Video Card: ATI Xpert 98 with 8MB AGP

Monitor: Relisys 17"

CD ROM: 45X

Network: Intel 10/100 Fast PCI

Tape Back Up: 12/24GB DAT drive

5 pieces of media

Mouse: Logitec 3 Button

Keyboard: Mutsumi 104 Key High Quality

Software: Seagate Back Up Exec Server Edition

O/S: Windows NT Server 4.0

Servers usually come configured with two central processing units. The need for a second processor is not as great as you might think. A second processor only benefits machines that require a constant processing of information. Computers conducting large amounts of calculations, running heavy graphic programs, managing large databases will show a significant performance increase with the addition of the second processor. Most small office and small offices network servers do not require two processors. The role of the server in smaller networks is typically to connect all workstations and share small amounts of information. Managers can make better use of financial resources by avoiding the second processor on their network. If you do manage a larger network however, it is advisable to add the second processor.

In the above configuration Intel's Pentium III 500 central processing unit (CPU) powers the server. Managers need to follow network standards and industry trends. Intel maintains a strong grasp on CPU technology with over an 80% share of the PC market. Software and hardware manufacturers' test and benchmark their products on computers powered by Intel CPUs. Intel now has several product lines for PC CPUs including the first generation product known as the Pentium, the second generation called the Pentium II and the third generation Pentium III processor. Each of these product lines builds on the technology of the previous generation. Depending on the amount of speed and power the server requires it is advisable to look at the price differences between these processors. If you manage a smaller network where the server is transferring small amounts of data, it is not necessary to purchase the fastest CPU possible. The server's processor speed has very little to do with the overall performance of the network. Again financial resources can be saved and better utilized on client machines requiring greater processor use.

Managers should not stray away from using non-Intel based PC servers. Using less expensive processors by Advanced Micro Devices and Cyrix could cause configuration nightmares because of hardware and software incompatibility.

The amount of memory installed on the server is more important than CPU speed to the overall server performance. When you access an application program or open a file the data is transferred from the hard drive or CD ROM and is held in memory while the processor conducts manipulations. Servers need to have enough memory to hold all the necessary data. A minimum of 128 MB of memory is standard on most servers. This limited amount of memory could slow down the server's performance by not allowing enough space to hold data. It is recommended taht you use a base of 256MBs or 512MBs of memory in your server. Larger network servers hold up to 4.0GBs of memory. This may be overkill for your network but you should not cut costs by limiting the server's memory.

The type of memory you use is also extremely important to the server's performance. At the beginning of 1998 Intel released their BX chipset on motherboards. This biggest advantage of this new chipset lies in the ability of memory to transfer data to the processor at 100 MHz. This is a 50% increase over the older TX and HX chipsets which allowed memory transfer speeds up to 66 MHz. Synchronous Dynamic Random Access Memory (SDRAM) is the memory required for today's fastest workstations and servers. This is a 168 pin memory module that offers the fastest performance for PC memory. SDRAM can come in a parity chip, known as Error Correcting Code (ECC), that tests the accuracy of data as it moves in and out of memory. ECC memory is not a necessity unless you are doing highly computational functions. Small office servers need not use ECC memory.

Hard Drives: SCSI vs. IDE

Hard drives offer the greatest degree of variability in the network server. They range in size from 2.1 GB to 36 GB. In addition, there is a fundamental choice between using two different interfaces: Integrated Drive Electronics (IDE) versus Small Computer Systems Interface (SCSI). IDE is a simple path for information stored on the hard drive to be processed using input/output technology. One of the drawbacks of input/ouput technology is it prevents the CPU from accomplishing any other tasks when it is accessing the hard drive for information. This limited multitasking ability greatly reduces the performance of the machine when transferring large data files.

To combat this problem, PC manufacturers turned to Apple for help. Apple first integrated SCSI technology into their Macintosh line of computers. The value of SCSI lies in the increased bandwidth and multi-tasking abilities it provides. SCSI accomplishes this multi-tasking function through Direct Memory Access (DMA) technology. DMA allows the CPU to work on other applications while information is being transferred from the hard drive. IDE hard drive manufacturers are joining the DMA bandwagon and introducing DMA technology into their newer drives.

SCSI come in a variety of configurations. SCSI drives can be either "Ultra", "Ultra Wide", or "Ultra 2". Ultra SCSI drives transfer data at a maximum of 20 MB/sec. Ultra Wide max out at 40 MB/sec and Ultra2 drives peak at 80MB/sec. This compares favorably to the 33MB/sec peak performance achieved by IDE drives.

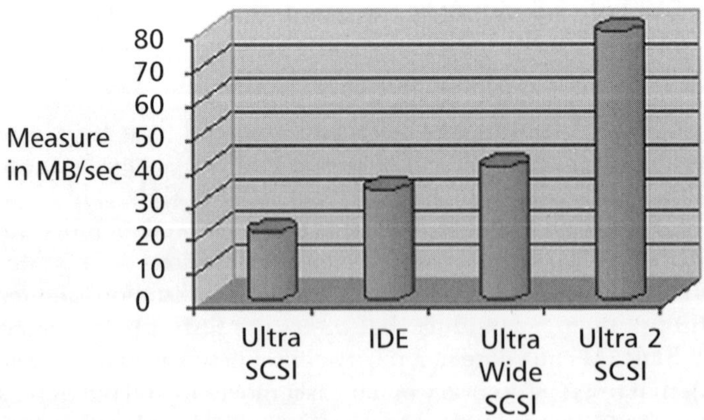

Figure 4.3: Data Transfer Rates

Managers should be aware of the performance of different hard drives. Data transfer speed is crucial to the performance of network servers. If the server is moving medium to large files Ultra 2 SCSI hard drives should be used. The time-savings allowed by faster data transfer is well within the additional cost for Ultra 2 technology.

Hard Disk Mirroring

If you go back to the PowerServe configuration, you will notice that there are two hard drives installed in the computer. The reason for this is not because of a need for greater hard drive space to store data. All of your data including files, programs, and operating system will reside on a single hard drive. The second hard drive is installed in order to ensure data back up. Windows NT Server 4.0 provides a very useful utility known as "mirroring". Mirroring allows two hard drives on the same system to store exactly the same information. If you have two mirrored hard drives in your server you essentially have two hard drives containing identical information. One hard drive is a mirror of the other. The reason why this is so useful is hard drives do fail. If one of the hard drives does fail, the mirrored drive can assume all of the functionality of the original drive. This translates to limited downtime for the server and increased network reliability.

Redundant Array of Inexpensive Disks (RAID)

If your data files are extremely large and important a Redundant Array of Inexpensive Disks (RAID) technology is the best option. RAID systems allow several hard drives to be connected together to act as a single drive. The computer recognizes this group of hard drives as just a single drive when it is accessed or written. RAID technology offers much greater fault tolerance than a single drive or two mirrored drives. A common term used in RAID is "striping".

Striping is "a method of connecting multiple drives into one logical storage unit. Striping involves partitioning each drive's storage space into stripes which may be as small as one sector (512 bytes) or as large as several megabytes. These stripes are then interleaved round-robin, so that the combined space is composed alternately of stripes from each drive. In effect, the storage space of the drives is shuffled like a deck of cards. The type of application environment, I/O or data intensive, determines whether large or small stripes should be used." (http://www.uni-mainz.de/~neuffer/scsi/what_is_raid.html) There exist six different RAID levels. Below is a brief summary of each level.

RAID-0 is the fastest and most efficient array type but offers no fault-tolerance.

RAID-1 is the array of choice for performance-critical, fault-tolerant environments. In addition, RAID-1 is the only choice for fault-tolerance if no more than two drives are desired.

RAID-2 is seldom used today because Error Correction is embedded in almost all modern disk drives.

RAID-3 can be used in data intensive or single-user environments which access long sequential records to speed up data transfer. However, RAID-3 does not allow multiple I/O operations to be overlapped and requires synchronized-spindle drives in order to avoid performance degradation with short records.

RAID-4 offers no advantages over RAID-5 and does not support multiple simultaneous write operations.

RAID-5 is the best choice in multi-user environments which are not write performance sensitive. However, at least three and more typically five drives are required for RAID-5 arrays.

In order to convert your server to a RAID system you will need a RAID controller and the software that comes with that controller. Mylex (http://www.mylex.com), DPT (http://www.dpt.com) and Adaptec (http://www.adaptec.com) offer excellent RAID controllers and plenty of technical support. Remember RAID is only used in networks with extremely large data files (10.0GB or larger).

Using Tape Backup Devices

To avoid a disastrous loss of data all network servers need to utilize a tape backup device. Backup devices along with backup software ensure that your network data is secure even if your server hardware fails. Installing a backup system is the single most important step in any network environment. Backup devices can be as simple as a floppy drive to the most complex Digital Analog Tape (DAT) device. Below is a list of several different backup devices along with their back up capacity and estimated cost.

Device	*Capacity*	*Cost*
3.5" Floppy Drive	1.44MB	$20.00
Iomega Zip Drive	100MB or 250MB	$90 to $165.00
CD Rewritable Drive	650MB	$400.00
Iomega Jazz Drive	2.0GB	$335.00
DAT Drive	Up to 24GB depending on model	$895.00

Using a floppy drive, zip drive, or CD Rewritable drive is not recommended because of the small capacity allowed by the media. Network managers need to evaluate the capacity of the data that needs to be stored, the frequency of back up, and the speed necessary to complete the backup. By analyzing these three factors managers can narrow down which backup device will match their network needs. DAT drives are the most reliable and offer the largest capacity for the price.

Simply installing any one of the devices is not enough. Network managers need to install software which will automatically backup the data periodically. Seagate's BackUp Exec is a perfect backup solution for small to medium size businesses. This software works with DAT drives of all capacity and speed. The key to the success of Seagate's product is the ease of use and flexibility it allows. Once this product is installed on the server it allows network managers to select the specific files that need to be backed up. Most small offices to mid size offices schedule a backup at the end of every week. All the manager needs to do is schedule the backup and install the tape cartridge.

Seagate's BackUp Exec offers the ability to backup not only data on the server, but data stored on client workstations. It is this flexibility which makes BackUp Exec such an effective weapon against network failures. Managers simply need to select the specific files on the server and the clients that they would like to be backed up. This is a very simple process that the backup software walks you through. This process is referred to as "backing up across the network." It is the most efficient way of backing up multiple systems but can be an extremely time consuming process that could cause significant increases in network traffic.

Managers need to analyze how quickly they would like this job to be done. Speed will be a determining factor in selecting the proper backup device. Today's DAT drives are considered the fastest backup devices available. There are slower devices that cost less money but it is recommended to use a device that matches your speed requirements.

Connecting Your Server to The Network

In order to connect your server to the network it needs to contain a Network Interface Card (NIC). The NIC allows a pathway for data transfer in and out of the server. It is a simple insert card that can either be pre-installed in the machine or added at a later date. Selecting the correct NIC can be extremely confusing unless you understand which type of connection you are using. There are two types of connections which depend on the cabling your network is using. The older type of connection is called BNC, 10Base2, or Coax.

BNC capable network cards have a cylinder terminal extending from the card. It is here that the cable is connected to your network card. This cable looks very similar to the round, stiff, black copper cable connected to your television. Copper allows data transfer speeds up to 10 MB/sec. This slow rate of data transfer, as well as the stiffness of the BNC cable, is limiting.

In order to combat the shortfalls of BNC cables, networks now use network cables based on twisted pair technology. Twisted pair, also referred to as 100BaseT, TPO, and RJ45, cable looks similar to telephone cable. It is much more flexible allowing the cable to be bent in any number of ways without destroying the integrity of the wire. More importantly these cables allow 100 MB/sec of data transfer. This provides for a significant increase in network performance.

There is no protruding cylinder connector similar to the one found in BNC network cards. Instead there is only a gap where the RJ45 cable can be inserted. Today's standard networks utilize 10/100 NICs but as a manager it is important for you to ask the building supervisor which type of cable is installed in the building.

Uninterruptible Power Supply (UPS)

All too often network managers assume that once they plug their server into the wall, the power will be supplied cleanly and at a constant level. Unfortunately this is not always the case. An uninterruptible power supply (UPS) is not a necessity to all networks but it should be added to networks in environments where there is a risk of power failure. The UPS acts as an automated external power supply that keeps a server running in the event of a power failure. Its job is to run the server for a short period of time and shut down the server in the proper fashion. If the power from the wall outlet fails, users are notified of the failure and the UPS takes over. American

Power Conversions (APC), http://www.apc.com, manufactures UPS for all sizes and varieties of networks including Windows based, Unix based, and Novell based. Choosing the correct UPS unit can be extremely confusing but it is important for managers to ask themselves the following questions:

1. How many components will the UPS need to support?
2. How much time is required to properly shut down the server and clients?
3. How will the UPS unit notify users of a power failure?

The simplest way to purchase the correct UPS is to contact your APC vendor with the answers to the above questions. They will tell you the minimum and maximum power requirements for your network.

Now your server hardware is almost completely configured. We will add the standard devices such a floppy drive, a CD ROM, a keyboard and mouse. The video card and monitor do not need to be anything special because most servers are not used as a general workstation with higher video requirements. Most servers come with a 15" SVGA monitor and a very simple video card with a maximum of 4MB of memory.

Selecting Server Software

Now that you have selected the network operating system and hardware required for the server, it is time to choose the software required to reach your networking goals. There are three basic server functions mentioned earlier. The first is file and print sharing. This is the very basic element of what a server can accomplish. All forms of network operating systems allow this functionality. No additional software is required besides the network operating system.

Most networks want to accomplish much more than this basic level of networking. Database sharing and communications ability are the driving force behind server based networks. Companies such as Microsoft, Oracle, and Netscape specialize in software suites that allow the functionality your network requires. Let's say you are designing your network to support a large database of customer contacts. As a manager you would like to store this database on the server and allow all of your users to access it. The solution is a product like Microsoft's Back Office Server. It provides an easy solution to managing customer databases.

Configuring The Client Workstation

In much the same manner that you selected hardware for the network server, the client workstations should be configured based on the software requirements of the machine. The client operating systems is much less robust than the network operating system. Microsoft remains the standard of choice for desktop OS. Microsoft Windows 2000 and Windows XP Workstation come with user friendly graphical user interfaces (GUI). More importantly these two OSs comprise 96% of the total client operating systems in the world today. This means that the majority of software programs are written specifically for Microsoft protocols. Apple's Mac OS and Linux are available options for client operating systems. However the lack of available software makes it extremely difficult to accomplish your tasks using either of these two operating systems.

Listed below is a typical client workstation offered by PSSC Labs.

Case: MidTower with 250 Watt UL Power Supply

Motherboard: Single Processor with Intel 440BX chipset

CPU: Intel Pentium III 550MHz with 512k pb Cache

Memory: 128MB SDRAM PC 100

Floppy: 3.5" 1.44 drive

Hard Drive: 8.4GB EIDE 33

HD Controller: (2) EIDE on motherboard

Video Card: Diamond Viper 550 with 16MB AGP

Monitor: Viewsonic 19" SVGA

CD ROM: 45X

Network: Intel 10/100 Fast PCI

Tape Back Up: Iomega Zip 250MB

Mouse: Logitec 3 Button

Keyboard: Mutsumi 104 Key High Quality

O/S: Windows NT 4.0 Workstation

Note that the basic differences between the hardware on the client workstations and the server. Client workstations are often relied upon to do more processor intensive work. For that reason the CPU in client computers should be equal to or more powerful than the CPU found on the server. This is not always necessary, however. Most client applications such as Adobe Photoshop and Macromedia Director are extremely processor intensive.

The amount of memory required on the client is usually much less than required on the server. The reason is client machines have less data stored in memory. Client workstations typically do not require high speed SCSI hard drives. A standard IDE hard drive is usually sufficient. However substituting a SCSI hard drive on the client may significantly increase performance if the client workstation is working with very large data files.

Video performance on client workstations is more crucial than on the server. The reason is client machines are required for everyday work functions. Individuals working on client machines prefer the largest monitor possible combined with a video card that is fast enough to smoothly run today's memory intensive programs. A 17" monitor and a video card with at least 8 MB of memory should be your office standard.

Even though you have configured your network server to back up data from client workstations it is still necessary to install some form of tape backup device on your client machines. An Iomega Zip drive or a CD Rewritable drive are perfect solutions for client machines that need to back up small amounts of data. Remember backing up your data as much as possible is essential to prevent the headaches caused by system and network failures.

All of the other hardware components including floppy drive, CD ROM drive, network card, keyboard and mouse should be almost exact replicas of the devices installed on your server. This hardware conformity will make it easier to work with machines in case of problems.

Connecting the Server and Clients

HUBS

Now you have configured the hardware and software necessary for the server and all client workstations and assigned IP addresses the only thing left to do is connect the machines together. In a simple peer-to-peer network this connection is accomplished via a network card and a cable. The same components are necessary on a server based network but an additional piece of hardware is required. This piece of hardware is a rectangular box known as a network hub. Over the page is a picture of several typical network hubs.

Figure 4.4: Network Hubs

There are some considerations you need to address before purchasing the hub. First, you need to assess how many machines will be installed on the network. Most hubs today come with ports to connect anywhere from 8 to 24 computers. If you require more ports you can purchase a "stackable" hub. Stackable refers to the ability to connect hubs together, thereby doubling or tripling the total number of accessible hub ports.

The next question is how fast do you need to transfer data. Networks are full of speed bottlenecks. One of the major bottlenecks is the network hub. Inexpensive hubs allow you to transfer data at a maximum of 10 MB/sec. Transferring large amounts of data on these hubs can be extremely time consuming. There are hubs that transfer data at 100MB/sec. Although these hubs are significantly more expensive, they can save large amounts of time because of the increase in data transfer speed. But what if you have a network consisting of some machines with 10MB/sec network cards and others with 100MB/sec network cards? Here you have to be extremely careful in selecting the proper hub. Certain hubs are said to be "switching" hubs meaning they allow 10MB/sec and 100MB/sec connections to the network. However this can be very misleading. Switching hubs will default to the slowest connection speed on the network. This means that even if you have machines using with 100MB/sec connections they will only run at 10MB/sec when they are installed on a network with machines running at 10MB/sec. In this case, the solution is to use a "switching hub". This type of hub will simultaneously allow data transfer speeds of 100MB/sec and 10MB/sec. These hubs cost significantly more money but again the time savings needs to be considered when purchasing this equipment.

Troubleshooting the Network

Maintaining the network is one of the most difficult jobs for network managers. The key here is to expect the unexpected. Network managers need to be aware of some of the basic problems that arise in networks of all sizes. The following hardware and software troubleshooting hints can help you defuse most of your basic network problems quickly and efficiently.

When you break down the computer to its most basic hardware components there are a finite number of hardware failures and conflicts that can occur. Managers must first be able to identify the symptoms of hardware problems in order to solve the problem. The computer has surprisingly few moving parts. Only the various drives and cooling fans in the computer possess items with mechanical components. All of the other components including memory, central processing units, motherboards, and insert cards are comprised of stationary circuit boards and silicon chips. This is important because typically only the items with moving parts can actually degrade with usage. Hard drives, floppy drives, CD ROMs, and tape back up devices will fail as time goes along. The stationary components usually will work for the life of the machine. Below is a detailed examination of each component, the symptoms of hardware failure, and the corrective action that should be taken.

Hard Drives

These are the components that most often fail in a system. PSSC Labs estimates that 10% of hard drives fail within the first year of a computer's operation. An additional 15% fail the second year and 15% more fail during the third year. This is extremely unfortunate considering this is the place where all of the data is stored. For this reason it is always necessary to have some form of backup including mirroring , RAID, or backup device installed in all servers and workstations.

Symptoms of failure: When a hard drive dies the machine will power on. It will go through the process of counting memory and installing the Basic Input Output System (BIOS). Once the BIOS is ready to run the OS residing on the hard drive, the machine will simply stop. The monitor will display a message that reads:

"No Hard Disk Found"

Solution: Shut the computer down. Remove the case from the machine and detach the hard drive from its housing. Disconnect all of the power connections to everything except the hard drive. While holding your ear as close as possible to the hard drive, turn the computer's power back on. If you hear no noise coming from the hard drive it is "dead". Completely remove the drive and contact your system administrator as soon as possible. If you would like to solve the problem yourself all you need to do is contact the hard drive manufacturer and obtain a Return Material Authorization (RMA) number. Certain manufacturers, like Samsung and Maxtor offer advanced swap policies on their drives where the manufacturer will send you a new drive before you send them the defective product. This is the most efficient technique manufacturers use to replace damaged products. One manufacturer to avoid because of lengthy product replacement time is Western Digital. Whenever possible avoid buying Western Digital components.

The data on the drive is now lost. There do exist companies specializing in recovering data from dead hard drives. This will cost you thousands of dollars and the effectiveness of this operation is questionable at best.

Floppy Drives and CD ROMs

These devices are typically more reliable than hard drives. However, because they are comprised of moving parts there does exist a chance of failure.

Symptoms: A floppy device failure will become apparent when you turn the machine on. On boot up the BIOS will check for the presence of a floppy drive. If this BIOS check fails a message will appear that reads "No floppy device present." A CD ROM failure does not make itself known until the machine enters Windows. CD ROM failure is apparent when you insert a piece of media into the drive and Windows cannot access the information. The CD ROM will continually spin without ever being able to read the media.

Solution: The solution is to follow the same steps as for replacing the hard drive. Floppy drives cost $25.00 and CD ROMs $60.00. It is advisable to keep a few extra CD ROMs and Floppy drives available in case of failure. Once you replace the drive you can send the defective drive back to the manufacturer.

Video Cards and Monitors

Video cards and monitors work in conjunction with one another. If one of the two devices fails, the other will be affected. It is important to realize this when diagnosing video issues.

Symptoms: Wavy lines, fuzzy picture, or a black screen are all symptoms of a monitor or video card failure. There is no way to determine which device, the video card or the monitor, is the cause of the problem until you follow the simple steps set out below.

Solution: While the machine is running, unplug the power and VGA cable on the back of the monitor. Find a replacement monitor on a nearby desk. Attach this new monitor to the machine and plug in the power cable. If the picture returns to normal the monitor you originally removed is bad. You will need to send the monitor back to the manufacturer and wait for the company to repair or replace their product. Monitor manufacturers are notoriously slow to return a working monitor back to you. The process could take up to six weeks.

If when you attached the replacement monitor the problem still persists you can conclude that the video card is defective. You will have to remove the card from the machine and send it back for repairs. If you do not have a spare video card the machine is essentially useless. Video cards can be as inexpensive as $25.00 for an AGP card with 4MB of memory.

Memory

Luckily memory will either work right away or it will not work at all. For that reason memory manufacturers offer a lifetime guarantee on their products.

Symptoms: The symptom of a memory problem is continual "crashing" of the operating system. This crash is immediately followed by the "blue screen". The blue screen is exactly what is sounds like. The monitor will only display a blue background as the computer tries to count the memory. Memory problems are probably the most difficult to diagnose because this blue screen can be caused by any number of defects including a faulty motherboard. If you receive many error messages called "page faults" you can narrow down the cause of the problem to memory. Page faults do not always appear but if they do you have bad memory.

Solution: Remove the case. Detach all of the peripheral devices including CD ROM, sound card, and network card. We are going to use the bare components of the computer. The only items necessary are the power supply, motherboard, CPU, memory, video card, and hard drive. Remove all of the memory except for one piece. Turn the machine on and see if the crashing persists. Continue to add memory one piece at a time until you have narrowed down which piece is the cause of the problems. Now you can return this piece and rebuild the machine to its working condition.

CPU

In ten years PSSC Labs has seen only one defective CPU from Intel. PSSC Labs has built thousands of workstations in that same period. It is safe to assume that the CPU is not going to go bad.

References

1. http://www.cis.ohio-state.edu/hypertext/faq/usenet/firewalls-faq/faq.htm
2. http://www.hsrl.rutgers.edu/ug/unix_history.html/unix stuff
3. http://www.redhat.com
4. http://www.pssclabs.com
5. Interview with Jason Valdry
6. Interview with Mike Holst
7. http://www.microsoft.com
8. *Networking Essentials*, Microsoft Press, 1997

5
Fault Tolerant Networks

Introduction to High Availability

When computer networking was in its infancy, networks were accurately viewed as a cost center, or a luxury that was not critical to the success or failure of the business. In the past, only access to mainframe data and applications was considered to be mission-critical. Early networks provided access to company information as a method of sharing that information, including a method of communicating with other network-enabled coworkers or associates.

Since that time, businesses have come to view networks as critical success factors in today's competitive landscape. A number of client/server data applications have become even more critical for efficient business operations. These applications include Enterprise Resource Planning (ERP) applications, new generations of specialized functional business applications, web-based information access for intranets, extranets and the Internet, and e-commerce. As businesses mold their operational procedures around these applications, the availability of the servers and networks on which these applications reside becomes an absolute necessity for normal business operations. In extreme instances, such as for web-based retailers, the network is more than just the computer; the network is virtually the entire business infrastructure. Many successful organizations even view the network and technology as a competitive differentiator that can be used to gain an advantage over their competitors by reducing time to market, increasing efficiency, and even to attract and retain talented employees.

The need for high availability networks will be further accelerated with the emergence of multiservice or "converged" networks supporting data, voice, and video applications over a common IP-based packet infrastructure. As a single infrastructure for all existing applications and communications services, and as the platform for new "converged" applications (e.g., Internet Call Centers and Unified Messaging), the multiservice network will be far more mission critical than any of the single-service networks it encompasses. Therefore, high availability networking must be considered one of the fundamental building blocks for multiservice networks, equal in importance to switched internetworking, quality of service (QOS), policy-based management, and directory-enabled networking.

While the range of networked applications that have become indispensable or "mission critical" varies considerably across industries and organizations, there are some common themes that apply to a broad base of organizations:

- Electronic mail has become the preferred channel of intra-enterprise and extra-enterprise communications. Email can be easily prioritized, where voice mail messages cannot. Email also allows the same message to reach a broad audience, produces a record of the communication, supports automated responses, and allows the enclosure of multimedia attachments. Email, in conjunction with enterprise telephony, provides a highly redundant communications system.
- Enterprise Resource Planning (ERP) systems have been deployed to automate and streamline the supply chains of many large enterprises in manufacturing, retail, and service-related industries. In these organizations, the ERP database has become the most important repository of up-to-date information required for critical business decisions. As ERP technology continues to mature, it is being customized for a wider range of industries and simplified for use by smaller businesses.
- A new generation of networked business applications has emerged focusing on improving the efficiency of key enterprise functional units such as sales, customer service, and human resources. Increasingly these applications are being interlinked among themselves and through ERP systems to achieve Enterprise Application Integration (EAI).
- The web browser and the search engine have become the universal user interface to business critical information throughout the Internet and the intranet. Dedicated terminal networks and 3270 terminal emulation solutions for access to mainframe applications are both being retired in favor of web browser access.

- The rapid growth of the Internet has made the external web site an important channel for both customer communication and product distribution. The web site conducts automated e-commerce and delivers investor information, customer support, product information, and marketing and public relations material. The external web site is itself mission-critical and increases the criticality of the intranet upon which it is dependent for up-to-date content.

While these trends will undoubtedly continue in the future, even more dramatic requirements for network availability will result from the emergence of multiservice or "converged" networks. In the multiservice model today's data networks will evolve to become a single network infrastructure based primarily on the IP protocol and capable of supporting traditional data applications plus a wide range of real-time and on-demand, multimedia communications services and applications. As more and more services are consolidated on a single infrastructure, the availability of the network will become critical to the entire employee population, regardless of job responsibility. For example, with email and voice on the same infrastructure, the redundancy between these two systems is lost unless the network itself ensures availability. And as the email experience shows, communications services tend to become mission-critical even if their delivery mechanisms are not truly real-time.

Among the emerging communication services are applications in the following categories:

- Streaming live video is a real-time, IP multicast application that has the potential to revolutionize corporate communications by allowing all employees to participate in corporate events without travel or disruption of normal work schedules.
- Video-on-demand, or video playback, is a store-and-forward unicast application that can deliver just-in-time distance training and other stored content (such as recorded live multicasts) without the expense of travel or the cost of recording media such as CD-ROMs and video tapes.
- Multimedia conferencing is a real-time, unicast or multicast application that can greatly improve collaboration among remote workgroups, while avoiding the cost and work disruption of travel to meetings.
- Voice over IP (VoIP), in which voice signals are digitized and compressed for transmission over IP networks, offers the possibility of replacing the traditional PBX with LAN telephony services based

on many of the same technologies as multimedia conferencing. In addition, VoIP gateway can allow voice or fax transmissions to bypass the PSTN toll network via either the intranet WAN or the Internet via ISPs with appropriate levels of service for voice traffic.

High availability is a critical requirement for virtually every industry in the business world today. A computer system failure quickly results in the failure of a business to operate, and every minute of downtime means a minute of lost revenue, productivity and profit. While avoiding downtime is not a new concept for businesses, the importance of high availability is compounded by business strategies such as e-business, enterprise resource planning (ERP) and business intelligence. With these highly integrated ERP and supply-chain environments, an entire enterprise's business may depend on one server. In addition, the explosive growth of the Internet has given rise to businesses that operate across multiple time zones and must be available 24 hours a day 365 days a year. Standish Group estimates that a minute of downtime typically costs approximately $13,000 per minute for ERP environments and $10,000 per minute for electronic commerce environments.

Failures can have a tremendous impact on the following:

- Revenues (direct loss, compensatory payments, lost future revenues, billing losses, investment losses, etc.)
- Financial performance (revenue recognition, cash flow, lost discounts, payment guarantees, credit rating, stock price, etc.)
- Productivity (number of employees impacted x hours out x burdened hourly rate)
- Damaged reputation (customers, suppliers, financial markets, banks, business partners, etc.)
- Other expenses (temporary employees, equipment rental, overtime costs, extra shipping costs, travel expenses, etc.)

In computer hardware and software, availability refers to the overall "uptime" of the system (or specific features of the system). For example, a personal computer may be deemed "available" for use if its operating system is booted and running. The standard way to define levels or "classes" of availability in a computer network system is a "scale of nines." System availability is often measured as the percentage of time a system can be used for its intended purposes. For example, 99% uptime translates to two nines of availability, 99.9% uptime to three nines, and so on.

Table 5.1 illustrates the meaning of this scale. It expresses each level in terms of the maximum amount of downtime per (non leap) year that could be tolerated to meet the uptime requirement. It also lists a few examples of the type of systems being built that commonly meet these requirements. Making hardware and applications available 99.999 per cent of the time is no trivial matter. Even ascending from 98 percent to 99 per cent uptime significantly increases hardware and software costs. Going for that fifth nine can result in an exponential cost increase.

Table 5.1: Availability Levels

Class	Uptime	Downtime per Year (maximum)	Examples
1	90.0%	36d 12h	Personal clients, experimental systems
2	99.0%	87h 36m	Entry-level business systems
3	99.9%	8h 46m	Top Internet Service Providers, mainstream business systems
4	99.99%	52m 33s	High-end business systems, data centers
5	99.999%	5m 15s	Carrier-grade telephony, health systems, banking
6	99.9999%		31.5s Military defense systems

For practical system analysis, outages should include both planned and unplanned. Unplanned outages range from software or hardware faults to operator or administrator errors and also include the most extreme form of unplanned outages – the natural disaster. Planned outages are due to such scheduled activities as services, upgrades and reconfiguration. Planned outages typically can account for significantly more downtime than unplanned outages. As such, planned outages should not be overlooked in system availability analysis.

When talking about availability levels, note that the overall window or time frame involved (weeks, months, years, etc.) should be specified to give the strongest meaning. A product that achieves 99.9% uptime over a period of one or more years has proven itself to a much greater degree than one whose availability has only been measured for a few weeks.

Given the range of availability options, it is helpful to break them into four categories:

Type of availability	Percentage of system uptime
Standard	99.0 to 99.5%
High	99.9 to 99.99%
Five nines	99.99 to 99.999%
Continuous operations	100%

One approach to building highly available networks is to use extremely fault tolerant network devices throughout the network, as shown in Figure 5.1. To achieve high availability end-to-end, the fault tolerance of each device is optimized. This is achieved by providing redundant backup within the device for each of its key components. For example, a highly fault tolerant switch might be configured with redundant power supplies, cooling fans, switch fabrics, and switch processors, plus it would have provision for redundant links via interfaces that support multi-linked connections. With this degree of internal redundancy, it is possible to achieve calculated device-level Mean Time Between Failure (MTBF) of up to 100,000 hours. Assuming a Mean Time to Repair (MTTR) of one hour, this MTBF would correspond to a theoretical device-level availability of 99.999%.

If a production network is designed with highly fault tolerant devices, but with a single point of failure as shown in Figure 5.1, the failure mechanisms of software configuration, electrical power, environmental control, and human operators, are likely to be the dominant causes of network downtime. Each failure that does occur will require human intervention for diagnosis and repair, resulting in a typical downtime (or Mean Time To Repair) of 2-4 hours. Thus, with single points of failure in the network design, the additional burden of highly fault tolerant devices may produce little measurable benefit in terms of the overall network system availability actually observed.

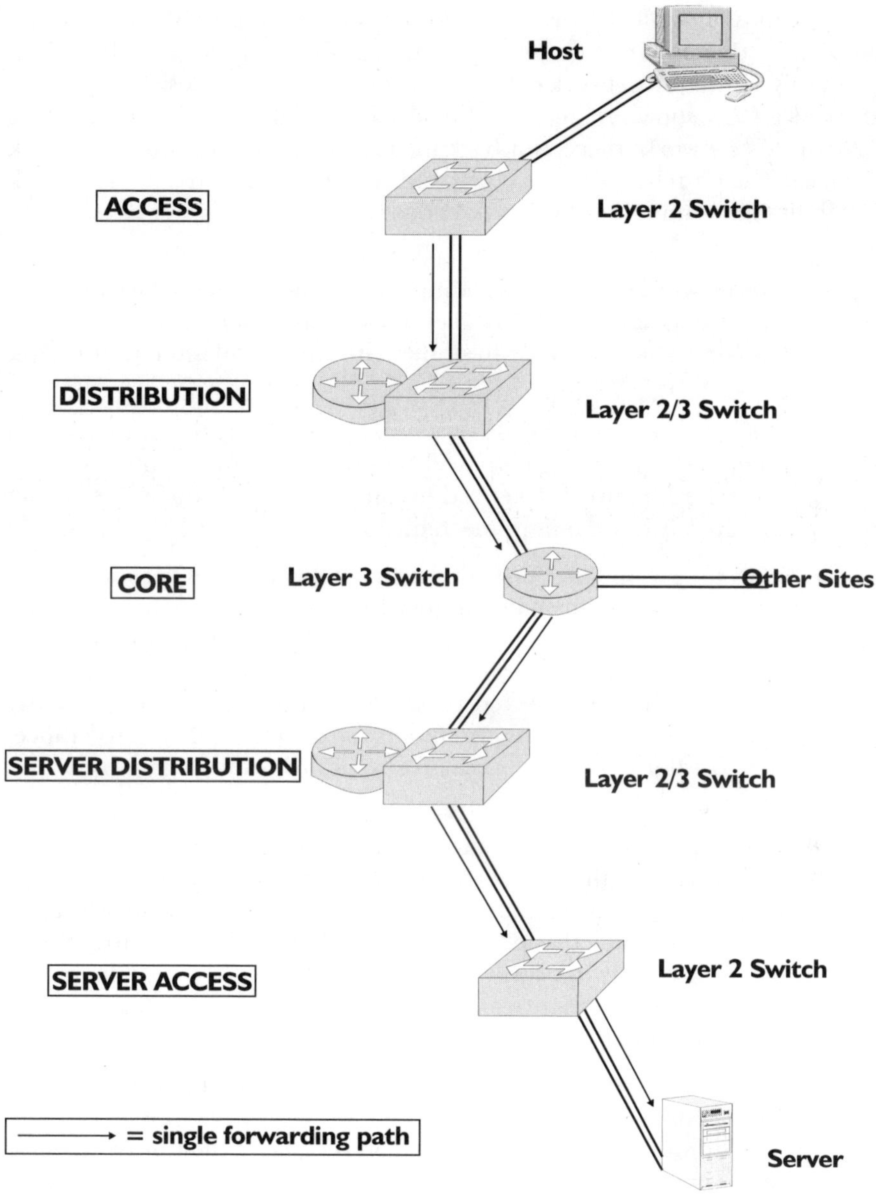

Figure 5.1: Office Network with Fault Tolerant Devices and Single Points of Failure

Another way to build highly available networks is to provide most of the reliability through redundancy in the network topology rather than primarily within the network devices themselves. With this design approach, the office LAN shown in Figure 5.1 would be modified as shown in Figure 5.2. In this network, there is a backup for every link and every network device in the path between the client and server. This approach to network reliability has a number of potential advantages:

- The network elements providing redundancy need not be co-located with the primary network elements. This reduces the probability that problems with the physical environment will interrupt service.
- Problems with software bugs/upgrades or configuration errors/changes can often be dealt with separately in the primary and secondary forwarding paths without completely interrupting service. Therefore, network-level redundancy can also reduce the impact of non-hardware failure mechanisms.
- With the redundancy provided by the network, each network device no longer needs to be configured for the ultimate in standalone fault tolerance. Device-level fault tolerance can be concentrated in the Core and Distribution Layers of the network where a hardware failure would be expected to affect a larger number of users. By partially relaxing the requirements for device-level fault tolerance, the cost per network device is reduced, to some degree offsetting the requirement for more devices.
- With appropriate resiliency features, plus careful design and configuration, the traffic load between the respective layers of the network topology (e.g., Access Layer to Distribution Layer) can be shared between the primary and secondary forwarding paths. Therefore, network-level redundancy can also provide increased aggregate performance and capacity, which in turn helps to reduce the incremental cost of a redundant network.
- Redundant networks can be configured to automatically fail-over from primary to secondary facilities without operator intervention. The duration of service interruption is equal to the time it takes for fail-over to occur. Fail-over times as low as a few seconds are possible.

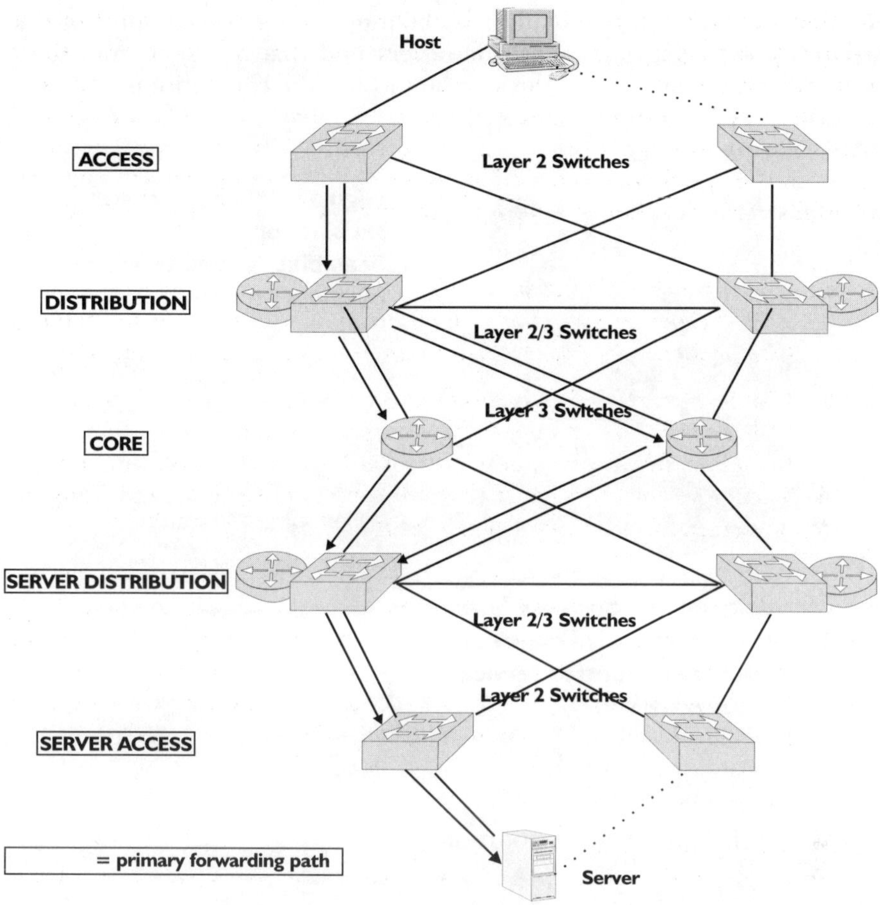

Figure 5.2: Redundant Office Network with No Single Points of Failure

IBM, one of the world's largest server companies, offers robust server solutions through clustering as a means to attain high availability. S/390 clustering, called Parallel Sysplex technology, uses a form of clustering where all servers appear to the business application as a single server. The S/390 servers support resource sharing and high-performance, multisystem, read/write data sharing. Data sharing and workload balancing enable work to be directed to available servers "on the fly" and permit servers to be dynamically added to the cluster without requiring costly downtime and providing 99.999 per cent availability. Data sharing and workload balancing enable availability – in the event of a planned or unplanned outage, hardware or software, work can be redirected with little effect on end users. S/390 Parallel Sysplex clustering is the business solution for helping to ensure that applications are available through any downtime event and that revenues from sales and other business opportunities are not lost to the competition.

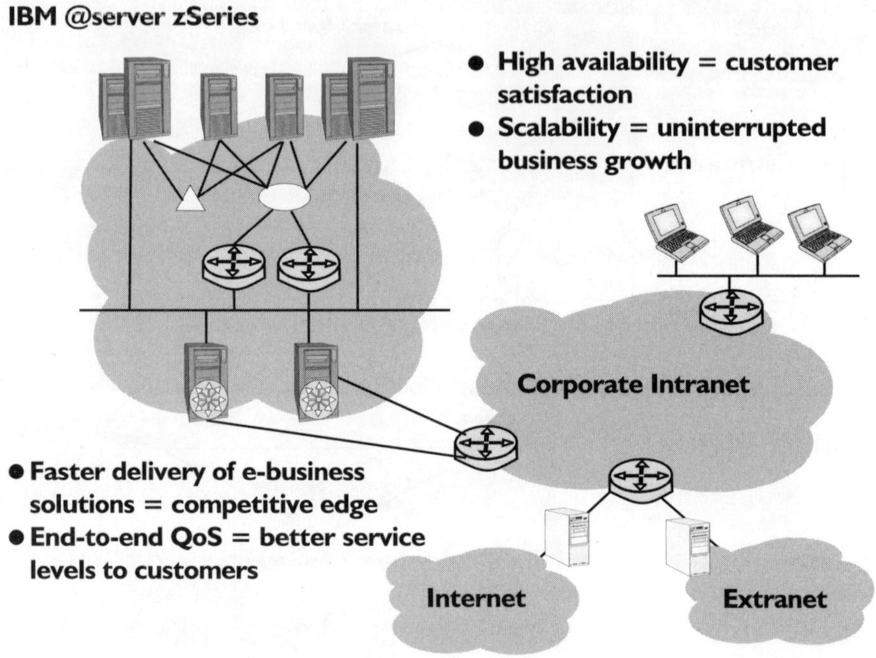

Figure 5.3: IBM e-business Solutions

Another significant and unique advantage of using S/390 Parallel Sysplex technology is the ability to perform hardware and software maintenance and installations in a non-disruptive manner. Through data sharing and dynamic workload management, servers can be dynamically removed from or added to cluster allowing installation and maintenance activities to be performed while the remaining systems continue to process work. Furthermore, by adhering to IBM's software and hardware coexistence policy, software and/or hardware upgrades can be introduced one system at a time. The ability to perform rolling hardware and software maintenance in a non-disruptive manner allows business to implement critical business function and react to rapid growth without affecting customer availability.

System availability boils down to finding faults rapidly – isolating those faults and switching to back-up without ever impacting active calls. The architecture must have partitioning such that one achieves non-interference electronically, mechanically, and through software partitioning so the fault will not impact the availability of the rest of the system. Non-stop systems must be designed to detect a fault. Fault-tolerant, redundant hardware and software will prevent active services from being dropped if an electronic or software component fails at any level or dimension.

One way to provide fault tolerant hardware and software is to monitor the health of the system by installing systems management software. Systems management software such as Desktop Management Interface (DMI) and Common Information Model (CIM) provide significant levels of system information. The fundamentals of hardware management include monitoring system voltage, fan speed, and thermal conditions, and examining the system hardware for specific types of faults. When the systems management software or firmware detects such faults, it can take a couple of different actions. In some cases – for example, when the software detects transient voltage drops or spikes – the software can only generate alerts. In other cases – for example, when the software detects a bad fan – the software can not only alert the network administrator but also transfer the cooling load to other fans in the box.

The next step up the reliability path is the add-on, hardware-based systems management card. IBM integrates this hardware – the Advanced System Management Processor – into its high-end servers. Compaq offers a couple of versions of its Remote Insight Board. Dell's implementation is the Remote Assistant Card. Any vendor selling enterprise-level server hardware offers similar technology. These boards and add-on processors build on the basic capabilities of the systems' standard management tools. These server-management cards often include the ability to dial in to a down server and run diagnostics or reboot the system.

After you decide on a systems management approach, the next implementation should be on hardware redundancy. Redundant, hot-swappable power supplies are standard on any server that claims to offer increased levels of reliability or availability. Most server vendors also offer some form of memory protection that can prevent soft errors from causing a server failure. An additional Error-Correcting Code (ECC) memory protection can prevent hard errors from faulting an entire memory DIMM. A soft error, typically caused by electrical surges, goes away when memory is refreshed, whereas a hard error is bad memory that you must replace. Vendors such as Compaq, offer additional memory protection schemes that provide enhanced protection from hard memory errors. IBM's Chipkill technology, when incorporated into the system motherboard, enables the use of nonproprietary memory and gives you additional protection from hard memory errors.

It is important to distinguish between fault tolerance and high availability. Both fault-tolerant and high-availability configurations are designed to maximize application and system availability. Both use backup resources – like mirrored servers and disks – as a fundamental approach for recovering from a failure. But while the goal of high-availability configurations is to help firms recover quickly from a crash, that is not good enough for fault tolerance – or for continuous availability, which seeks to eliminate the recovery time completely. The emphasis is on application takeover and not merely fail-over to a new server.

With fault tolerance, you are taking about five minutes or less of downtime a year. With high availability, you have a fail-over time which can vary quite a bit depending on the configuration. Fault tolerance is about true redundancy at a physical level where any component can fail and nobody knows about it even for a second. As a result, all systems in a fault-tolerant configuration feature a high degree of built-in hardware redundancy, serviceability and remote management capabilities. All main hardware elements – such as CPUs, memory and disks – are duplicated. Replicated components perform the same instructions at the same time, so even if one were to fail, the application would keep running without change. As a result, administrators can add capacity, run routine maintenance or remove failed components without taking the whole system down.

The principles that guide the design of practical fault-tolerant networks are simply put:

- Design and implement all-important components in a fully redundant fashion with the capacity to continue processing in the event of failure.
- Use fault resilient components to minimize component failure.
- Deploy network nodes in a matrix topology with robust recovery. No dependence on single point of failure links.
- Insist on industry standards for all components in order to ensure protection of investment and interoperability.
- Instrument all components so they can be managed.

There are three levels at which fault tolerance can be applied. Traditionally, fault tolerance has been used to compensate for faults in computing resources (hardware). By managing extra hardware resources, the computer subsystem increases its ability to continue operation. Hardware fault tolerance measures include redundant communications, replicated processors, additional memory, and redundant power/energy supplies.

Hardware fault tolerance was particularly important in the early days of computing, when the time between machine failures was measured in minutes.

A second level of fault tolerance recognizes that a fault tolerant hardware platform does not, in itself, guarantee high availability to the system user. It is still important to structure the computer software to compensate for faults such as changes in program or data structures due to transients or design errors. This is software fault tolerance. Mechanisms such as checkpoint/restart, recovery blocks and multiple-version programs are often used at this level.

At a third level, the computer subsystem may provide functions that compensate for failures in other system facilities that are not computer-based. This is system fault tolerance. For example, software can detect and compensate for failures in sensors. Measures at this level are usually application-specific. It is important that fault tolerance measures at all levels be compatible.

Backbone Network

For fault tolerant facilities we need a reliable, self-healing backbone. The factors involved in local backbone design are isolation, alternate paths, and robust media. Isolation means dedicating a LAN segment exclusively to communications among routers. Only routers or equivalent switches can participate in these links. This means that end-user segments, servers, and gateways are deliberately excluded. The local backbone is the final line of defense, the citadel. You want to defend it from problematic devices and protocols.

Alternate paths as it applies to the local backbone mean at least two links. As shown in Figure 5.4, there is not a single point of failure in the high availability design. Redundant network paths are provided by having two ISPs giving Internet access to the corporate network. For example, if the network path to ISP 1 is unavailable, ISP 2 can automatically take over and provide all the network access. For the intranet design, having multiple Cisco LocalDirector and Catalyst 6500 provides redundant network devices. Other redundant network devices can include clustered web servers and databases for high availability.

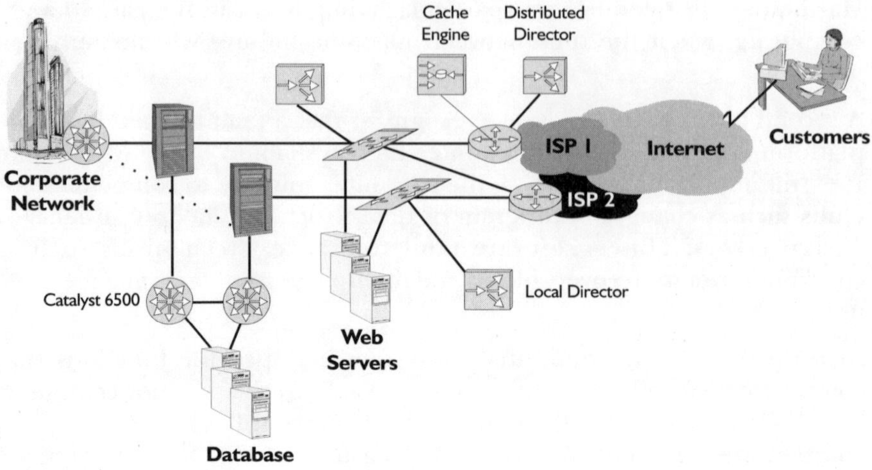

Figure 5.4: High Availability Design

Robust media means over-engineering. FDDI is a mainstream solution for local backbones. Each FDDI ring is actually two separate rings. The protocol allows traffic to alternate to the backup ring to bypass failures. Even with a robust link between the routers, a second backup is advisable.

The need for high availability has led the industry to focus on eliminating any single point of failure in their network. A very common single point of failure is the firewall server. Many times firewalls will have a limited traffic load capacity, thus having a single firewall may lead to interruption of connectivity when upgrades or service need to be completed. One solution to this problem would be implementing firewall load balancers with multiple firewall servers.

Load balancers are hardware or software devices that control the flow of packets to the machines for which they are balancing the load. Load balancers can automatically detect when a server becomes unstable or unavailable and control the flow of traffic accordingly. This type of flow management also enables growth in the environment, by allowing more servers to be added transparently, as needed to meet growth requirements.

Excite Inc., a 24-hour portal and commerce site, designed its site's infrastructure to include higher levels of redundancy, load-balancing, and additional connectivity, and higher switching capacity to make sure that each visitor's experience remains consistently rewarding. It uses Cisco's LocalDirector hardware/software devices to direct requests for content to an available server for processing. For example, if one server is not functioning or is occupied with other user requests, LocalDirector will route the call to an available server. In a typical application of the Cisco LocalDirector product, application servers are load-balanced by LocalDirector and appear as a single virtual IP address. As shown in Figure 5.5, a Web-hosting site can implement several application server clusters to support different applications such as Web, file transfer, and mail. LocalDirector front-ends these servers and, using port addresses to recognize different applications, redirects the traffic to the different servers based on their applications.

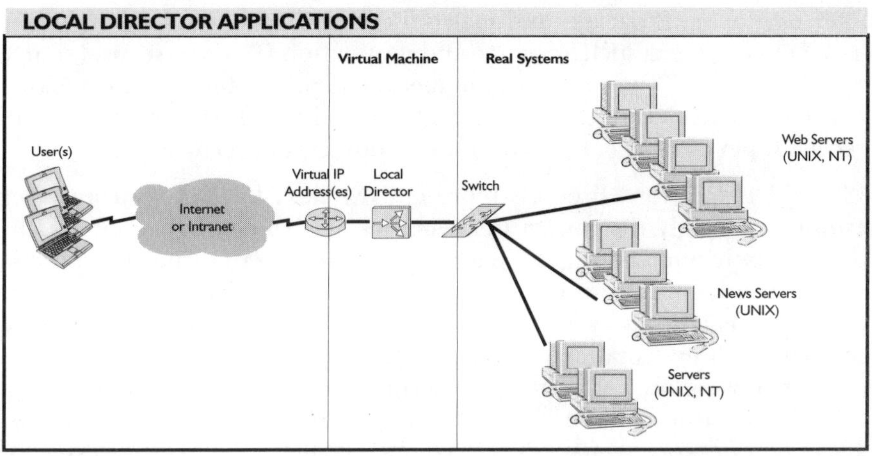

Figure 5.5: Cisco's LocalDirector

At the Internet access layer, the Excite Web site connects to multiple ISP's points of presence (POPs) via Cisco 7500s and Cisco 12000 series Gigabit Switch Routers (GSRs). Connecting to multiple ISP POPs over several DS3 (45 Mbps) and OC-3 (155 Mbps) links adds diversity and redundancy to the network design. For instance, if one ISP's network experiences a fault, traffic to and from the site will simply be rerouted across other segments of the Internet.

Firewalls

There are many vendors and types of firewalls on the market that perform a variety of functions, but traditionally firewalls can be segmented into three categories. These categories define a firewall by its functionality. The categories are:

1. Packet Filters
2. Stateful Inspection
3. Application-Level Gateway (Proxy)

"Packet Filter" technologies are typically not employed in modern firewalls, but are commonly found in routers and Layer 3 switching devices. A packet filter examines IP packets and makes a decision to accept or deny traffic based upon criteria such as source and destination IP addresses and source and destination TCP/UDP port numbers. The packet filter is only looking at a packet header, not the data contained within. It therefore provides excellent performance, but only a minimum level of security.

"Stateful Inspection" firewalls represent the next level of security. They examine not only the packet headers, but also some of the contents by transparently intercepting the packet and reviewing the data. The criteria that is used for the accept/deny decision is based upon the header information (as in a packet filter) and the data contained within. The firewall compares this information with known data types to ensure that a packet is safe for passing on to the internal (secure) network. A stateful inspection firewall can also provide Network Address Translation (NAT) services, allowing a network to have private (non-Internet routable) addresses for additional security.

The third category of firewalls, "Application-Level Gateway (Proxy)" firewalls, accepts and initiates connections for the LAN clients. This type of firewall allows for the highest level of security, as the firewall can inspect all of the data within a packet. Consequently, the application-level gateway also has more processing overhead, as it must initiate two sessions for each actual session (one to the client, one to the host).

No matter which category of firewall is chosen, an added layer of complexity is brought to light when there are multiple firewalls. The multiple firewalls can either be synchronous or asynchronous. Synchronous firewalls allow either directional traffic to flow through multiple firewalls by exchanging connection information. Because of this sharing, the connection does not need to be revalidated each time a different firewall is used. Asynchronous firewalls do not exchange information, and the connection must be revalidated each time a new firewall is used. When multiple active firewalls are installed, it is important to understand its function in the network and the possible consequences.

Firewalls load balancers can help alleviate some of these issues. Load balancers can be hardware or software based. Software solutions typically involve additional software to be installed on each of the servers in the cluster. An unknown amount of resources could be used by the installation of another piece of software. Resource allocation might also be affected by the running of multiple software products on each of the servers. Alternately, a hardware solution is a physical box that handles the flow and directing of traffic between the clients and the servers. Because the solution is a standalone machine, it would typically allow for operating system independence and for a heterogeneous server environment. However, hardware solutions are limited in the information they have about the servers. Information like CPU usage and memory utilization is difficult, if not impossible, to obtain without software counterparts loaded on each server.

Although both the software and hardware solutions have independent disadvantages, they do offer several advantages when implemented. Load balancers overcome the limited traffic load capacity on a firewall machine. They also make it easier to add firewall servers to the environment without adding complexity. If configured correctly, they can eliminate a single point of failure at the firewall level. Load balancers ensure the highest degree of availability by implementing redundancy and fault tolerance.

Server Availability

Server availability can be approached in two ways – lost data or lost time. Step one is to make the servers fault resilient. Several vendors sell fault resilient servers with basic levels of fault tolerance including ECC memory, RAID arrays, and multiple NIC cards. Every vendor from AST to Zenith has some type of resilient server. Other pertinent features to look for in a fault tolerant system are:

- Hardware-based reboot and power on/off for remote support of hung servers.
- OS support for changes in configuration and drivers without requiring a reboot.
- A UPS interface for a graceful shutdown in the event of a major power outage.
- Hot-swappable drives, both RAID and otherwise.
- Realistic backup and restore. Will the system be usable during backups and restores, or will performance or file locks reduce usability?

- Remote management of everything.

The way to prevent lost data is to record it simultaneously to multiple media. Use either disk mirroring or RAID.

System Clustering

Both planned and unplanned hardware and software failures can be reduced or eliminated through clustering. Many enterprise customers now use clustering technology to provide greater availability and scalability for their high-end mission-critical applications such as customer order entry. Clusters are essentially multiple servers that are used as a single unified resource. By inter-connecting multiple servers many times with access to shared resources, the resulting cluster is able to deliver greater availability. This increase in availability is due to the ability of the other systems in a cluster to assume the workload of a failing system resource. Likewise, maintaining access to data is a key element for high availability. If your server is running but it cannot reach critical data, your business may cease to function. Properly designed cluster solutions can also reduce or eliminate planned outages due to hardware or software upgrades by allowing selected systems to be taken off-line for nondisruptive upgrades. Robust cluster solutions depend on strong data sharing models, the ability to have systems co-exist at different software release levels, the ability to dynamically vary systems off-line, as well as robust recovery models that do not interrupt end users.

A clustering example is the point-of-sales system in a retail operation. It provides critical ongoing access to the store's database of products, codes, names, and prices. If the point-of-sale system fails, cashiers cannot log sales and the operation loses money, customers, and its reputation for quality service.

As shown in Figure 5.6, clustering technology can deliver high system availability. The clustering solution would allow a pair of servers to access the multi-port storage devices (Disk Array) on which the database resides. In the event of a server failure on Server 1, the workload is automatically moved to Server 2 and end-users are switched over to the new server – with no operator intervention and minimal downtime, as shown in Figure 5.7. The addition of clustering technology means that the overall system remains on line.

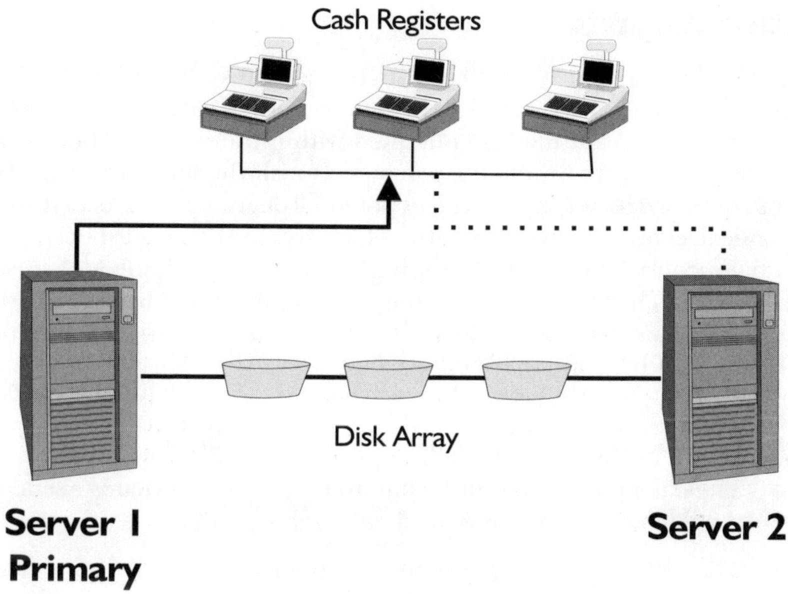

Figure 5.6: Retail Branch Before Failure

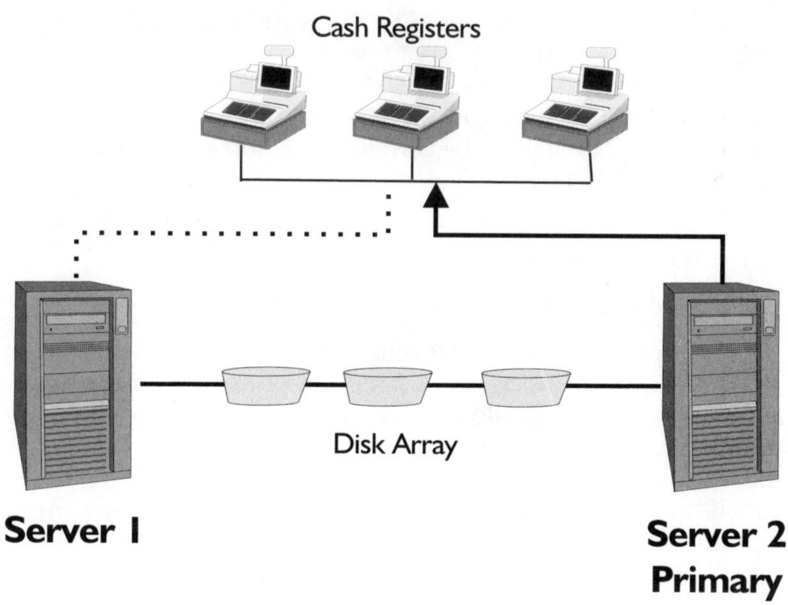

Figure 5.7: Retail Branch During Failure

Data Integrity

Data integrity is defined as information that adheres to a strict standard of value and completeness. That is, the data is both accurate and the entire population of relevant data is contained within the data warehouse. The credibility of the data warehouse solely rests with the integrity of its data. The problem of data integrity involves potential degradation of data through mechanical error (an overlap with reliability) and more especially the potential inconsistencies arising during updates or simultaneous access to a shared file. There are other numerous factors that can cause a data warehouse project to fail. The problem may be due to any one or a combination of the following items:

- Data corruption or lack of data within the originating source systems.
- The data migration routine fails to move the data that is needed by the end user community or data warehouse.
- The data cleansing process does not resolve data inconsistency issues or correctly verify selected information.
- The data transformation process is not properly constructed or contains incorrect transformation rules.
- The data loading process into the warehouse is not properly constructed or fails.

If a security breach occurs, knowing that data has not been altered inappropriately can be vital. To be certain of the integrity of stored data, you need to know:

1. When data is changed
2. Who made changes
3. The details of the actual data change

Knowing the first two things allows you to determine if changes were valid; you can examine each change to see if the person or process making any particular change did so appropriately. The third piece of information may not always be available, but if full change logs are maintained, they will also allow controlled rollbacks of inappropriate changes. If a criminal prosecution were likely, knowing these elements would be crucial.

Logs should be maintained at all levels. Operating system logs can track user logins, file access, and privilege use. Database logs (actual data logs, not the built-in transaction logs) can be constructed to automatically log user manipulation of data; a common technique is to use database triggers that fire on inserts, updates, and deletes. Using triggers ensures that activity is logged regardless of data access method. This is especially important, since a malicious user may not use the "normal" methods of changing information.

Logs need to be actively and regularly analyzed to find unexpected actions. In addition to providing an additional layer of intrusion detection, well-tuned log analysis can pinpoint application or process flaws that result in data changes that are inconsistent with established business rules (examples might be invalid data values in certain columns, users with unexpected access privileges, or even patterns of data usage that were unanticipated in application design).

Suggestions for maintaining data integrity within the operating system, applications and data files include:

1. Always have current, reliable backups of all data and code. This is important for both possible recovery, as well as comparison if necessary.
2. Implement proper code change management techniques. Good techniques facilitate verification, auditing, and rollback procedures.
3. Maintain appropriate audit logs. Auditing should occur at all levels: operating system, application code, application execution, and data files or databases.
4. Develop procedures to perform routine data validation of coded documents. This includes appropriate code generation as well as verification.
5. Integrate security policies with general policies, including personnel processing and training.
6. Train staff in validation and incident response procedures.

Introduction to RAID

RAID (Redundant Array of Inexpensive Disks) is a method whereby information is spread across several disks, using techniques such as disk striping (RAID Level 0) and disk mirroring (RAID Level 1) to achieve redundancy, lower latency and/or higher bandwidth for reading and/or writing, and data recoverability. Different RAID levels offer dramatic differences in performance, data availability, and data integrity depending on the specific I/O environment. RAID can be implemented in software or hardware or any combination of both. There is no single RAID level that is perfect for all users.

Commonly used RAID designations run from RAID 0-10, with RAID 0, RAID 1, RAID 3, and RAID 5 being the most readily available (and used).

- RAID 0 is striping with no redundancy.
- RAID 1 is disk mirroring.
- RAID 3 uses striping and requires an extra drive for parity, used to recover a failed drive.
- RAID 5 is similar to RAID 3 but spreads the parity information over all of the drives in the set.

Table 5.2: Common Flavors of RAID

RAID Type	Read/Write Performance	Fault Tolerance	Hard Disks Required
RAID 0 Striping	High/High	Low	2
RAID 1 Mirroring	Medium/Low	High	2
RAID 1+0 Striped Mirrored Array	High/Medium	High	4
RAID 4 and 5 Striped Array w/ Parity	High/Medium	Medium	3

As shown in Table 5.2, to implement RAID at least two hard disks are required. They should be identical in size and performance. Either IDE or SCSI hard disks can be used, but IDE hard disks are substantially cheaper. A hard disk controller capable of RAID such as a hardware RAID can be implemented. But an operating system which supports RAID in software can also be used.

RAID 0: Striping

RAID level 0 refers to striping data across multiple disks without parity or any redundant information. It requires a minimum of two drives. An array configured to RAID level 0 is an array configured for performance, but at the expense of fault tolerance or data integrity. Striping reduces the level of data availability since a disk failure will cause the entire array to be inaccessible. For example, if one of the drives in the RAID 0 fails for any reason, there is no way of retrieving the lost data as can be done in other RAID implementations. As such, it is usually implemented in conjunction with RAID 1 to get resilience.

The reason RAID 0 is a performance-enhancing configuration is that striping enables the array to access data from multiple drives at the same time. In other words, since the data is spread out across a number of drives in the array, it can be accessed faster because it is not bottled up on a single drive. This is especially beneficial for retrieving very large files, since they can be spread out effectively across multiple drives and accessed as if it were the size of any of the fragments it is organized into on the data stripes.

RAID 1: Shadowing/Mirroring/Duplexing

RAID level 1 refers to maintaining duplicate sets of all data on separate disk drives. It is done to ensure data reliability or a high degree of fault tolerance. Of the RAID levels, level 1 provides the highest data availability since two complete copies of all information are maintained. In addition, read performance may be enhanced if the array controller allows simultaneous reads from both members of a mirrored pair. During writes, there will be a minor performance penalty when compared to writing to a single disk.

In a RAID level 1 configuration, the RAID management software instructs the subsystem's controller to store data redundantly across a number of the drives (mirrored set) in the array. Although striping is not used within a single mirrored drive pair, multiple RAID 1 arrays can be striped together to create a single large array consisting of pairs of mirrored drives. All writes must go to both drives of a mirrored pair so that the information on the drives is kept identical. In other words, the same data is copied and stored on different discs (or "mirrored") to ensure that, should a drive fail, the data is available somewhere else within the array. In fact, all but one of the drives in a mirrored set could fail and the data stored to the RAID 1 subsystem would remain intact.

Peer-to-Peer RAID 1: Mirroring over Networks

Peer-to-peer RAID level 1 refers to maintaining duplicate sets of all data on separate disk drives, on separate hosts or servers. Of the RAID levels, this level provides the highest data availability since two complete copies of all information are maintained. During writes, there will be a minor performance penalty when compared to writing to a single disk.

RAID 3: Striping and Parity

In RAID level 3, data is striped across a set of disks. This RAID level is really an adaptation of RAID level 0 that sacrifices some capacity, for the same number of drives, but achieves a high level of data integrity or fault tolerance. It takes advantage of RAID level 0's data striping methods, except that data is striped across all but one of the drives in the array. This drive is used to store parity information that is used to maintain data integrity across all drives in the subsystem. The parity drive itself is divided up into stripes, and each parity drive stripe is used to store parity information for the corresponding data stripes dispersed throughout the array. With RAID 3, data chunks are much smaller than the average I/O size and the disk spindles are synchronized to enhance throughput in transfer rate intensive environments. RAID 3 is well suited for CAD/CAM or imaging type applications. Since parity is used, a RAID 3 stripe set can withstand a single disk failure without losing data or access to data.

RAID 5: Striping and Parity

In RAID level 5, both parity and data are striped across a set of disks. At least three drives are required for RAID 5 arrays. Data chunks are much larger than the average I/O size. Disks are able to satisfy requests independently, which provides high read performance in a request rate intensive environment. Since parity information is used, a RAID 5 stripe can withstand a single disk failure without losing data or access to data.

Unfortunately, the write performance of RAID 5 is poor. Each write requires four independent disk accesses to be completed. First old data and parity are read off of separate disks. Next the new parity is calculated. Finally, the new data and parity are written to separate disks. Many array vendors use write caching to compensate for the poor write performance of RAID 5.

Table 5.3: Advantages and Disadvantages of RAID Levels

RAID Levels	Advantages	Disadvantages
RAID 0 – Striping	- High performance - No cost penalty – all storage is usable	- Significantly reduced data availability – no fault-tolerance
RAID 1 – Mirroring	- Excellent data availability - Higher read performance than a single disk	- Used to be expensive – requires twice the desired disk space
RAID 1 – Mirroring over Networks	- Excellent data availability - Improved system uptime rating	- Used to be expensive – requires at least 2 disks on each host - Moderately slower write performance
RAID 3 – Striping & Parity	- Good data availability - High performance for transfer rate intensive applications - Cost effective – only 1 extra disk is required for parity	- Can satisfy only 1 I/O request at a time - Poor small, random I/O performance - Complicated
RAID 5 – Striping & Parity	- Average data availability - Cost effective – only 1 extra disk is required	- Poor write performance - No performance gain in data transfer rate intensive applications - Complexity - Requires special hardware

Software RAID vs. Hardware RAID

To make informed disk array purchase decisions, it is important to look beyond a discussion of the different RAID levels and also understand the differences between software and hardware-based RAID implementations. Software-based arrays are typically the least expensive to implement. Conversely, the true cost of this implementation can increase dramatically if the demand they place on the server CPU necessitates an upgrade to maintain acceptable network performance.

Software-based RAID implementations are either operating system-based, or they are application programs that run on the server. Most operating system RAID implementations provide support for RAID 1, and to a lesser degree, some operating systems also provide support for RAID 5. All array operations and management functions are controlled by the array software running on the host CPU. Just like any other application, software-based arrays occupy host system memory, consume CPU cycles and are operating system dependent. By contending with other applications that are running concurrently for host CPU cycles and memory, software-based arrays degrade overall server performance.

Most hardware-based arrays are implemented directly on a host-based RAID adapter and tightly couple the array functions with the disk interface. Additionally, this design allows all of the array operations and management to be off-loaded from the host CPU and instead, be executed locally on an embedded-processor. Since the host CPU can execute user applications while the array adapter's processor simultaneously executes the array functions, the result is true hardware multitasking. Hardware arrays also do not occupy any host system memory, nor are they operating system dependent. Different hardware RAID adapters offer some combination of RAID configuration support options including RAID 0, RAID 1, RAID 3, and RAID 5.

The bottom line is that hardware RAID, while initially more expensive than software-based RAID, is the best selection of overall system performance. In software-based RAID, overall server performance is bottlenecked. It uses the host CPU for parity calculations and forces the operating system to perform all of the additional disk I/O. In a hardware-based RAID, the hardware-based RAID adapter processes all of the additional I/O, thereby off-loading the host CPU and freeing it to more adequately handle user applications and operating system functions.

Sample Network Configuration

Data Return delivers Internet, intranet and extranet application infrastructure services using Cisco switches and routers and Microsoft Windows 2000 hosting technology, as shown in Figure 5.8. Its strategy is to provide specialized servers, a highly available global network infrastructure, and clustered hosting solutions to companies seeking to outsource management of business-critical, Internet-based applications. Data Return delivers scalability and reliability to its global customers in two ways. First it addresses reliability through extensive use of clustering technologies for application-specific solutions based on Microsoft Windows 2000 technology. This reliability extends worldwide by delivering its web hosting services across a global network based on Cisco infrastructure.

To achieve its goal of end-to-end high availability, all of Data Return's data centers are designed using a redundant, fault-tolerant Cisco router and Catalyst switch architecture to ensure there is no single point of failure. Relying on the Catalyst 2924XL Fast EtherChannel Uplink feature to eliminate bandwidth constraints between the switch and Compaq servers, the Data Return architecture ensures continuous customer access to application services. Virtual local-area networks (VLANs) are used to segment individual customer's servers into separate "logical" networks. By taking full advantage of VLANs with independent spanning trees, they eliminate the risk of single spanning tree failure from taking down the entire network or multiple subnets.

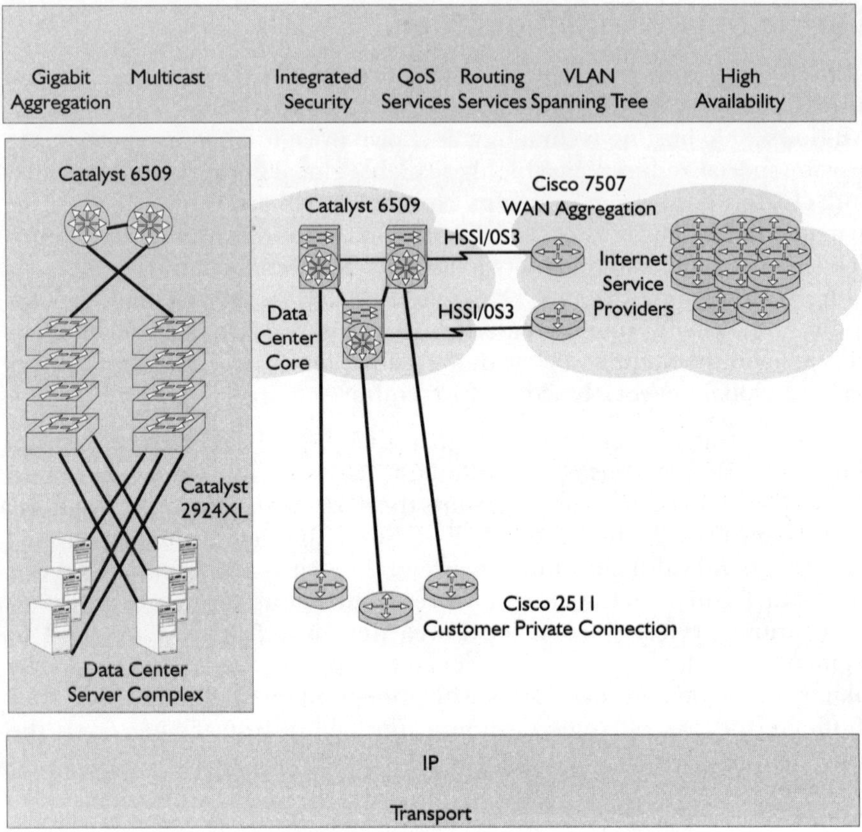

Figure 5.8: Data Return Network Infrastructure

At the core, Catalyst 6509 switches with dual Layer 3 switching processors provide continuous high-speed access to the Internet. These are connected to Cisco 7507 routers providing interconnectivity to upstream service providers. The Cisco 7507 redundant Route Switch Processor (RSP) and power supplies give Data Return maximum fault-tolerant support for connectivity via DS3 to their Internet Service Providers (ISPs).

Data Return uses Microsoft Windows 2000 Server as its operating system equipped for high-end, clustered hosting solutions. It has deployed Microsoft solutions on more than 100 Compaq ProLiant servers to support its customers all over the world. Data Return's high-performance hosting service is optimized for applications that use Windows 2000 Advanced Server, Microsoft Site Server Commerce Edition, Active Server Pages, and Microsoft SQL Server 7.0. It offers a range of cluster hosting solutions that provide linear scalability to accommodate high traffic loading.

6
LAN Switching

Aspects of LAN switching must be considered by the reader to achieve maximum benefit of use. Of special interest in effectiveness terms are routing, switching modes, protocols, and results achieved. Switching may involve various layers.

- Switching Modes
- Switching and Routing
- Multilayer Switching
- Routing Protocols
- Switch Performance

LAN and WAN Technology

Computers are powerful tools that enable users to store and process large amounts of data quickly. Practically every business, no matter how small, uses computers to handle bookkeeping, track inventory, and store documents. By any name, a confluence of computers, communication technologies, and demographics is transforming the way any enterprise conducts itself and carries out its organizational mandate. A business that ignores it will fall hopelessly to the rear in the global race for the competitive edge.

At the heart of the transformation is information. The generation and movement of information have been profitable by those companies that have taken up the technological challenge posed by the myriad machines that automate so much of our lives. A business survives and thrives on information, for it is essential within the organization and for information exchanges with suppliers, customers, and regulators. Moreover, the information needs to be consistent, accessible, and at the right location. So fundamental is information communications technology to business success that it is emerging as the foundation of a new strategy now taking shape in the U.S. business – management structures to gain competitive advantage.

We are unquestionably dependent on computers and the communications devices and services that connect them. The number of computers and terminals at work in the world today is over 100 million. It constitutes a critical mass. The overwhelming need of organizations and their workers now is connectivity, for integration, for ease of access to information. The technology that is making much of this possible is networking.

As businesses grow, they often need several people to input and process data simultaneously. For this to be beneficial, those people must be able to share the data each person enters. Networking computers becomes beneficial in this situation. Networks are simply a "group of computers connected by cable or other media so they can share information" (Nash 5). Networks make sharing data much easier and more efficient for users. The network systems help users not only to be more productive because several people can enter data at the same time, but they can also evaluate and process the shared data. So, one person can handle accounts payable, another can handle accounts receivable, and someone else process the profit – and those statements. When companies combine specialized software with a computer network, they can track and utilize information to help make the business run smoothly. Communications technology helps companies overcome those kinds of basic organizational difficulties. Good networks make "geographically dispersed companies more manageable; they help top – heavy companies trim down middle management. They also help companies break down barriers between divisions."

Many types of networks exist, but the most common types of networks are Local Area Networks (LANs), and Wide Area Networks (WANs). In a LAN, computers are connected together within a "local" area (for example, an office). In a WAN, computers are farther apart and are connected via telephone/communication lines, radio waves, or other means of connection.

As mentioned, "the number of computers in use worldwide, according to the International Data Corporation, is in excess of 100 million." Moreover, because of the expanding memory and processing power of these computers, users can put the machines to work on new kinds of applications and functions. Accordingly, the pressure from the users of these systems for ways to communicate among all these machines is irresistible. It is changing the way vendors think and the way in which automation products and services are sold. A network is no longer just a group of computers in one office or even one large building. Networks are constantly being connected to each other to form larger internets (not Internet). An internet is a large network made up of connected smaller networks. Local area networks (LANs), and wide area networks (WANs) are examples of communications networks. As been mentioned, a "communications network is a facility that interconnects a number of devices and provides a means for transmitting data from one attached device to another."

Networks are usually classified using three properties: Topology, Protocol, and Architecture.

> Topology specifies the geometric arrangement of the network. Common topologies are bus, ring, and star. Protocol specifies a common set of rules and signals the computers on the network use to communicate. Most networks use Ethernet, but some networks may use IBM's Token Ring protocol. We recommend Ethernet for office networking. Architecture refers to one of the two major types of network architecture: Peer-to-peer or client/server. In a Peer-to-Peer networking configuration, there is no server, and computers simply connect with each other in a workgroup to share files, printers, and Internet access. This is most commonly found in home office configurations, and is only practical for workgroups of dozen or less computers. In a client/server network, there is usually an NT Domain Controller, which all of the computers log on to. This server can provide various services, including centrally routed Internet Access, mail (including e-mail), file sharing, and printer access, as well as ensuring security across the network. This is most commonly found in corporate configurations, where network security is essential (Nash).

Originally, only medium to large-sized businesses could afford the cost of networking hardware. In the last decade, prices have rapidly dropped as new technology has developed, and it is now possible to grab a basic 4-PC 10Base-T home networking kit for under $150. The affordable price and added convenience of having a LAN has made it commonplace to see networks in many offices. Each PC in a LAN is able to access shared files and devices anywhere on the LAN. This makes the sharing of expensive devices, such as laser printers or large removable storage drives a cost-effective alternative to purchasing a device for every user.

If a user decides to use Ethernet for his or her LAN, they can choose from "10Base-T (10 million bits per second, or 10 Mbps) or 100Base-T (100 million bits per second, or 100 Mbps), that is usually referred to as *Fast Ethernet*." 10Base-T is used mostly for small office networks, because "it offers both affordability (around $20-30 for a network card, and $80 for an 8-port hub) and decent performance. For large businesses, a 100Base-T solution may be required if there are a large number of PCs connected to the network, or if large amounts of data are frequently transferred.

There are three different types of cabling for Ethernet networks. "Unshielded Twisted Pair (UTP) is the most popular type of cabling, and the one we recommend because of its wide availability and low price." The other two types are coaxial and Shielded Twisted Pair (STP). "STP cable provides more shielding against outside Electromagnetic Interference (EMI) than UTP, but is more expensive." Coaxial uses much thicker and more expensive cable, and must be run in a ring configuration, from PC to PC, unlike Twisted Pair, which is run directly from each computer to a port on the hub, making wiring more convenient in most cases.

Ethernet networks require a Hub, which is "a hardware device that all PCs on a network are connected to by cabling." The hub manages receiving and transmitting data from networked devices. Hubs come in many different port configurations, but they will probably need a 4, 8, or 12 port hub, depending on the number of PCs the user wants to connect together. Each port supports a single 10Base-T connection from a PC or peripheral. If the user is using Coaxial cable, he or she needs to find a hub with one Coaxial port and minimal Twisted Pair ports. There are both Manageable Hubs, which "allow advanced configuration of Hub properties via a software package, and Standard Hubs, which are cheaper, and usually used for small office networks."

The sizes of networks are generally categorized into three different groups. One type of network that has become increasingly common is the local-area network (LAN).

Local area network or LAN is the smallest network size. Indeed, the LAN is to be found in virtually all medium-sized and large office buildings. As the number and power of computing devices have grown, so have the number and capacity of LANs to be found in an office. For higher-speed data support, networks that rely on a shared transmission medium and the use of packet transmission are needed. These are referred to as local-area networks (LANs).

Some characteristics of a LAN are "high speed, small error counts, and inexpensive price" (Nash 30). Since LANs are contained in small areas, high-speed cable can be used. Also, since the installed media is usually high quality, few to no errors are generated on the network. Prices of LAN equipment are fairly low. Network cards for individual computers can be found for as little as $30 each.

Although standards have been developed that reduce somewhat the number of types of LANs, there are still half a dozen general types of local-area networks to choose from. Furthermore, many offices need more than one such network, with the attendant problems of interconnecting and managing a diverse collection of networks, computers, and terminals. Local networks are used to interconnect equipment within a single building or cluster of buildings. Some local networks rely on circuit-switching technology; these include "digital private branch exchange (PBXs) and digital data switches. This technology is suited primarily to telephone and terminal support." For higher-speed data support, networks that rely on a shared transmission medium and the use of packet transmission are needed.

Whereas wide-area networks may be public or private, LANs usually are owned by the organization that is using the network to interconnect equipment. LANs have much greater capacity than wide-area networks to carry what is generally a greater internal communications load. A simple example of a LAN that highlights some of its characteristics is that there would be several workstations along with a server. All of the devices are attached to a shared transmission medium. All other devices attached to the same network can receive a transmission from any one device. As a result, much of the effort in LAN development has been in the development of high-speed LANs, with data rates of 100 Mbps or more.

The key to the development of the LAN market is the availability of a low-cost interface. The cost to connect equipment to a LAN must be much less than the cost of the equipment alone. This requirement, plus the complexity of the LAN logic, dictates a solution based on the use of chips and very large scale integration.

Traditionally, networks have been classified as either LANs or WANs. Wide-area networks cover a "large geographical area, often require the crossing of public rights-of-way, and often rely at least in part on circuits provided by a common carrier. Typically, a WAN consists of a number of interconnected switching nodes. A transmission from any one device is routed through these internal nodes to the destination device specified."

Traditionally, WANs have provided only relatively modest capacity to subscribers. For data attachment, "either to a data network or to a telephone network by means of a modem, data rates of 9600 bps or even less have been common." A recent development in WANs in this range of performance has been the development of the T1 service, which provides network services at rates up to 1.544 Mbps. The continuing development of practical optical fiber facilities has led to the standardization of much higher data rates for WANs, and these services are becoming more widely available. These high-speed WANs provide user "connections in the tens and hundreds of Mbps, using transmission techniques known as *frame relay* and *asynchronous transfer* mode (ATM)."

As with wide-area networks, a local-area network is a communications network that interconnects a variety of devices and provides a means for information exchange among those devices. There are several key distinctions between LANs and WANs:

The scope of the LAN is small, typically a single building or a cluster of buildings. This difference in geographic scope leads to different technical solutions, as we shall see. In particular, most LAN designs involve the shared use of a transmission medium rather than an interconnected set of network switching nodes.

It is usually the case that the LAN is owned by the same organization that owns the attached devices. For WANs, this is less often the case, or at least a significant fraction of the network assets are not owned. This has two implications. First, care must be taken in the choice of LAN, since there may be a substantial capital investment (compared to dial-up or leased charges for wide-area networks) for both purchase and maintenance. Second, the network management responsibility for a local network falls solely on the user. The internal data rates of LANs are typically much greater than those of wide-area networks (Stallings and Slyke).

Traditional LANs have provided data rates in a range from about 1 to 20 Mbps. These data rates, though substantial, "have become increasingly inadequate with the proliferation of devices, the growth in multimedia applications, and the increased use of the client-server architecture." As a result, much of the effort in LAN development has been in the development of high-speed LANs, with data rates of 100 Mbps or more.

For transmission of data (including voice, image, and video, as well as ordinary data such as numerical and text), beyond a local area, communication is typically achieved by transmitting data from source to destination through a network of intermediate switching nodes. The purpose of the switching nodes is to "provide a switching facility that will move the data from node to node until they reach their destination. The end devices that wish to communicate may be referred to as stations. The stations may be computers, terminals, telephones, or other communicating devices."

Two quite different technologies are used in wide-area switched networks – circuit switching and packet switching. These two technologies differ in the way the nodes switch information from one link to another on the way from source to destination. Circuit switching is the dominant technology for both voice and data communications today and will remain so for the foreseeable future. Communication via circuit switching implies that there is a dedicated communication path between two stations. That path is a connected sequence of links between network nodes. On each physical link, a channel is dedicated to the connection. The most common example of circuit switching is the telephone network.

Circuit switching is used in public telephone networks and is the basis for private networks built on leased lines and using on-site circuit switches. Circuit switching was developed to handle voice traffic but can also handle digital data, although the latter use is often inefficient. With circuit switching, a dedicated path is established between two stations for communications. Switching and transmission resources within the network are reserved for the exclusive use of the circuit for the duration of the connection. The connection is transparent. In terms of performance, there is a delay prior to signal transfer for call establishment. However, once the circuit is established, the network is effectively transparent to the users. Information is transmitted at a fixed data rate with no delay other than the propagation delay through the transmission links. The delay at each is negligible.

An important characteristic of a circuit-switching device is whether it is blocking or nonblocking. Blocking occurs when the network is unable to connect two stations because all possible paths between them are already in use. A blocking network is one in which such blocking is possible. Hence a nonblocking network permits all stations to be connected (in pairs) at once and grants all possible connection requests as long as the called party is free.

Around 1970, research began on a new form of architecture for long-distance digital data communications – packet switching. Packet switching is employed to provide an efficient means of using the shared facilities in a data communications network. Although the technology of packet switching has evolved substantially since that time it is remarkable that the basic technology of packet switching is fundamentally the same today as it was in the early 1970s networks, and packet switching remains one of the few effective technologies for long-distance data communications. The two newest WAN technologies, frame relay and ATM, are essentially variations on the basic packet-switching approach.

With packet switching, data are transmitted in short blocks called packets. Each packet contains some portion of the user data plus control information needed for proper functioning of the network. Each packet contains a portion (or all for a short message) of the user's data plus some control information. The control information, at a minimum, includes the information that the network requires being able to route the packet through the network and deliver it to the intended destination.

Packet switching adds several new alternatives for wide-area networking in addition to those that can be provided using circuit-switching technology. Just as there are public and private circuit-switching networks, there are public and private packet-switching networks. For data traffic, the number of wide-area networking choices is even broader. Roughly the alternatives are:

- Public packet-switching networks. These are available to be shared by a number of separate subscriber communities. The technology may also be employed to build a private packet-switching network. There are a number of such networks in the United States and at least one in most industrialized countries. Typically, the user must lease a line from the user's computing equipment to the nearest packet-switching node.
- Private packet-switching networks. In this case the user owns or leases the packet-switching nodes, which are generally located with the user's data processing equipment.
- Private leased lines. Dedicated lines can be used between sites. No switching is involved, so a leased line is needed between any pair of sites that wish to exchange data.
- Public circuit-switching networks. With the use of modems or switched digital service, the user can employ dial-up telephone lines for data communications (Stallings and Slyke).

The choice between circuit and packet switching depends on a host of considerations, including cost, performance, reliability, and flexibility. Both technologies will continue to be important in wide-area networking.

Beyond the traditional alternatives for packet-switching and circuit-switching services, a wide-area networking alternative is the integrated services digital network (ISDN) which provides both opportunities and challenges for the manager. The ISDN service is based on the concept of providing a set of channels at a single interface. The B channel, at 64 kbps, is the principal channel used for circuit switching, packet switching, and dedicated (leased) circuits. The D channel is used for control signaling (call setup) and may also carry some data. For residential users and small businesses, a basic access service of two B channels and one D channel is adequate.

As the speed and number of local-area networks (LANs) continue their relentless growth, increasing demand is placed on wide-area packet-switching networks to support the tremendous throughput generated by the LANs. In the early days of wide-area networking, "X.25 was designed to support direct connection of terminals and computers over long distances. As LANs have come to play an increasing role in the local environment,

X.25 is being recognized as an inadequate tool for wide-area networking." There are three high-speed WAN services available, (1) Frame relay, (2) switched multimegabit data service (SMDS), (3) asynchronous transfer mode (ATM), also known as cell relay. In addition, there is a follow-on specification for a second-generation ISDN, known as broadband ISDN (B-ISDN), that is based on the use of ATM.

The most widely available such service is frame relay. Frame relay was developed initially as part of ISDN but is offered by a wide variety of providers for both public and private network configurations. Frame relay makes "use of variable-sized packets, called frames, and a processing scheme that is considerably simpler than X.25." Data rates of up to 2 Mbps are readily achievable. Frame relay is designed to eliminate much of the overhead that X.25 imposes on end-user systems and on the packet-switching network. The advantage of frame relay is that we have streamlined the communications process. The protocol functionality required at the user-network interface is reduced, as is the internal network processing.

Switched multimegabit data service (SMDS) is a high-speed switched data service developed by Bellcore on behalf of the seven regional Bell operating companies (RBOCs). It was first introduced in late 1991 and is now widely available in the United States. It is also offered now by a number of public carriers in other countries. The switched multimegabit data service (SMDS) "offers data rates of up to 45 Mbps and makes use of fixed-size packets called cells. The cell size for SMDS is the same as that used for ATM and B-ISDN. It is intended that SMDS will provide a transition path to ATM and B-ISDN."

As with frame relay and ATM, SMDS defines a service at the user-network interface. Unlike frame relay and ATM, which are connection-oriented services, SMDS is a connectionless, or datagram, type of service. It shares with frame relay and ATM an emphasis on streamlined processing to achieve high data rates. As with other public switched services, access to the networks is by means of a dedicated line. Internally, SMDS packets are switched from source to destination through one or more SMDS switches. These switches are connected by high-speed trunks, such as DSI or SONET transmission.

Asynchronous transfer mode (ATM) is even more streamlined than frame relay and SMDS and is intended to provide capacity in the hundreds of Mbps and eventually in the Gbps. ATM technology is finding use in a wide variety of offerings for both wide-area and local-area networking. In addition, ATM is the transmission technology used in broadband ISDN, which may be viewed as a package ATM offering that provides a variety of standardized services.

As the speed and number of local-area networks (LANs) continue their relentless growth, increasing demand is placed on wide-area packet-switching networks to support the tremendous throughput generated by these LANs. In the early days of wide-area networking, X.25 was designed to support direct connection of terminals and computers over computers over long distances. At speeds up to 64 kbps or so, X.25 copes well with these demands. As LANs have come to play an increasing role in the local environment, X.25, with its substantial overhead, is being recognized as an inadequate tool for wide-area networking. This has led to increasing interest in frame relay, which is designed to support access speeds up to 2 Mbps. But now, even the streamlined design of frame relay is faltering in the face of a requirement for wide-area access speeds in the tens and hundreds of megabits per second.

As what has been discussed, there are four factors, which have made a solid understanding of the fundamentals of data communications essential to business and information management. One of the factors is the increasing use of data processing equipment. As the cost of computer hardware has dropped, data processing equipment has become an increasingly important and pervasive part of the office, factory, and engineering environment. The second factor is the increasing use of distributed systems. The dropping hardware costs have resulted in the increasing use of small systems, including minicomputers, microcomputer workstations, and personal computers. These systems are distributed throughout a business and must be interconnected to exchange messages, share files, and share resources, such as printers.

The third factor is the increasing diversity of networking options. The emergence of a broad range of local area network (LAN) standards plus the evolution of LAN technology have led to a broad, overlapping range of products for local-area communications. Similarly, the planning for the next generation of telephone equipment and networks, and the evolution of new transmission and networking technologies have led to a broad, overlapping range of options for long-distance communications. The fourth factor is the sudden emergence of the Internet and the World Wide Web. In a very short time, the Internet and especially the World Wide Web, have attracted millions of business and personal users. No business can ignore the potential of this enormous facility.

Now, technological developments and the widespread acceptance of standards are transforming the ways in which information is used to support the business function. In addition to the traditional communications requirements for voice and data (meaning text and numerical data), there is now the need to deal with pictorial images and video information. These four types of information (voice, data, image, and video) are essential to the survival of any business in today's competitive international environment. What is needed is a treatment not just of data communications but of information communications for the business environment.

Information communications and computer networking have become essential to the functioning of today's business, large and small. Furthermore, they have become a major and growing cost to organizations. Management and staff need a thorough understanding of information communications in order to assess needs; plan for the introduction of products, services, and systems; and manage the systems and technical personnel that operate them. This understanding must comprise (1) Technology – the underlying technology of information communications facilities, networking systems, and communications software. (2) Architecture – the way in which hardware, software, and services can be organized to provide computer and terminal interconnection. (3) Applications – how information communications and networking systems can meet the requirements of today's businesses.

Networking and Client/Server Standards

How is all of this client/server technology connected together? To answer this question, we must look at what makes up a network. A physical network is a collection of communication links, cables, routers, switching equipment, personal computers with network cards in them, and the transport stacks or protocols that help them all communicate effectively. Networks can be divided into two main categories: Local Area Networks (LAN) or Wide Area Networks (WAN). A LAN covers short distances like a building. A WAN covers extended geographical distances. A corporation is said to have a Wide Area Network when this corporation has two local area networks connected together by a fast digital line that goes through the public Internet. Examples of this fast digital lines are frame-relay, ATM, XDSL, or any of many digital lines. For example, an airport retail company has many shops in different airports throughout the United States and perhaps overseas. Their headquarters might be located in Long Beach. The network located in the main headquarters will be described as the main LAN. The networks created at each one of the stores are described as the remote offices or remote LANs. When you connect this remote LANs through a high-speed digital line to the main headquarters' LAN, you are said to have a Wide Area Network (see Figure 6.1).

Your basic LAN will have the following similar connections: a PC client with an Ethernet card connected through copper wire, usually category five cabling, to a central Ethernet switch. All of your PCs in the same building will connect to an Ethernet switch. These Ethernet switches can be cascaded together to accommodate all of your computers. Your computers will need to speak a common language or protocol to be able to communicate with each other. The most common used protocol is the Transmission Control Protocol/ Internet Protocol (TCP/IP). It is a non-proprietary protocol that is used not only in LAN but is the protocol of choice in the Internet. Routers are another device used both at the LAN level and the WAN level. In the LAN, the router is used to segment different areas and keep traffic local. For example, you might notice that the Human Resources Department personnel are always sharing files mainly between themselves. You might also notice that this is also the case for your other departments. In order to keep traffic low, you will want to add a router between each of these departments and the backbone. Another solution is to purchase switches that can create virtual networks. A router is most commonly used when you have a remote LAN connecting to your main LAN. The router is the main connection between the high-speed digital line and your local Ethernet switch. You need to have another router on the other side in order to complete the connection. Routers are capable of speaking various languages. It is not uncommon for companies to be running multiple protocols on their networks. Some of the most common protocols are TCP/IP, IPX/SPX, and Appletalk. IPX/SPX is a proprietary protocol developed by Novell. Appletalk is also a proprietary protocol developed by Macintosh. Some of the most common high-speed WAN links are Frame-Relay, Asynchronous Transfer Mode (ATM), and Switched Multimegabit Data Services (SMDS). T3's maximum speed is 45 megabit per second, ATM's maximum speed is 2.4 gigabit per second, and SMDS's maximum speed is 45 megabit per second. Out of these three technologies, frame-relay is the most commonly used. In summary, client/server configurations are connected by wires, switching equipment, routers, and high-speed lines. They are governed by protocols that are the equivalent to languages. The connectivity speed is decided by the amount of bandwidth that will be utilized.

Testing

This is a revolution and as with any other revolution there will be pitfalls to watch out for. It is imperative that we continue on forward with this technology since it is the technology of the future. If you want to be an integral part of these new standards, you must participate in programs like the beta testers and attend various meetings where the standards are being discussed and written. As with any new technology, testing is an integral part of the design. Test your new products thoroughly before you implement them. You should have a testing lab that is not connected on your network and that closely resembles your actual production servers. Here you can test all of the latest products and test to see how well they would do in your network. Make sure that every little oddity has been tested before you implement it on your actual production servers. A good example of thorough testing is the case of the Los Angeles County Office of Education. They are in the midst of implementing a new Peoplesoft powered ERP system. In order not to take any chances, they are running the new ERP system parallel with their existing production servers for an entire year before going with a full production implementation. This speaks greatly to how thorough you must be before implementing a major client/server technology. It is recommended you do as much testing as possible before production implementation.

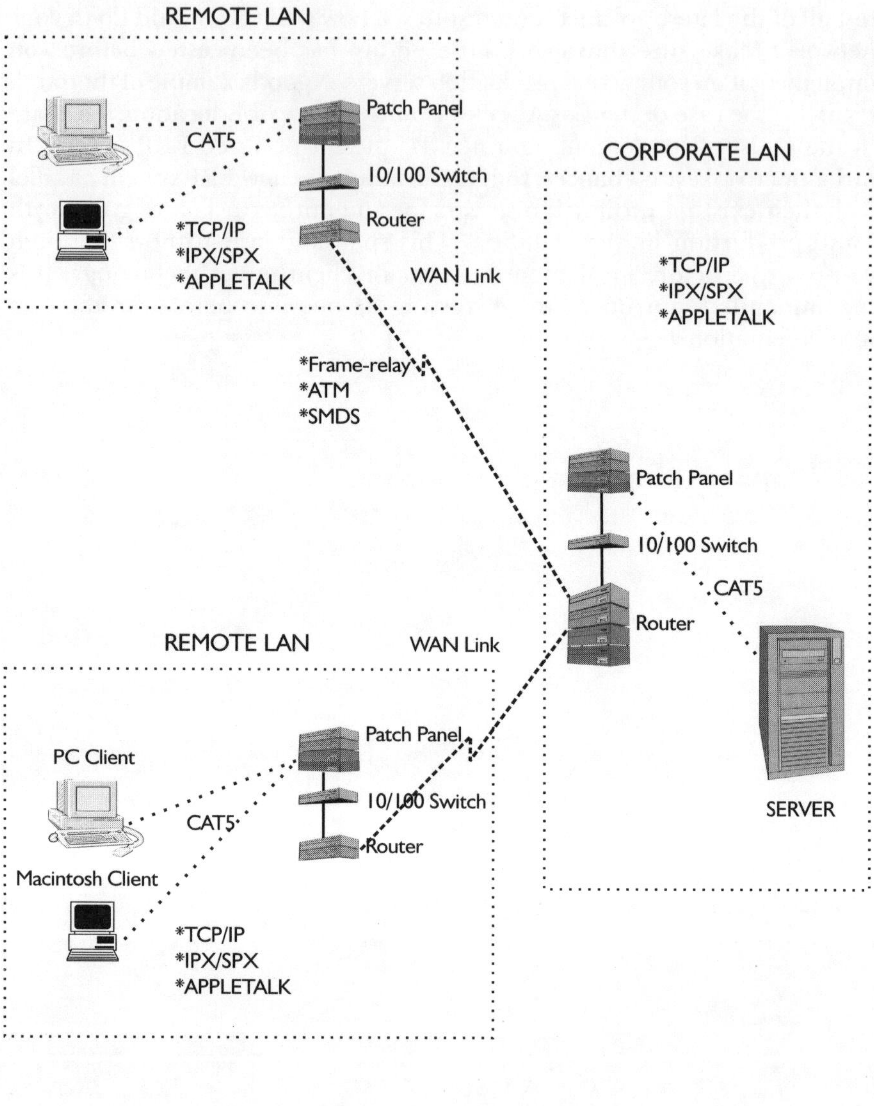

Figure 6.1: Typical Wide Area Network

References

Nash, Jason. *Networking Essentials MCSE.* Chicago: IDG Books Worldwide, Inc., 1998.

Stallings, William, and, Slyke Richard. *Business Data Communications*, 3rd ed.: New Jersey: Prentice-Hall, Inc., 1998 .

"Computer Networking." Online Internet.

http://www.geocities.com/ResearchTriangle/8892/networkings.htm. December 14, 1999.

"Computer Network Technology." Online Internet.

http://www.geocities.com/ResearchTriangel/6838/privacy.htm. December 14, 1999.

"Networking." Online Internet.

http://src.doc.ic.ac.uk/bySubject/Networking.html. December 14, 1999.

7

Data Storage and Management[1]

Assessing Storage Requirements

The average computer user is often guided by processing speed as the primary criteria for judging a system's worth. Unfortunately, this often renders the storage capacity as a neglected specification. This short-sightedness often catches up with everyone; in fact the first hardware upgrade made after the purchase of any computer system is some type of storage upgrade.

Storage requirements vary greatly based on the type of business. In addition to the actual storage of data, one should consider how stored data is managed. For example, a library has vastly different storage needs than a medical center. Some companies just need to back-up their data records and frequent access is not essential, while other companies may rely on continuous access to data. The advent of the Internet has resulted in many companies approaching the "paperless office." Storage must be viewed from several angles so that the organization's needs are properly addressed.

Understanding Data Storage

We need to understand some technical aspects of computers and how they work with data before we can address different storage modalities. Computers store data in a binary format; that is, data is stored as a series of the numbers 0 and 1. A group of 0's and 1's defines a character. The vast majority of computers today use an international standard known as the

American Standard for Coded Information Interchange (ASCII). ASCII consists of 256 characters, including the 26 letters of the English language and other Latin characters, numerals from 0 to 9, characters found on a typical *qwerty* (typewriter) keyboard and other special characters. In order for the binary system to be able to represent all ASCII characters the minimum number of binary digits needed is 2x2x2x2x2x2x2x2 = 2^8 = 256. Each binary digit is a *bit* (from *bi*nary digi*t*). Eight bits are used to represent one ASCII character.

Eight bits of information are also referred to as one *byte*. When the letter "A" is typed in your computer, you have just taken up 1 byte of information. 1,024 bytes are referred to as a kilobyte(Kb). The vast majority of storage needs will be defined either by kilobytes or multiples of the kilobyte, such as a megabyte (1,024 kilobytes) or a gigabyte (1024 megabytes).

Storage Devices and Media

Various methods may be employed to actually store data. The term data storage is very broad. It encompasses a wide range of devices and media, including:

- RAM
- ROM
- punch cards
- chads
- floppies
- diskettes
- solid state disks
- hard drives
- magneto-optical disks
- CD-ROMs
- rewritable CDs
- DVD
- DVD RAM

Memory

Most computers have memory in various forms. Memory that is directly on the computers main board is referred to as either Read Only Memory (ROM) or Random Access Memory (RAM). ROM is a fixed memory system. It is used to permanently store data that will essentially never change. A good example of ROM memory is the game cartridge used by many video game consoles such as the Nintendo. The user purchasing the game cartridge is unable to modify the program stored on the cartridge. ROM memory is virtually indestructible. ROM memory also does not require any outside power source to maintain the information. ROM memory is extremely fast.

RAM memory is used to hold data temporarily. RAM is the memory a computer uses to store a particular program while that program is being used. Once the user has finished using the program, the information is cleared from RAM. RAM memory is faster than other memory storage solutions, though not as fast as ROM. The disadvantage with RAM memory is that it looses all its data once its source of power is eliminated.

Another memory system inside computers is the Erasable Programmable ROM, also known as EPROM. This is similar to ROM but it can be erased and changed. EPROM are used in situations where the user may need to change memory on a limited basis, for example, some hand held information devices (like the palm pilot) employ an EPROM to store important phone numbers, so that even if the battery dies, the information is still available. EPROMS, however, can succumb to erasure if they are exposed to ultraviolet light. There have also been reports of EPROMS not making it through older security X-RAY machines at airports.

While it may seem that this type of storage would suffice, the main problem is its cost. Chip based memory costs more than other types of memory storage primarily due to the high rate of rejection in the manufacturing process.

Another problem with chip based memory is transportability and compatibility. Since these types of memory work with a given computer system, another brand of computer may not easily accept the memory. Many companies use proprietary connectors and different voltages, which results in incompatibilities. Due to the cost of ROM and RAM, these storage methods were employed on a large scale only in universities and corporations with significant resources.

The earliest storage system was punch cards. In fact, punch card technology was used in player pianos as early as the 1800's, where holes on a roll of paper would be used to allow a piano to strike keys in a predefined sequence, resulting in music. Ironically it is exactly the same pre-industrial age technology that was used by some voting districts in the 2000 United States election that resulted in the count fiasco. It is evident that this storage medium, while simple, is limited by its mechanical nature and is prone to errors. In addition, it is vulnerable to damage since most punch cards are made out of paper products. Punch cards can also occupy large amounts of space.

As a result of this need for compatibility and transportability the floppy disk drive came into being. Consisting of a metallic coated Mylar plastic disc in a protective cover the discs proved to be quite popular. Although slow and limited in storage by today's standards, the floppy disk is present on most personal computers sold today. Originally started as an 8 inch disk, later expanded to 5.25 inches and finally to a 3.5 inch size, floppy disks can be given from one user to another easily. Floppy disks are also very inexpensive. Unfortunately, this great storage medium is prone to damage easily. Placement of a floppy disk in varying climatic conditions or near magnetic fields can severely disrupt or destroy the data recorded onto them. Floppy disks have now been delegated mainly to back up small documents or for transporting files between computers.

Although introduced in 1956, affordable hard drives with capacities greater then 10 megabytes came in the 1980s and were the answer to larger storage requirements. The hard drive consists of a solid metal platter (the medium where the actual data is stored) encased in a dust free container. The reading head rests just above platter at a distance less than that of a human hair. The platter spins at extremely high speeds (as much as 10,000 revolutions per minute). In addition to holding much more than the standard floppy, the hard drive has an extremely fast access time to the data which it is storing. First generation hard drives were large and unmovable. Hard drives available today are small, and in some cases portable or removable.

Hard drives record data in a very interesting way. The platter is divided into tracks, cylinders, and sectors. This creates a systematic recording structure. A track can be thought of as a ring around some point on the platter. Data is recorded to the outermost track first. As the hard drive gets filled with data, it makes its way toward the center of the platter. When data is erased in the middle of a platter, then new data will be written in the new empty space. This technique eventually results is data being stored all over the platter area. While this may seem disorganized, it is actually advantageous. Should a hard drive fail for any reason (also known as a hard disk "crash"), and if data were recorded sequentially, one may loose extremely large amounts of data. However, with dispersal over a large area, you will usually lose only a small fraction of the data. For example, if you were to have a list of 200 people, and the hard drive crashed, you may still be able to extract data for 190 people since the damaged area would contain only a small fraction of the data you are trying to retrieve. Hard drives are susceptible to temperature variations and G-forces (falling on the floor). In fact, platters have been shown to expand and contract. Moreover, any metallic disc spinning at 10,000 rotations per minute can generate a lot of heat. However, increased sophistication in electronics and changes is physical design have greatly eliminated these problems.

While the hard drive sounds like the panacea for our storage needs, there is a hitch. The drives themselves may store and send data at fast speeds, but can your computer keep up with the speed of the hard drive? Hard drives interface with your computer by essentially 2 standards. The *Advanced Technology Attachment* (formerly known as Integrated Disk Electronics) or *ATA* interface and the *Small Computer Systems Interface* (*SCSI*, pronounced "skuzzy"). The ATA interface dominates the computer industry with over 90% using this interface. SCSI is used primarily in high end workstations and servers. SCSI drives have traditionally been used in "data" sensitive environments because they have a proven track record of extremely high speeds, fast transfer rates, and huge storage capacities. This results in improved performance. However, ATA drives are catching up and for the average use do not make much of a difference. SCSI drives and interfaces are of greater cost as well.

Mid to large size businesses use synchronized groups of hard disks known as drive arrays. This allows data to be spread out over several drives to prevent a massive loss of data and to facilitate restoration of data, if necessary. This system is known as *Redundant Array of Inexpensive Disks* or *RAID*. When one records data on their computer system, the data are sent to multiple drives on the RAID system simultaneously. If one drive fails for any reason, another has the capability to take over the system with no "downtime". Until recently, RAID was available only to SCSI drives and was used extensively in server environments.

To provide ease in transport of data from one computer to another, removable hard drive technology began in the mid 90s. Essentially similar to a hard drive, the platter was encased in a dust free plastic case which the user could eject out of the drive reader unit and then walk over to another computer and use. While not as fast as a standard hard disk drive, the benefits are obvious. However, several competing companies manufacture these drives, resulting in proprietary technologies that are incompatible with each other.

Another portable storage system is the magneto-optical or "floptical" drive. In the early days of hard drive data storage, many thought that this technology would be the next evolution in portable storage. A crystalline metal alloy on a platter changes shape when a magnet together with a laser are employed to record a byte of information on the disc. This medium is extremely durable and is not effected by light, magnetic interference or G-forces. However, the capacity of these drive units became overshadowed by removable hard drive technology and MO drives are now used only in some specialized applications (one example of MO technology is the SONY minidisc).

The next evolution in disc storage was the Compact Disc, Recordable and Rewritable. Compact disks store data on one physical track. This track spirals from the center to the outer edge of the disk. The current maximum storage capacity of a CD is about 800 megabytes.

CDs have high transfer rates, but slow access times. This means that the CD takes times to find the initial bit of data, but once it knows where it is, it sends that data almost as quickly as a hard drive. CD-ROMs are inexpensive. It is easy to transport and can be read by millions of computers. In terms of data storage, 800 megabytes is not that large, but considering the cost per megabyte, it is currently the most cost effective storage format. When considering blank CD-ROMs for data storage, the GOLD reflective layer discs should be selected. Gold discs cost slightly more but will out last silver and other dyes.

Backup

Rapid recovery of data after a disruption is essential in every organization. Data records are stored on all types of media, including electronic media and hard copies. Backing up data poses unique challenges for most organizations due to its continually evolving nature.

The backup strategy should consider recovery not only of data files, but also hardware, software, and documentation. Without a backup strategy, it would be time consuming and expensive to reconstruct lost information.

While software and hardware backup is essential in an organization, software and hardware tends to be relatively static. Unlike data, software and hardware are not changing or being modified every second. In an online, real-time system, each transaction modifies some element of the data.

The backup strategy must consider the transitional nature of data. Procedures must be in place to ensure that the most recent data records are backed up.

Backup Policy

A backup policy should be established for the short-term as well as the long-term backup needs of the organization.

The backup should consider the procedures to be followed. It should indicate which procedures take precedence over others. It should indicate which procedures are to be performed sequentially or simultaneously. A flowchart or diagram can greatly simplify the recovery process.

Identifying Critical Data Records

Data includes not only data files and databases, but also source documents and output documents. Critical records must be recovered quickly following a disruption. The following should be considered:

- What documents and records are likely to be needed first?
- Where are the critical records stored?
- What special equipment, if any, or other resources will be needed to restore or recover the vital records?
- Once the critical data records have been recovered, where will these records be stored and how will they be protected?

Backup Facilities

Both onsite and offsite backup must be planned for hardware, software, data and documentation. The advantages of onsite backup include convenience and immediate access. Generally, a fire-resistant safe on the premises will suffice for onsite backup. It should be used to store the most current copy of the backup files. Onsite backup is useful when there is a need to recover from a minor disruption, such as the malfunction of hardware equipment or the corruption of a data file.

However, for more significant disruptions, such as a disaster or catastrophe, onsite backup protection may be inadequate. In a disaster, such as a fire, an earthquake, or a tornado, there is a significant probability that whatever will affect the primary hardware, software or data files, will also affect the backup hardware, software or data files. Offsite backup facilities are essential in recovering from a significant disruption.

Offsite facilities may be located nearby for convenience or far away for greater protection. A disruption, such as a fire, may not affect your nearby backup facility. However, a widespread disaster such as a hurricane, a flood, or an earthquake may damage your nearby offsite backup facilities.

A balance must be struck between convenience and protection. One compromise is to have both a local offsite facility and a remote offsite facility.

A fire-resistant safe in a building within a few miles of your primary facilities can serve as a local offsite facility. This will provide reasonable protection against disasters such as a fire. A local offsite facility offers the convenience of accessing data files on a daily basis. Files may be kept in this location for a few days.

A remote off-site facility provides the greatest protection. The greater the distance, the greater the protection. The remote off-site facility should be at least several miles from the primary facility. Ideally, it should be at least a hundred miles from the primary facilities. The remote offsite facility should have a fire-resistant storage area. It may be used to house data files for several weeks or longer.

Several generations of backup files should be kept. An organization should have procedures in place to move data files between different backup facilities. The logistics of moving files to different facilities can become complex very quickly. It is essential to ensure that data files are correctly backed up and transported.

The organization should make an attempt to utilize standard technology. Recovery in a disaster can be expedited if the organization does not use unique hardware or software. Hardware equipment and software should be kept current to make it easier to replace equipment and software after a disruption.

Hot sites are capable of becoming fully operational on a short notice, often in a matter of hours. Warm sites can be made operational in a matter of days or weeks. Cold sites typically have only a skeleton structure in place; it may take anywhere from several days or weeks to even months to get a cold site operational.

Hot sites are suitable for critical applications. If a delay in computer processing will cause significant financial loss, an organization should consider using a hot site. Hot sites are used when even a partial degradation in processing is considered unacceptable.

The fast recovery time of a hot site comes at a significant price. It is essential all hardware and software at the hot site be completely compatible with the primary processing site. Whenever equipment or software is updated at the primary site, it must be updated at the hot site.

Warm sites tend to be less expensive than hot sites. The tradeoff is that warm sites require more time to become fully operational. An organization may be able to use older or less expensive equipment at its warm site. More expensive equipment and software may be purchased or leased if necessary after a serious disruption. Warm sites are suitable when a partial degradation in processing output is considered acceptable.

Cold sites are suitable for longer term computing needs. A cold site may not necessarily have any hardware equipment or software. A cold site may consist of nothing more than a basic skeleton facility. The cold site provides the infrastructure for a computer facility, such as cables for voice and data transmissions, air conditioning/heating units, humidity controls, electrical wiring, flood and fire protection systems, etc.

The three types of backup sites do not have to be mutually exclusive. It often makes sense to utilize three or two types of sites in contingency planning. A hot site may be used for initial short-term processing until a warm site or cold site is ready to meet long-term processing needs.

Identifying the best recovery strategy is a complex issue. It requires a thorough technical analysis to identify the cost of hardware equipment and software applications. Cost-benefit analysis should be performed for each strategy option. The cost-benefit analysis should consider not only the direct cost of hardware and software, but also the indirect costs, such as wiring, air conditioning, and fire prevention systems.

Backup Techniques

Several techniques are frequently used to backup data. One approach is to differentiate between larger and smaller data files. For instance:

- Smaller files, which are not continuously updated, may be duplicated in their entirety on magnetic or optical media.
- Larger files, especially those that are frequently updated, require an incremental backup strategy. Generally, it is neither practical nor cost effective to backup the entire data files.

In a batch processing system, it is common to update the original master file with transactional changes to create a new master file. By keeping copies of the original master files and the transaction files for two or three generations, a new master file can be recreated if necessary.

Real-time, online processing systems require a different strategy. The continuously changing nature of data files necessitates the use of techniques such as the duplicate logging of transactions or keeping before and after images of master records. Copies of individual records may have to be kept before updating. Database management systems typically contain specialized backup procedures for recovery of data in online, real-time systems.

Remote Backup

Remote backup using the Internet is available from several companies. Data files are automatically backed up on a daily basis. Data are stored in an encrypted format. Remote backup using the Internet is inexpensive. It is especially useful for small businesses that may not be keeping backup copies of their data on a regular basis at an offsite location.

Another more expensive option is to transmit a copy of each transaction to a remotely located computer facility. Mirror processing not only transmits the data, but also updates a backup duplicate database, always keeping the backup current.

Network Data Management Protocol

Network Data Management Protocol (NDMP), *www.ndmp.org*, is an open standard protocol for enterprise wide backup of heterogeneous network-attached storage. It was jointly developed by Network Appliance (NetApp) & IntelliGuard Software.

An NDMP server can provide two services to either a *data* server or a *tape* server. The server either reads from disk or tape and produces an NDMP data stream (in a specified format) or reads an NDMP data stream and writes to disk or tape, as necessary.

Testing

The backup recovery system should be tested on a regular basis. Testing will help identify problem areas that should be fixed. Testing also provides a way to train employees for an emergency. The guidelines for testing the recovery system should be defined within the backup policy plan. The testing of the plan should be documented.

Recovering Data

Caution must be exercised when recovering data from backup files. It is advisable to always duplicate the original backup files before attempting recovery to minimize the likelihood of accidentally damaging the backup files. Controls should be used to ensure the integrity of the data in the recovery process.

Data Compression

Data compression involves storing data in a format that requires less space than normal storage. Compression saves an organization time and money. Files can be saved and retrieved at a faster speed. It is quicker to transmit compressed files versus uncompressed files. Less storage space is also required. Disadvantages of compression includes possible loss of performance and reliability. Time is needed to decompress data before it may be used. The compression/decompression algorithm may make the data less reliable. There are two basic types of data compression methods: the *lossless* compression and the *lossy* compression.

Lossless Compression

Lossless compression may be used to compress virtually any type of data. Its primary advantage it that it compresses and decompresses data without the loss of any information. The data restored from compression and the original data are exactly the same. It relies upon a variant of *Huffman Encoding*. It is based upon the mathematical theory of probability distribution. The compression rate varies depending upon *entropy*, or the uniqueness of information stored in the data. LZ78 compression involves the comparison of blocks of data against a predefined dictionary. A reference to that entry is used to compress data each time a match occurs. The effectiveness of the compression method depends upon the dictionary used. Data for faxes typically uses the LZ78 compression. PKZIP also uses the LZ78 compression with its own dictionary. PKZIP is a very popular lossless compression format.

Lossy Compression

Lossy compression is frequently used when slight distortions in data are acceptable. Lossy compression is much more effective when slight changes in data will be unnoticeable. Lossy compression is predominantly used in multimedia, such audio and video compression. JPEG (Joint Photographic Experts Group) format is frequently used on the Internet for graphic images. JPEG relies upon the lossy compression technique.

[1] This chapter was coauthored by Dr. Pradeep Nagachandran, Computer Technology consultant.

8

Java Applications and Applets

Java is an object oriented language that has become immensely popular in recent years. Java's popularity has been due to several factors, including its ability to run not only on desk top and local area networks, but also over the Internet and the World Wide Web. Java's pre-built class libraries allow for rapid application development (RAD), with graphical user interfaces (GUI). It is simple to add multimedia capabilities of graphics, images, animation, audio, and video to Java applications and applets.

Java is capable of running on a variety of computers and operating systems without modifying the source code. Java is often used when portability among different platforms is of prime importance. Java's capabilities include automatic garbage collection, excellent string handling capabilities, dynamic arrays, applet and application generation, and extensive standard packages or class libraries. Java lacks pointers. Pointers have been a source of numerous bugs in C/C++ applications. Note that Java does offer the ability to integrate existing C and C++ code. However, this limits the Java program to a single platform and portability is sacrificed.

Java's code is interpreted and hence somewhat slower than compiled code. Java also does not offer multiple inheritance capability like C++. It does offer the use of interfaces to provide some functionality similar to multiple inheritance.

Client/Server programming in a networked environment may be performed using most programming languages. Presently, among the most popular are Java and C/C++. C/C++ is a general purpose language which is used extensively for business and engineering applications. C/C++ is very fast and is often used when speed is of paramount importance.

Applications and Applets

An *application* is a program that executes using the Java interpreter. An *applet* is a Java program that runs using a web browser such as Microsoft Internet Explorer or Netscape Communicator. Java applets can be embedded in Hypertext Markup Language (HTML) documents. The ability to create applets was a major factor in Java's initial popularity and wide acceptance.

Java Development Kit

This chapter is based on Sun's Java 2 platform (Java Development Kit (JDK) 1.2). The JDK is available without charge from Sun's web site at: http://java.sun.com. When installing the JDK, it is important to ensure that the *PATH* variable is set properly for your computer system. The three JDK tools that are most frequently used are:

- **javac**: The javac command is used to compile Java source code into Java bytecode. The bytecode is used by the Java virtual machine to run the program.
- **java:** The java command is used to run the application in the platform's implementation of the Java virtual machine.
- **appletviewer:** The appletviewer is used to read HTML files and execute Java applets embedded within them. A web browser may be used instead of the appletviewer. However, it is generally easier to use the appletviewer during application development.

Running a Java Application

The Java source code may be created using any text editor. The Java source code must always be saved in a file with a .java extension. For example, the programmer may create a program called MyProgram using a text editor and save it in the following file: MyProgram.java

This program can be compiled (coverted to Java bytecode) by issuing the following command:

- javac MyProgram.java

  Assuming no errors are generated in the compilation process, Java creates another file with the same name, but with .class extention. In other words, the following file is automatically generated:

- MyProgram.class

  To execute the bytecode program in the .class file using the Java interpreter, the following command is issued:

- java MyProgram

  Note that Java will execute the file MyProgram.class. However, the .class extension is not included when issuing the command.

Program Structure

Java is an object-oriented language. A Java program always consists of one or more classes. A class is a template that specifies or defines an object. A class normally consists of both *data* and *methods*. Each class typically consists of a separate file. The file name must be the same as the class name and have the .java extension.

Creating a Simple Java Application

Listing 1 shows the source code for a simple application that outputs "Hello World!" on the computer screen. Listing 1 may be typed using any text editor. Do not type the line numbers; the line numbers are there for reference purpose only. The text file should have the same name as the class name, along with the ".java" filename extension. Therefore, the class HelloWorld should be saved in a file with the following name:

- HelloWorld.java

  This file should be compiled into bytecode by issuing the following command:

- javac HelloWorld.java

  If any errors are encountered, check for typing errors. Java is case-sensitive; uppercase and lowercase letters are different. Everything should be typed exactly as in the listing. If the compilation is successful, the program may be executed by issuing the following command:

- java HelloWorld

Listing 1

```
1. class HelloWorld
{
    public static void main (String[] arguments)
    {
     System.out.println("Hello World!");
    }
}
```

Output

```
Hello World!
```

Line 1 begins the class definition of **HelloWorld**. Every Java program consists of one or more classes. A *left brace*, {, (line 2) begins the body of every class definition. A corresponding *right brace*, }, (line 7) must end each class definition. Lines 3 to 6 are indented. Such white-space (blank lines, space characters and tab characters) is ignored by the Java compiler. However, white-space is used frequently to make the programs easier to read.

Line 3 is a critical part of every Java application. Java applications automatically begin executing at **main**. The parentheses after **main** indicate that **main** is a *method*. There must be one and only one **main** method in an application and the program execution begins at it. The other keywords in line 3, **public**, **static**, and **void** have specific meanings; however, we will defer their discussion until later. For now, it is important to remember that they must be included in every Java application program.

The left and right braces at lines 4 and 6, respectively, are used to define the body of a method. The real work of this program is done by line 5. It is used to output the text to computer screen. The entire line 5 is called a *statement*. Every statement in Java must end with a semicolon(;).

Creating a Simple Java Applet

Listing 2 contains the code for a simple Java applet that outputs (or draws a string) to the screen.

Listing 2

```
2. import javax.swing.JApplet;
import java.awt.Graphics;
public class HelloWorldApp extends JApplet
{
    public void paint (Graphics g)
    {
     g.drawString ("Hello World Applet!", 50, 50);
    }
}
```

Output

```
Hello World Applet!
```

Listing 2 should be saved under the filename: **HelloWorldApp.java**. It should be compiled using the following command:

- **javac HelloWorldApp.java**

  If there are no syntax errors, the resulting bytecodes will be stored in the file **HelloWorldApp.class**. Unlike Java applications, applets can only run within a web browser (or appletviewer). Web browsers require HTML files, and a Java applet must be embedded within the HTML file. HTML files typically end with the "**.html**" or "**.htm**" file name extension. A complete discussion of HTML is beyond the scope of this chapter. However, Listing 3 shows how an applet can be embedded within a HTML file. The HTML file may be created using any text editor or dedicated HTML software.

Listing 3

```
<html>
<applet code = "HelloWorldApp.class" width=200
height=50>
</applet>
</html>
```

The code in Listing 3 should be saved in a file with an "**.html**" filename extension as follows:

- **HelloWorldApp.html**

Object Oriented Programming

Java is an object-oriented language. Everything in an object oriented language utilizes "objects." An object is made up of data and methods that operate on the data. Problems are solved using objects that model reality. An object's *behavior* is defined by its *methods*. Methods are used to manipulate the instance variables of an objected. Methods may also create new state; that is, a new object may be created using an object's methods.

Every object that a program uses must have a corresponding class definition. An **instance** of a class is used to refer to an existing object of the class.

Object oriented languages have the following characteristics:

- Encapsulation
- Inheritance
- Polymorphism
- Dynamic Binding

Encapsulation refers to hiding **data** or **methods**. This is typically accomplished by specifying an item as **private**, **protected**, or **public**. By defining an item (data or methods) as private, no object other than the class creating the object can access a **private** datum or method. **Public** access is at the other extreme. It allows any class to access a public item. One can ensure that only valid values are set for data by using a combination of private data variables and public methods.

Polymorphism means that different objects are capable of reacting differently to the same message. For example, a "draw" message sent to a *triangle* will be processed differently than when the same "draw" message is sent to a *square* object.

Inheritance makes it possible to reuse code. New classes and behavior can be based on existing classes to obtain code re-use and code organization.

Dynamic binding provides flexibility in writing a program. With dynamic binding, it is not necessary to know an object's specific type at the time of writing the program.

With object oriented programming, each object is defined in its own class. Using data and method hiding, we create independent classes that may be used repeatedly in other applications.

Comments

Comments are notes in a program. Comments are ignored by the Java compiler. Comments are used to explain some fact about the program. Java supports three styles of comments:

- All text between /* and */ is ignored by the Java compiler.
- All text following // is ignored by the Java compiler.
- All text between /** and */ is ignored by the Java compiler. This type of comment is used to create automatic documentation in Java. The extra asterisk (*) identifies it as an automatic documentation comment.

Data Variables and Calculations

Information in a Java program is saved in *variables*. A variable is a location in the computer's memory. Variables must be declared with both a *name* and a *type* before being used in a program.

Naming Variables

The *name* chosen for a variable (or anything else that needs to be named in Java, such as methods or classes) is called an *identifier*. The identifier:

- may be of any length
- must *begin* with a letter, an underscore (_) or a dollar sign ($)
- cannot begin with a number
- the *remainder* of the identifier may consist of any characters other than those used as operators by Java (such as +, -, *, /, %)
- cannot be a *keyword* (keywords are words that are an integral part of the Java language, such as *public*, *class*, and *static*)

- cannot be constants (such as the number *45* or alphabetic constants like *true* or *false*)

Naming Conventions

Java is case-sensitive; uppercase and lowercase characters are treated differently. Variables should be given a meaningful name. This helps in self-documenting the source code. By convention, variables are given a meaningful name by joining several words together as follows:

- the first letter of the variable name is lowercase
- each successive word in the variable name begins with a capital letter
- all other letters are lowercase

Examples of variable names following the standard convention include:

- ageOfEmployee
- dateOfBirth
- creditLimit
- lastName

Variable Types

The *type* of a variable determines the type of data that may be stored in that variable. Java has eight basic data types, known as *primitive* data types. In addition, Java variables may also reference objects; objects will be discussed in a later section. Two primitive data types are used to store *boolean* and *char* data types. Four primitive data types are used to store integer data: *byte*, *short*, *int*, and *long*. Two of the primitive data types are for floating point numbers: *float* and *double*. Floating point numbers are used to store numeric values which are not integers.

Casting and Computations

Certain types of computations will not work correctly without an *explicit cast*. Some types of casting is done automatically and some types require the programmer to explicitly convert the data types. Without explicit casting, the user may get undesirable results.

An automatic cast is applied when there is no possibility of losing information. In other words, an automatic conversion will be made when going from:

- byte → short → int → long → float → double

An explicit cast is required when one wishes to go from **double** to **float** or **long**.

Strings

In Java, a string is an object rather than a basic (primitive) data type. One of Java's strengths is its excellent string handling capability. Java provides numerous methods to modify, combine and compare strings. String literal consist of a series of characters inside double quotation marks. Special character *escape codes* may be included in the string. Character escape codes are used to perform certain actions. For example, including a **\n** in a string forces a new line or including a **\t** inserts a tab space. String arithmetic may be performed using the "**+**" operator to concatenate two strings. Two strings may be concatenated using the "+" operator.

Care should be taken when "adding" two string values. The program can sometimes yield an unexpected result. For example, if one string contains "2" and another string contains "3," the result is "23," not "5." When two strings are added, the values **2** and **3** are concatenated; the expected arithmetic operation is not performed. Therefore, the result is **23** and not **5**.

Arrays

An array is a group of contiguous memory locations with the same name and data type. Arrays are used in the exact same way as you use a single variable of the same data type. Array variables of type **int** may be *declared* using either of the following statements:

- int [] age;
- int age[];

To declare multiple arrays in the same line, the former statement is preferable to the second statement. If we wish to declare multiple array variables of type **String**, either of the following statements may be used:

- String [] firstName, lastName, titleOfJob;
- String firstName[], lastName[], titleOfJob[];

The two statements above are equivalent. Both create three arrays of type **String**. After the array has been declared, it needs to be *defined* before it may be used:

- age = new int[10]; // Defines an array of Ten integers

Arrays may be *declared* and *defined* in a single statement as follows:

- int[] age = new int[10];

Note that no memory is allocated when the array is *declared*. Memory is allocated only when the array has been *defined*. Once the array has been defined, array *elements* may be referenced. Each variable in the array is called an array element. A particular array element may be referenced using its *index* value. The first element will have an index of 0, the second element will have an index of 1, the third element will have an index of 2, and so on. The index value must be of type **int**. If type **long** is used, it must be explicitly cast as type **int**. Expressions using type **short** and type **byte** produce a result of type **int**, and may be used as an index without an explicit cast.

Arrays may be *declared*, *defined*, and *initialized* in one step as follows:

- int age[] = {3, 1, 5, 9, 12, 2, 2, 5, 7, 10};

This statement creates an array of ten elements and initializes them with the values between the braces. The size of the array is determined by the number of initial values.

Exceptions

Exceptions arise when something serious goes wrong in a program. Exceptions are generally used to signal an error. Exceptions can be handled using the **try/catch** combination.

- Code that may give rise to one or more exceptions is enclosed in a **try** block.

- Code designed to handle the exception is enclosed in a **catch** block.

For example, if we attempt to divide a number by zero, an error will be generated. Without enclosing the code in the try/catch block, the program will not execute and generate a "/ by zero" error message. Enclosing the code in the try/catch block allows us to execute the program. When the error is generated, it is caught by the catch block and a programmer provided message is given to the user. It is possible for a single try block to have multiple catch blocks to catch different types of errors.

Java Class Library

The Java Development Kit (JDK) contains a large number of standard *classes* grouped into *packages*. The classes in these packages allow a programmer to use code written by other programmers. The use of pre-written code makes software development easier and faster. The reuse of classes reduces the likelihood of bugs or errors in software. The code in the standard classes has been optimized, and enhances the performance of newly developed software.

Classes must be **imported** before they can be used. In Listing 2, when we introduced Java applets, we **imported** two important classes:

- the **JApplet** *class* is part of **javax.swing** *package*
- the **Graphics** *class* is part of **java.awt** *package*

We will be using the **javax.swing** and **java.awt** packages extensively. The **javax.swing** package contains components for building graphical user interfaces (GUI). The **java.awt** provides classes necessary to build window applications. Whenever possible, programmers should use classes from the **swing** package rather than the **awt** package. Some of the other commonly used packages include:

- the classes in **java.lang** package support basic Java language features and are always available automatically
- **java.util** package contains various classes for managing data
- input and output operations are performed using classes in **java.io**
- **java.awt.event** package contains classes that are used to implement event handling in windowed applications

Logic and Loops

Java provides six relational operators for comparing two data values. A basic statement for making comparisons and decision making is the **if** statement. The **if** statement takes following form:

- **if (*expression*)**

  statement;

where, if the value of the *expression* is true, the statement executes. Enclosing statements in braces can allow multiple statements to execute if the expression is true. Several levels of nested **if-else** statements may be created using the **else if** statements.

The *switch/case* statements can be used to do the work of multiple *if-else* statements. The *break* statement tells the program to stop evaluating the cases.

Loops allow you to execute a statement repeatedly. Loops are used frequently in almost every program. The *for* loop statement consists of the following format:

- **for (*initialization*; *condition*; *increment*)**

{

// **one or more statement;**

}

The **while** loop is similar to the **for** loop. It executes as long as a given logical expression is *true*. When the expression becomes *false*, the program exits from the loop. The *while* loop statement consists of the following format:

- **while (*condition*)**

{

// **one or more statement;**

}

Both **for** and **while** loops can be used interchangeably and it is a matter of personal preference. Typically, **for** loops are used when the number of loops is known or determinable in advance, whereas **while** loops are used when the number of loops is not known and depends upon some condition becoming true (such as the user entering a value to exit from the loop).

The **do while** loop is a variation of the **while** loop. The *do while* loop statement consists of the following format:

- **do**

  {

  // one or more statement;

  } while (*condition*);

Like the **while** loop, the **do while** loop executes as long as a given logical expression is *true*. However, unlike the **while** loop, the **do while** loop does not exit immediately when the expression becomes *false*. In other words, the use of the **do while** loop guarantees that the loop will execute at least once, even if the expression is *false*.

Java offers three ways to terminate or skip a loop:

- The **break** statement may be used to terminate a loop immediately
- The **continue** statement is used to skip a loop iteration
- For nested loops, the **labeled continue** statement may be used to stop the execution of the inner loop and continue with the next iteration of the outer loop.

Conclusion

The basic concepts of structured programming in Java were illustrated in this chapter. Using the components discussed, users can create simple Java applications or applets.

The "Example of Coding" consists of a single program based on the object oriented paradigm. It illustrates the concepts discussed in this chapter.

Example of Coding

Listing A-1

```
// PhoneBookMain.java
import javax.swing.JOptionPane;
public class PhoneBookMain
{
 public static void main(String[] args)
 {
 String output="";
// ask user for the total number of names
 NumberOfNames namesArray = new NumberOfNames();
// determines the array size
 int totalNumberOfNames = namesArray.getArraySize();
//create an array of object InputNames to hold the maximum number of names
 InputNames name [] = new InputNames[totalNumberOfNames];
//create an array of object InputNames to hold the maximum number of names
 InputPhone phone [] = new InputPhone[totalNumberOfNames];
//loop to entr data into the array
 int counter=0;
```

```
do
{
name[counter] = new InputNames();
phone[counter] = new InputPhone();
output = output + name[counter].getName() + "\t"
+ phone[counter].getPhoneNumber() + "\n";
counter++;
} while (counter<totalNumberOfNames);
System.out.println(output);
System.exit(0);
}
}
```

Listing A-2

```
// NumberOfNames.java
import javax.swing.JOptionPane;
public class NumberOfNames
{
private int numberOfNamesInArray;
public NumberOfNames()
{
ArraySizeValid arraySize;
do
{
```

```
String arraySizeInput = JOptionPane.showInputDialog
(“Maximum number of names you wish to enter?”);
arraySize = new ArraySizeValid(arraySizeInput);
if (arraySize.getArraySizeValid())
numberOfNamesInArray = arraySize.getArraySize();
} while (!arraySize.getArraySizeValid());

}

public int getArraySize()
{
return numberOfNamesInArray;
}
}
```

Listing A-3

```
// ArraySizeValid.java
// Determines if the array size is valid.
import javax.swing.JOptionPane;

public class ArraySizeValid
{
private int arraySize;
private boolean isArraySizeValid;

public ArraySizeValid(String arraySize)
{
StringToInteger string1 = new
```

```
   StringToInteger(arraySize);
if (string1.getIsInteger())
if (string1.getIntegerValue()>=1)
{
this.arraySize = string1.getIntegerValue();
isArraySizeValid = true;
}
else
{
isArraySizeValid = false;
JOptionPane.showMessageDialog(null,
"You must enter a value greater than Zero!",
"Invalid Array Size",
   JOptionPane.ERROR_MESSAGE);
}
}

public int getArraySize()
{
return arraySize;
}

public boolean getArraySizeValid()
{
     return isArraySizeValid;
     }

   }
```

Listing A-4

```
   // StringToInteger.java
```

```

import javax.swing.JOptionPane;

public class StringToInteger
{
 private String stringNumber;
 private int integerNumber;
 private boolean isInteger;

 public StringToInteger(String stringNumber)
 {
 this.stringNumber = stringNumber;
 isInteger = true;
 try
 {
 integerNumber = Integer.parseInt(stringNumber);
 }
 catch (NumberFormatException nfe)
 {
 isInteger = false;
 JOptionPane.showMessageDialog(null, "You must enter a numeric integer value",
 "Invalid Number Format", JOptionPane.ERROR_MESSAGE);
 }
 }
 public boolean getIsInteger()
 {
 return isInteger;
```

```
    }
    public int getIntegerValue()
    {
    return integerNumber;
    }
  }
```

Listing A-5

```
  // InputNames.java

  import javax.swing.JOptionPane;

  public class InputNames
  {
    private String fullName;

    public InputNames()
    {
    String firstNameInput = JOptionPane.showInputDialog
    ("Enter First Name: ");
    String lastNameInput = JOptionPane.showInputDialog
    ("Enter Last Name: ");

    String firstName = capFirstLetterOfName(firstNameInput);
    String lastName = capFirstLetterOfName(lastNameInput);

    fullName = firstName + " " + lastName;

```

```
    }
    private String capFirstLetterOfName(String name)
    {
    String trimName = name.trim();
    String lowerCaseName = trimName.toLowerCase();
    String firstLetter = lowerCaseName.substring(0,1);
    String capitalizedFirstLetter = firstLetter.toUpperCase();
    String remainingLetters = lowerCaseName.substring(1,lowerCaseName.length());
    String capitalizedName = capitalizedFirstLetter + remainingLetters;
    return capitalizedName;
    }

    public String getName()
    {
    return fullName;
    }
}
```

Listing A-6

```
//InputPhone.java
import javax.swing.JOptionPane;

public class InputPhone
{
 private String areaCode, phoneNumber, fullPhoneNumber;

```

```
public InputPhone()
{
NumberValid validateAreaCode, validatePhone;
do
{
String areaCodeInput = JOptionPane.showInputDialog
("Enter Area Code: ");
validateAreaCode = new NumberValid(areaCodeInput,3);
if (validateAreaCode.getIsNumberValid())
areaCode = areaCodeInput;
} while (!validateAreaCode.getIsNumberValid());

do
{
String phoneInput = JOptionPane.showInputDialog
("Enter Phone Number: (Enter \"555-1234\" as \"5551234\"");
validatePhone = new NumberValid(phoneInput,7);
if (validatePhone.getIsNumberValid())
{
String partOne = phoneInput.substring(0,3);
String partTwo = phoneInput.substring(3,7);
phoneNumber = partOne + "-" + partTwo;
}
```

```
    } while (!validatePhone.getIsNumberValid());

    }

    public String getPhoneNumber()
    {
    fullPhoneNumber = "(" + areaCode + ") " + phoneNumber;
    return fullPhoneNumber;
    }
  }
```

Listing A-7

```
  // NumberValid.java
  public class NumberValid
  {
    private boolean isNumberValid;

    public NumberValid(String numberString, int stringLength)
    {
    LengthOfString string1 = new LengthOfString(numberString, stringLength);
    StringToInteger string2 = new StringToInteger(numberString);
    if (string1.getStringLengthValid() & string2.getIsInteger())
    isNumberValid = true;
    else
    isNumberValid = false;
    }
```

```

    public boolean getIsNumberValid()
    {
    return isNumberValid;
    }
   }
```

Listing A-8

```
   // LengthOfString.java

   import javax.swing.JOptionPane;

   public class LengthOfString
   {
    private boolean stringLengthValid;

    public LengthOfString(String input, int length)
    {
    if (input.length() != length)
    {
    JOptionPane.showMessageDialog(null, "The value entered must consist of " +
    length + " numbers!", "Invalid Length of Value",
    JOptionPane.ERROR_MESSAGE);
    stringLengthValid = false;
    }
    else
    stringLengthValid = true;
    }

```

```
    public boolean getStringLengthValid()
    {
    return stringLengthValid;
    }

}
```

9

Client Server Technologies[1]

This chapter discusses client server architecture, applications, and network operating system. The features and uses of clients and servers are presented. The types of servers are enumerated. There is a discussion of middleware and groupware. Distributed computing applications and transaction processing are addressed.

Systems administration is managing client/server computers including the network operating system. A computer can be a server, client, or both. Client/server application software handles data and processing. Client/server refers to the sharing and interaction of information between client and server computers connected by a network. Mobile agents distribute code and data across clients and servers. Mobile code consists of a sequence of executable instructions.

A client/server network is a local area network (LAN) with one or more computers acting as the host for various applications. LAN acts as the infrastructure for client/server applications. The LAN relays service requests from the client to the server and carries the resulting data back to the computer. A LAN is a network of computers located near each other, for instance in the same building. There is a sharing of resources. LANs make up an important part of the overall enterprise network infrastructure serving as the backbone linking many distributed components of the client/server information system.

The objective of the client/server structure is to optimize the utilization of computer resources. It serves as a means for varying computer devices to perform together, each conducting its appropriate responsibilities. Lower cost clients (typically PCs) are able to share more costly servers.

In a client/server environment, different computers specialize with different tasks (server computers) and any other computers (client computers) on the network can request services such as file loading, database processing, Web hosting, and printing.

Client/server network operating systems are designed to satisfy network requests, provide security, share resources, store files, and logically organize data for fast retrieval.

The client/server system enables managers and workers to access voluminous access points to obtain data. The client/server system provides many tools to process information. Large databases can be accessed immediately and fully throughout the organization by staff.

Key elements that may be used across a variety of clients and servers are processing, data management, presentation, and business logic. These components are linked by the enterprise network. Dissimilar platforms are transparently interoperable by adding translating middleware. The design of client/server systems mandates comprehending the interfaces between the various layers of hardware and software required for inter-operation.

SOCKS formulates a proxy data channel to the application server on behalf of the application client. SOCKS audits screens and filters all data going between the application client and server. SOCKS controls traffic communications.

Vertical software assures compatibility and direction of protocols to assure the adequate operation of software and hardware within a single client or server. It relates to the transparency between different software or hardware layers within a particular client or server.

Horizontal software compatibility relates to the transparency between similar software among different clients and servers.

In a calendar access protocol, users can mix and match different clients and servers. An enterprise network connects the client to the server. It is a transportation system of the client/server architecture. It assists in cooperation between distributed processors and databases.

A storage area network is shared by multiple application servers or computing clusters.

Portability refers to the ability for client/server applications to be developed in one computing environment and deployed in others. Scalability is a distributed processing system's ability to add clients and/or servers without negatively impacting the performance of the system.

Hypertext Transfer Protocol (HTTP) is the client/server protocol on the World Wide Web. HTTP is a request/response model for messages between client and server.

Testing in a client/server system is when data from all sites is transmitted to assure the proper working of the network system including computers, software, interfaces, devices, and personnel.

Benefits of client/server networks are:

- Ability to accommodate a vast number of users.
- Ability to share costly equipment and devices.
- Centralized control and security.
- Central storage of files.
- Ability to work from the same data set.
- Sharing of resources.

Drawbacks of client/server networks are:

- Costly network operating system software.
- Costly hardware.
- Need to hire a network administrator.

Client Server Architecture

Architecture is a specification to determine the building of something and how it will function. It includes the protocols and interfaces enabling the working together and transmission among components. Architecture refers to how one segregates the client/server application into operational units that can be assigned either to the client or to one or more servers. The typical functional units are the user interface, business logic, and the shared data. There are many ways to initiate a client/server system. Some of these processes may apply to downsizing, rightsizing, upsizing, and smartsizing. Architectures assist us in describing the shape and structural elements that may be used as building blocks to develop more complex client/server systems. There are many variations of multitier architectures depending on how an application is split and the middleware used to communicate between the tiers. Client/server architecture is usually referred to as 2-tier, 3-tier, and n-tier.

Client/server architecture includes reference to the overall organization of software categories within a representative client workstation including information management, presentation, application, and network operating system.

A key benefit of client/server architecture is the ability to incorporate hardware and software from multiple vendors into a transparently interoperable information system.

Businesses require flexible, integrated, comprehensive, and responsive applications to support the full range of business processes.

In client/server systems, PCs are the "clients" which are connected to specialized, powerful "servers" (e.g., databases, mainframes, communication devices), that are shared through local and world-wide networks. In this structure, telecommunication standards exist enabling interconnection of various kinds of hardware or software. It should be noted that the client/server architecture is used to support the Internet and the intranets.

An activity may be conducted only on the client, only on the server, or split between the client and server. In the client/server model, servers furnish the clients with information and processing like a centralized relationship. However, clients by themselves may form a peer-to-peer network structure. The prevalent situation is to have multiple, specialized servers meeting the data and communication dictates of a single client.

Client/server information systems reflect the interaction of presentation, application, and data activities across multiple servers. The segregation of the functions of presentation, application, and data requires sophisticated communications between cooperating computers via a network (usually a LAN). The network is comprised of both software (network operating systems) and hardware (network interface cards, hubs).

A client/server infrastructure is comprised of many different kinds of servers and the various networks connecting those servers to each other as well as to the clients served. The following applies to client/server infrastructure:

Local area network administration concentrates on the management of clients and servers and the operating system and network operating system that makes the clients and servers communicate properly. Subprocesses include client/server management, distributed application management, fault tolerance and availability management, configuration management, desktop management, and help desk assistance. The functionality of systems administration includes monitoring the servers, tracking hardware and software inventory, license metering, tracking LAN and server activity, asset management, and software distribution.

Enterprise network management emphasizes the hardware, software, network services, and media required to link and manage distributed clients and servers throughout the company. It may be considered internetwork administration between LANs. Subprocesses involve local area network (LAN) and wide area network (WAN) management, internetworks, Internet management, remote access management, and internetworking device management. The tools that may be used include network baselining, network auditing, network modeling and simulation, and network analyses and remote monitoring. Careful planning is required in integrating tools and protocols into a cohesive system. The functionality of enterprise network management includes carefully evaluating and managing internetwork technology including bridges, routers, hubs, and switches as well as monitoring and managing WAN links.

Figure 9.1 presents a client/server configuration.

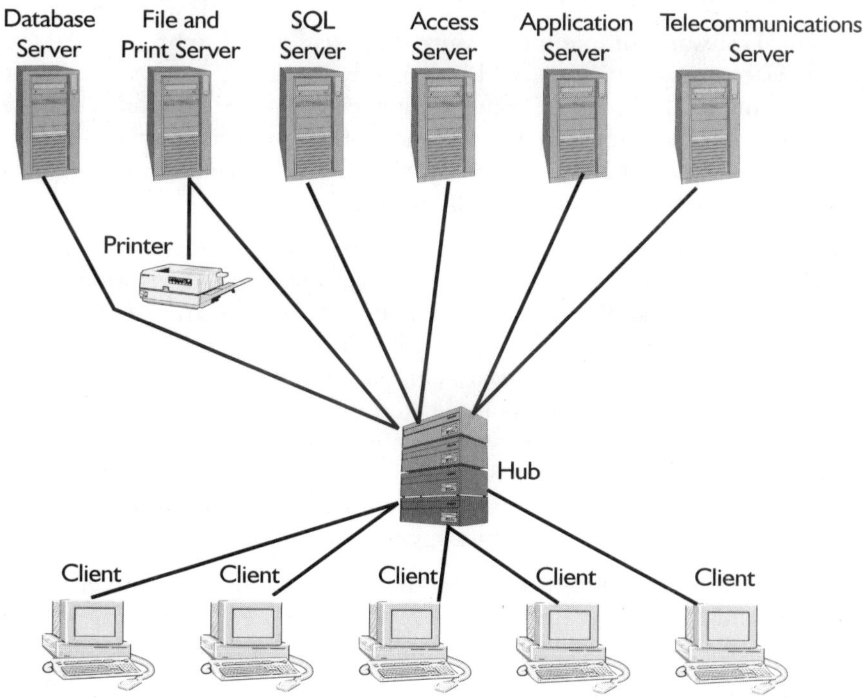

Figure 9.1: Client/server Configuration

The basic structure of the client/server architecture is a client device and a server device that are distinguishable, but interact. The client process has solution-specific logic and provides the interface between the user and the rest of the application system. The server process acts as a software engine managing shared resources like databases, printers, modems, or high-powered processors. The front-end task and back-end task have different requirements for computing resources, such as memory, capacities, disk speeds, processor speeds, and input/output devices. The environment is usually heterogeneous. The hardware platform and operating system of the client and server are not typically the same. Client and server processes communicate through a well-defined set of standard application program interfaces. A key attribute of client/server systems is scalability which may be in horizontal or vertical terms. Horizontal scaling refers to adding or removing client workstations with just a slight performance effect. Vertical scaling means changing into a larger and faster server or multi-servers. Note that multi-servers are common in a client/server structure. Servers are most likely distributed graphically. Some servers will be situated remotely and linked to geographically dispersed clients through a wide area network.

Vertical software compatibility assures the existence of compatible protocols for system and hardware within a client or server to assure sufficient operation. Vertical software compatibility relates to the transparency between different layers of software or hardware within a particular client or server. Vertical software compatibility is achieved by adjacent software layers supporting common compatibility protocols.

Horizontal software compatibility relates to the transparency between similar software layers among different clients and servers. Horizontal software compatibility usually applies to having different software of the same category to interoperate transparently between clients, servers, or clients and servers. Horizontal software compatibility between different types of software or computers is a complicated undertaking. Horizontal software compatibility is typically derived by software referred to as middleware. In some situations, horizontal software compatibility applies to transparent interoperability between multiple software pieces in the same software company category layer within a single client or server.

In explaining the architectural arrangement of the client/server system a number of questions requiring answers include:

- What function will peer networks serve?
- How is the application split among server and client?
- What activities will be conducted by the server?
- What operations will the client do?
- What size business is the client/server architecture suitable for?

Figure 9.2 presents a typical client/server architecture.

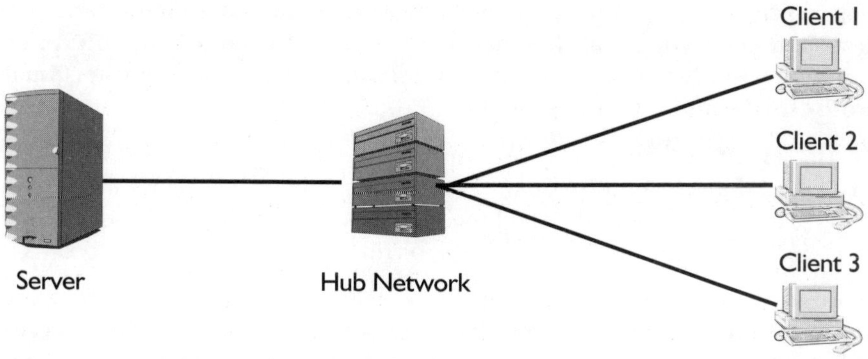

Figure 9.2: Client/Server Architecture

Server	Network	Client Workstations
• Data storage • Information requested by an application program • Data retrieval	• Communication of data request • Transportation of actual data	• Storage of data • Data retrieval • Presenting data to user • Application program request for data • Transportation of actual data • Transportation of information requests

Enterprise Network

An enterprise network is the transportation system of the client/server architecture which in addition to middleware acts as transparent cooperation of distributed databases and processors. Enterprise network management emphasizes the hardware, software, media, and network services to link and manage distributed client and server computers in the business.

The objective of the enterprise network is to furnish the integration and transparent interoperability of the client/server architecture. The enterprise network typically consists of video conferencing traffic, and host-terminal traffic in a sound integrated manner.

In the event the company is comprised of widely disbursed regional branches, than the enterprise network will usually include WAN links. The enterprise network typically combines network mechanisms and connections of the categories of either WAN, LAN, or internetwork (e.g., LAN to LAN).

The enterprise network serves a key role in managing client/server information systems. Because the distributed nature of a client/server information system is comprised of a host of widely situated processors, there is only one location by which all traffic passes which is the enterprise network. Hence, the enterprise network is the only reasonable form of managing many shared resources of the client/server information system. The enterprise network is the backbone of the client/server information system. In effect, the network is the computer.

An integrated client/server management system not only includes an enterprise network management system but also provides the capabilities of transaction processing management, enterprise distributed processing management, desktop management, and database management.

Models

Some models of client/server information systems segregate the application layer into sublayers to emphasize the application layer's interfaces with the presentation and data layers as follows:

- Presentation logic. The part of the application program accountable for interfacing to the user interface.
- Business logic. The actual program rules of the application accountable for controlling program execution and enforcing business rules.
- Database logic. The part of the application program accountable for interfacing to the database management system.

Interprocess communication is a mechanism allowing multiple pieces of an application to communicate with one another. It enables separate pieces of an application to behave as if it were a single application process running on a single computer. Interprocess communications protocols facilitate the sending of messages and commands between processes and putting together difficult transactions involving the execution of multiple processes on multiple computers. Two widely used interprocess communications protocols are remote procedure calls (RPC) and message-oriented middleware (MOM). These IPC protocols are examples of middleware.

Agent Event Manager

An agent event manager is one of the three cooperating elements of the agent part of the client/agent/server architecture. The agent event manager coupled with a customer transaction handles forms of the intelligent agent. The intelligent agent is on the local server. Once the agent event manager receives a request from a mobile agent, it acts for that client in all communications with the local server until the initial client request is met. The objective of a client/agent/server architecture is to reduce the amount of mobile client to server network traffic. This can be achieved with as much intelligence as possible into the server-based agent so it can serve on behalf of the mobile client application.

Client/Server Network Operating System

Client/server network operating systems incorporate both standard client application software operation with advanced network operating system features for a wide range of network servers. Client network operating systems mandate sophisticated server software. Client/server network operating systems are more scalable relative to peer-to-peer network topologies and enable faster more stable operations. Client/server networks are also much more complex and costly than peer-to-peer network operating systems. Client/server network operating systems support many users and foster interaction with other network operating systems via gateways. The dedicated server provides a platform for additional system security and administrative features.

In a network operating system, server software makes a single-user computer into a multi-user one. A server has many users who can share the server's peripherals, disks, printers, and plotters but the network operating system does not allow multiple users to share its processor. In an intelligent network, clients use their own processor and software to process information provided by the server.

An up-to-date network operating system will support most common LAN media types – Ethernet and token ring. In terms of security, a network operating system must support disk quotas, encryption, managing intruders, and access control lists.

A client/server network operating system is depicted in Figure 9.3.

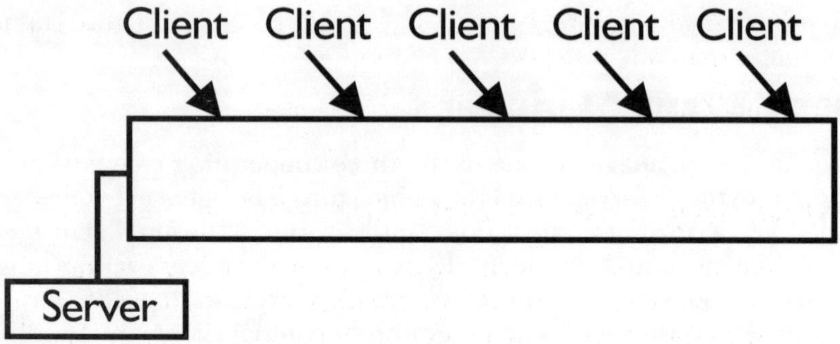

Figure 9.3: Client/Server Network Operating System

Client/server network operating systems provide the following:

- Peer-to-peer functionality.
- Client software and network drivers allowing a client to communicate with a compatible server. We refer to this as Media Access Control (MAC) protocols.
- Network redirectors to process application program interface (API) calls so as to support application program file systems.
- Conventional operating system capabilities.
- Client/server interoperability with other network operating systems.
- Network transport protocols packaging and transporting messages between clients and servers.
- Symmetrical multiprocessing enabling multiple simultaneous processing requests from clients. This is especially important for large client/server installations where clients use advanced multi-dimensional software applications.
- Multithreading enabling multiple subprocesses to execute synchronously.
- Support for file access including indexed files.

The following questions should be asked about the architecture of the network operating system:

- Is there support for multiple microprocessors?
- Can network operating system functions be partitioned to run on more than one processor at the same time?
- Is there support for multiprocessor systems?

Connecting Clients and Servers

In a simple peer-to-peer network, the connection of the clients and server is achieved by a cable and network card. In a server-based network, we also need a network hub (rectangular box). In selecting the appropriate hub, consideration should be given to how many computers there are. Typically, a hub has ports to accomodate up to about 24 computers. If more ports are needed, a "stackable" hub may be bought in which hubs are connected together possibly tripling the total amount of accessible hub ports.

A decision must be reached as to how quickly data must be transmitted. What bottlenecks exist in the network and how may they be overcome? An important bottleneck is the hub. Care must be exercised in choosing the best hub for the system. The authors suggest buying a 100 MB/Sec Switching hub because it enables both data transfer speeds of 100 MB/Sec and 10 MB/Sec. While these hubs are more costly, they result in faster connection speed.

Application gateway is the connection between requesting clients and servers. Before providing the requested information, the server confirms the legitimacy of the request.

Databases

A data file is a collection of related data records organized in a particular logical way such as a payroll record. A data element is a grouping of information items to be used as a unit in a particular case.

A database file is an accumulation of related records describing a subject by using a set of fields. A database is a collection of interrelated records of different types. The stored information is used to meet the needs of multiple users within the company and/or other companies accessing it. A database is an accumulation of corporate and other information. It is a collection of interrelated or independent stock items to be used in business applications.

A database management system is the software enabling the creation and maintenance of a database and the execution of software using the database. In data control, the manager monitors and tracks data going into and out of the system.

A data dictionary is a centalized place describing and explaining the data in the database such as meaning, interrelationships, format, and usage. It aids in planning, managing, and appraising the stored data. The information may be stored in a dictionary-like document or a text file.

A database management system is generalized software used to manage databases. The software enables users to define, create, modify, and maintain the database and furnishes controlled access to it. The software allows for the interaction of the user's application programs and the database. Information is organized in an accurate, reliable, and timely way enabling easy data retrieval.

In a homogeneous database management system, the same database management system is used by all sites. A database may be split into segments of similar (homogeneous) records. An example is customers with a predetermined range of income or age.

Database utilities are programs to fine-tune the database operation such as allocating resources, removing unused or redundant records, assigning user priority, and organizing information in a logical order.

Database clustering is a method in which data is stored in a logical manner. For example, employee data may be stored and retrieved sequentially in departmental order.

A network database consists primarily of a collection of record types which are linked (associated) in a database. No restriction exists on the number or direction of links between record types. It is possible to have more than one kind of link. The network database is similar to a hierarchical database except that each child can have more than one parent record. A child record is referred to as a "member" while a parent record is termed an "owner."

In database concurrency, there is a simultaneous access of database objects by more than one user or application.

A distributed database management system consists of one logical database split into fragments. The sites are linked by a communications network where interrelated data is shared. In a multi-database system, each site in the distributed database has its own autonomy. Users can access and share data without requiring database integration. Users can opt to handle their own databases without centralized control.

A data hierarchy is a data organization comprised of sets and subsets such that each subset of a set is of lower rank relative to the data of the set. A hierarchic database is created as a family tree of related record types. Lower-level records are subordinate to higher-level ones. There exists one "root" record type which may participate with one-to-many relationships with "child" record types. Each child may in turn have child record types. NOTE: A parent record may have more than one child, but a child always has only one parent. Figure 9.4 presents a hierarchic database.

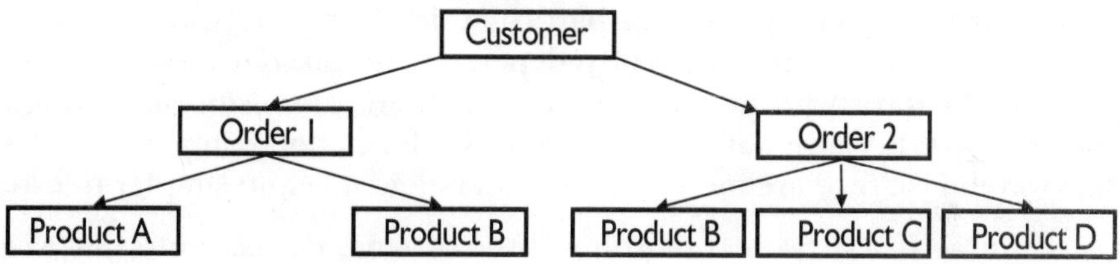

Figure 9.4: Hierarchic Database

In a relational database, data is represented in two-dimensional tables consisting of rows and columns. The tables are called "relations." The relational database relates data in different files through the use of a key field, or common data elements.

An object-oriented database (OOD) stores objects which are shared between different applications. OOD is comprised of a collection of classes. A class represents objects structured in a similar way. The object is defined by its properties such as a customer having the properties of name, number, address, and status. Examples of object classes are customers, orders, and merchandise items. An object consists of text, sound, images, and instructions on the action to be conducted on the data.

A temporal database is comprised of the time property in the underlying data and usually has validity intervals associated to each data component. The temporal database tracks previous versions of the database after it is updated.

A spatial database is geometric or geographic-related. It presents real-time access, query processing, and pattern recognition.

A multimedia database consists of abstract data such as video, audio, animation, image, hypertext, and graphics.

Knowledge discovery as applied to databases is the process of not only choosing and preparing data but also searching the information and interpreting the findings.

A data warehouse is a subject-oriented, integrated large corporate database furnishing information for management decision making.

Meta-data is a description of information included in the data warehouse such as origin, content, and usefulness. The data going into and out of the warehouse is described.

A data mart is a subset of a data warehouse furnishing business information, typically in summarized form, to a particular department or division within the company. The data mart can be independent or linked to the company's centralized data warehouse (containing large amounts of detailed information). Because data marts have less information compared to the data warehouse they are more readily understandable and simpler to use.

Data mining is the process of going through large databases to uncover useful relationships, trends, patterns, and correlations. We collect and evaluate significant data stored in data warehouses or data marts by using artificial intelligence and statistical methods. Data mining is used in making business decisions. The key functions associated with data mining are segmentation, highlighting deviations, prediction, and linking. Examples of data mining applications are to uncover fraud, portfolio selection, whether to give credit and how much, and customer profiling. Data mining examines data trends and patterns based on a classifiction or clustering model.

Domain Name System

The domain name system in a distributed database system stipulates name-to-address mapping for clients.

Domain name service is a network service in a Transmission Control Protocol/Internet Protocol (TCP/IP) network to convert host names to IP addresses. There may be a fully qualified domain name which is an address having the workstation name in addition to the domain name.

Each computer connected to the Internet has a unique Internet protocol address. One IP address consists of 32 bits and is divided into 8-bit segments separated by a period. The IP address identifies the computer on the Internet. Because the IP address is very difficult to remember, a domain name corresponding to each address is used. The domain name has two levels as follows:

- Top-level domain names are classifications of different purposes of usage. For example, "com" refers to commercial organizations, "gov" refers to government agencies, and "net" refers to an Internet service providers. Other top-level domain names include countries such as "US" for the United States, "cn" for China, and "UK" for United Kindom.
- Lower-level domain names are unique names identifying the specific servers. For example, IBM and Ford are lower-level domain names.

For example, www.ibm.com refers to ibm as a commercial business.

The combination of the top-level domain name and the lower-level domain name can thus identify the type of business and its country.

Unlike the IP address, there is no limit on the number of possible domain names. One server may have several domain names provided there exists a way to map between domain names and their associated IP addresses. Such a database is kept in the domain name system (DNS) server, or name server, accessible by a router. In other words, after a user types the domain name, it is transferred to a corresponding IP address and the server with that IP address is accessed.

Client

The client PC is a workstation. A workstation is a configuration of input-output equipment where an operational activity occurs. A client is a computer or software requesting services (e.g., file storage) from another computer (e.g., server) or program. A client computer requires only what is needed for local application software. Clients use network resources but do not furnish them. Client computers dedicate their processing capability to performing tasks for the end-user. A client can provide locally controlled computering directly to a user's desktop while continuing to share computing, devices, and data resources with other users. Clients want from servers information reliability and security in addition to fast and easy access to data. Clients are run by the network client operating system. Client workstations should be configured based on the software needs of the machine. The client operating system is much less robust than the network operating system. Microsoft Windows 2000, Windows XP and Windows NT workstations are used for almost all client operating systems.

The client process sends a message to a server asking the server to conduct a service. Client processes typically manage the user-interface part of the application, substantiate data entered by the users, dispatch requests to server processes and sometimes execute business logic. The client-based process is the front-end application that the users see and interact with. The client process also manages the local resources that the user interacts with such as the workstation CPU and peripherals.

The front-end has user interfaces formatting data requests or processing from the server, and formatting information obtained from the server for output to the user. In most cases, in a client/server environment, the client computer is responsible for presentation management and application logic. Application logic is the part of the software that determines what needs to be performed in a specified situation.

File synchronization software is software that can synchronize versions of files on desktop and laptop workstations. It is a feature in most client network operating systems.

Because client workstations usually do more processor-intensive work, the central processing unit (CPU) in the client computers should be at least equal to or more than the server's CPU. The memory needed on the client is typically much less than that needed on the server because the client machine has much less data stored in memory. A SCSI hard drive should be used on the client if it works on very voluminous data files.

Domains are a group of network and clients sharing secured information. To access the information approval is required from special servers referred to as domain controllers.

Remote access is the linking of remote PCs to local LANs. Remote client computing is a remote client PC operating as if it were locally attached to network resources. Remote control is a remote PC supplying input and output devices for the local client. The local client interacts with the local server and other local attached LAN resources.

Server

The main part (focal point) of the network is the server which centralizes information and keeps control of it. A server uses special software and operating system. A server is a large and powerful network computer sharing resources, services, and information (e.g., data storage) with clients and network users. Servers manage data flow. Servers may be mainframes, minicomputers, workstations, and occasionally high-end PCs. Subsystems are groups of servers. Local servers are servers executing in a separate process from their clients on the same machine and operating system. Servers can be programmed and equipped with special functions for optimal processing efficiency. A server requires a large hard drive and a fast CPU. In a large network, there are many servers for different functions such as for access, database, telecommunications, files, applications, SQL, and printing. In fact, multiple clients can be serviced at the same time. In most cases, servers are more powerful than individual clients.

After the network operating system server software is decided upon, suitable hardware can be bought and configured. In selecting a server, consider cost, memory, power, speed, and backup capabilities. NOTE: The server's amount of memory is more crucial than CPU speed to the overall server performance. In a smaller network where data transferred is limited, the fastest CPU is probably not needed. A large network should have a second processor for its server. Servers are typically configured with two CPUs. Computers having many significant computations, handling large databases, and running significant graphic software will have substantial additional performance when the second processor is added. However, small offices do not need a second processor for the server because the purpose of the server in smaller networks is usually to connect all workstations and share limited data.

Which hard drive should be bought for the network server? This will depend partly on how many gigabytes are needed. The choice between interfaces is typically either Integrated Drive Electronics (IDE) or Small Computer Systems Interface (SCSI). IDE is an easy path for data stored on the hard drive to be processed based on input/output technology. A deficiency, however, of input/output technology is it hinders the CPU from achieving other functions when it is accessing the hard drive for data. This restricted multitasking ability significantly lowers the machine's performance when voluminous data files are transferred. SCSI technology was first used by Apple. SCSI is attractive due to its multitasking abilities and increased bandwidth. SCSI achieves this multitasking function through Direct Memory Access (DMA) technology. With DMA, the CPU can work on other applications while data is being transferred from the hard drive. There are different configurations of SCSI that relate to how fast data is transferred and the ensuing performance of network servers. TIP: If the server is moving large files, Ultra 2 SCSI is the way to go. All of the company's data including operating system, programs and files will reside on a single hard drive. The second hard drive is installed to assure data backup.

A server's desired memory size depends on a number of factors including programs run, performance goals, type and number of users, purposes to be achieved, network operating system, and amount of activity. If memory is insufficient, processing problems and possibly crashes can occur.

The network server houses the network operating system, stores information gathered from different parts of the network, distributes data to clients, and backs up appropriate information.

Servers can support multiprocessing where many CPUs share processing activities. Scalability is possible by adding additional processors as required. In symmetric multiprocessing (SMP), server activities and results are distributed evenly over multiprocessors within one computer.

A network interface card (NIC) is used to connect the server to the network. The NIC provides a path for transferring information to and from the server. Interface definition language describes the functions and parameters that a server exports to its clients.

Most companies have voluminous databases requiring a server-based network. Such networks provide growth opportunities because of resource sharing and size. Security features of these networks are good. Through controlling resources and central administration, greater efficiency and effectiveness may be achieved.

A server process satisfies the client request to conduct a requested task. The server process usually receives requests from client processes, conducts database retrieval and updates, manages data integrity, and dispatches responses to client requests. On occasion, servers can perform common or complex business logic. The server-based process may run on another machine on the network. This server could be the host operating system or network file server; the server is then providing both file system services and application services. The server process acts as a software engine managing shared resources, such as communication links, printers, databases, or high-powered processes. The server process conducts the back-end functions common to similar applications. The back-end (processing engine) part of the server program has three major uses of retrieving and storing requested data, performing computations and application processing, and providing the necessary security and management activities. The back-end conducts all or most of the processing except the interface and formatting-associated functions conducted by the front-end. The back-end distributed applications represents most of the total processing.

The network operating system allows for the functioning of file-and-print sharing, which is a basic function. Software suites must be bought to accomplish the operations of communications and database sharing. For example, Microsoft's Back Office Server manages customer databases.

Server capacity planning is software managing the server including its planning and monitoring abilities. How is the server performing? How can the operation of the server be improved? The server network operating system selected should be based on its ability to best satisfy the required functions.

In evaluating the server, answer the following questions:

- Is the server being fully utilized? This can be determined by ascertaining the extent of processing of the CPU.
- Is the server's performance hindered by errors in the network?
- Is the memory of the server optimal?
- Can the server operating system handle the fastest chips?
- Can the cabling handle additional throughput?

Universal client capability is when a client workstation can interact with different network operating system servers. Universal data access system is middleware that takes client requests and reformats them for a server to comprehend.

Dedicated versus Nondedicated Servers

There are different kinds of network servers. The selection of a network topology governs if a server is dedicated or not, and the topology selection is governed by what the network is doing. A dedicated server is a central computer used only to manage network traffic and provide services for network users. A nondedicated server is a computer that is used as a client as well.

Dedicated servers have faster network performance because they do not take requests from both local users and network stations. Further, these machines are not vulnerable to crashes due to local users' errors. Dedicated servers are costly and cannot be disconnected from the network and used as stand-alone computers. Dedicated servers are easier to administer than nondedicated because all data is in one place. They are quicker because they do not have a local user to serve. However, it is more difficult to make resources available on an ad hoc basis, because setting up a server is more difficult and time-consuming. If a dedicated server malfunctions, all users are forced to cease working due to the fact that all resources are centralized. A dedicated file server is the main computer setting for the operation and applications on the network. It cannot be a user workstation.

Nondedicated servers provide for additional flexibility because users can make resources available on their computers as needed. However, a nondedicated server scheme mandates that the users are willing to take some administrative responsibility for their computers and requires LAN knowledge. In a nondedicated approach, it is more difficult to establish security, back up shared data, and set forth access rights. Nondedicated servers will also likely experience worse performance when used simultaneously as a server and workstation. Nondedicated servers involve a higher price-performance ratio for companies having infrequent use of the server as a local workstation.

Fragmentation

Fragmentation is the splitting up of parts of a table among multiple servers. Horizontal fragmentation is splitting a table into two or more groups of rows and storing such rows on separate servers. Vertical fragmentation is splitting a table into two or more groups of columns (or locations).

To inform workstations of what is available on the network, the authors recommend using name (browse) servers. These servers are given the responsibility of monitoring network services. As new services and data become available, the name servers are updated on a periodic basis. The browse servers are backed up to other servers in case the primary server goes down.

Clustering

Shared nothing clusters are clusters in which each node has its own memory space. Communication is through passing messages. Applications are rewritten or segmented so parts of the application are executed independently on multiple servers of the cluster.

In clustering, there is a group of servers to assure information will be retained and available if a server meltdown occurs. Clustering enhances server capabilities and results in fault tolerance. Multiple servers also make for quicker response time.

Diagnostic

Servers should be equipped with diagnostic programs and fault tolerance software. The software can track performance as well as identify potential problems and correct them. A server may log the following to keep track of these activities for management and control: number of logons, by whom, time used, and what files were accessed; applications used; incidence of errors; and performance statistics. If a component within the server malfunctions or fails, the server can still function and recovery can be expedited. Further, the server may be shut off in an orderly fashion without information loss. For security purposes, a secondary (failover) server may be connected to the primary server so if something happens to the primary one the secondary one continues to operate assuring that the system is still functioning.

Backup

There is nothing more crucial than having proper backup. In deriving a backup policy, consideration should be given to the speed needed for a full backup, data capacity to be stored, and the frequency of backup. The authors recommend DAT drives because of their reliability and the voluminous capacity they offer. The backup device used should be appropriate for your storage and speed requirements taking into account your budgeted funds. Examples of various backup devices are a floppy drive, zip drive, CD rewritable drive, jazz drive, and digital analog tape (DAT). The first three backups listed are not suggested for a network system because of their limited capacity. DAT drives are the fastest for backing up. Duplexing refers to the copying of hard drives.

Besides the backup device, backup software must be installed to automatically perform the operation on a periodic basis. In the case of a small-to medium-size business, Seagate BackUp Exec software may be suitable because it is compatible with all DAT drives, is simple to use, and is flexible. Many small to medium-size businesses conduct their backup weekly. Backup should be done more often as the importance of data increases. A differential backup is one in which the backup has been changed from the last complete backup. With multiple server backup, the server can operate without interruption so the servers are functioning productively with the additional backup feature. Seagate's BackUp Exec cannot only be used to backup the data on the server but also the data stored on client workstations. Thus, we are backing up across the network. While Seagate's BackUp Exec is efficient in backing up multiple systems, the drawback is that it is time-consuming and results in a lot of network traffic.

Mirroring duplicates data in two different drives or duplicates the information on two locations of the same drive. This will be a "life saver" if one hard drive fails because the mirrored drive can take over all of the functioning of the failed drive. This results in enhanced reliability and less downtime for the network. For example, mirroring is a utility present with the Windows NT Server.

If data files are extremely important and large (at least 10 GB) Redundant Array of Inexpensive Disks (RAID) should be taken advantage of: in this several hard drives are connected to serve as a single drive recognized by the computer. RAID protects against a failed disk. Under RAID, there is significantly more fault tolerance compared to a single drive or two mirrored drives. The approach used to connect multiple drives into one logical storage unit is referred to as "striping." Striping partitions each drive's storage space into stripes so the combined space consists of stripes from each drive. To convert your server to a RAID system, a RAID controller and the associated software must be acquired. One vendor of RAID is Adaptec (http://www.adaptec.com).

An uninterruptible power supply (UPS) should be installed in network environments where risk of power failure exists. The UPS will run the server for a short time period and shut it down in the proper way. A good vendor for various types and sizes of UPSs depending on the network type is American Power Conversions (http://www.apc.com). In selecting the appropriate UPS unit, consider the following questions:

- Manner in which the UPS unit will inform users of a power failure.
- Time needed to shut down clients and servers.
- Number of type of components supported by the UPS.
- Power requirements of the network.

Types of Servers

There are various types of servers in a network including file, database, transaction, application, communication, print, directory, message, and security.

File Server

A file is a set of related records considered as one unit. An example is an arrangement of sales orders. General files may include accounting, finance, or engineering data. A file control system is used to store and retrieve information. File interrogation software can review and appraise a computer file's content. In file management, we keep a file current by adding, deleting, or changing the data. We administor, control, and optimize the continued generation of file information on the network system. File management software designates, recognizes, and categorizes file data to make it easier and more useful when retrieving the information from storage. Files are processed and utilized as needed. If all or part of the network system fails, steps must be taken to reconstruct the lost or damaged files.

A file server is a central network server that makes files and applications available to users. The server PC can make its entire disk and specified directories available. Network users can read, write, retrieve, exchange, manage, and move information from the file server. File update synchronization assures that file users have the most current version. This process involves user tracking and stamping the date of update so the files are in chronological order. The file server's hard disk is an extension of each user's PC. File sharing servers control user access to information stored across the network. Client machines can access files located on the server for manipulation or printing by downloading the data from the server and storing the file on the client's local hard drive. File data service involves large bandwidth and can significantly slow down a network with many users. Conventional LAN computing allows users to share resources like data files and peripheral devices by moving them to a networked file server. The network administrator ascertains user privileges to access specified files. The file server stores network and application software for the network. For example, in a retail store the file server stores the inventory management software to share with other workstations. A common file server is Novell Net Ware. Novell's Net Ware is a proprietary operating system and answers requests from network computers expeditiously.

It is advisable to use file archiving which is backing up files and information on the servers to offline storage devices (e.g., otpical disks). It is also advisable to back up the information on the clients. Hence, all network files on servers and clients may be on central storage devices.

In conclusion, file servers are network computers with centralized disk storage which can be shared by client computers. File servers store documents, images, etc. for client access. They store the files required for the database management system and applications.

Transaction Server

In transaction servers, clients invoke remote procedures residing on servers which also contain an SQL database engine. There are procedural statements on the server to carry out a group of SQL statements (transactions), which either all succeed or fail as a unit. The applications based on transaction servers are referred to as online transaction processing (OLTP) and tend to be mission-essential applications which mandate immediate response, tight security, and a reliable database. There is minimal communication overhead because the exchange usually is comprised of a single request/reply (as distinguished from multiple SQL statements in database servers).

Database Server

A database server is a program running on a server allowing the manager to establish, access, manage, handle, and maintain a database. Database servers can store and retrieve key and/or sophisticated data. Information is distributed in an efficient manner especially with distributed databases located close to where the stored information must be accessed. A database server does things such as searching, sorting, and indexing. This reduces network traffic because fewer items are travelling between the server and user. Security is enhanced because all data is stored centrally. A database server uses its own processing power to find requested data instead of passing the records back to the client and then allowing the client to find its data, as is the situation with a file server. Thus, a database server is a more efficient process then a file server. Data sharing occurs when multiple applications access common data simultaneously.

Database servers typically use the UNIX operating system. An example of a database server is Microsoft Exchange Server.

In the client/server situation, database servers are usually standard query language (SQL) servers. SQL servers use SQL distributed databases as distinguished from network, hierarchical, or object databases. Client computers make SQL requests of the database server, which accesses the stored database to process each request or query and returns the result to the relevant client computer. After receiving the data, clients could process it and if appropriate submit changed or new data back to the server. Optimizer is part of the SQL server opting for the most efficient method for query processing. NOTE: Requested data must be managed because it may be in multiple servers.

Locking refers to the system used to safeguard that data changes are not performed by more than one user at the same time to a record or table. If this occurred, data would not be reliable.

Application Server

An application server runs application programs. An application server processes information for clients. Application servers enable clients to have similar abilities as a file-and-print server but the files are not downloaded from the server. Rather, clients have the ability to access and alter files stored just on the server. This is suitable for networks having large databases that must be shared among many clients. By using the resources on the server, users can share data, while security and management services, both of which are also based on the server, assure data integrity and security.

Object-oriented applications refer to software designed to achieve complicated tasks by merging smaller applications referred to as objects. Server objects provide support for sharing and concurrency. Object transportation monitors (OTMs) are application servers for distributed objects.

Workgroup applications include routing forms and documents among clients.

Specialized servers may be for particular applications.

Communication Server

A communication server enables the routing of e-mail messages and Internet access. The communication server provides a gateway for incoming and outgoing information. The server manages such communication devices as modems.

Print Server

A print server allows clients and network users to share printers. The print server may be in the form of hardware, software, or a mix. Printing services are managed and controlled. There is simultaneous and multiple access to printing resources. Data may be printed from the computer or data terminal. Jobs are sent to the printer in an organized manner based on either priority or time sequence. By sharing access to printers overall cost is reduced and there is versatile usage.

Directory Service Servers

A directory service server aids users in finding, storing, and retrieving information on the network. It assists in user and group organization, resource sharing, and security.

Message Server

A message server furnishes message services. Data may be in the form of text, graphics, audio, and video. Message services can transmit data over the network. There must be coordination between documents, applications, and users. Message services may be in the form of e-mail, directory services, object-oriented applications, and workgroup applications.

Gateway Server

A gateway server can be used to communicate and translate data into a comprehendable format.

Groupware Server

A groupware server furnishes collaboration tools.

Remote Node Server

A remote node server controls remote access to local area network connected resources and acts as a gateway to those resources. Application services are provided. Remote procedure call enables programs running on the client to use the services of a program running on the server. It is middleware which works like a locally run application program that activates and calls a subroutine stored outside of the application program itself, but usually in a subroutine library on the same computer. Remote servers execute a separate process on a remote machine and perhaps on a different operating system.

Web Application Server

A Web application server returns documents to clients who ask for them. It augments standard HTTP servers with server-side component frameworks. It is a server storing HTML documents accessible on the Web.

Proxy Server

A proxy server is a computer placed between the Web server and a Web browser (client) to filter requests and improve performance. The proxy server intercepts messages going into and out of the network. A proxy server undertakes a single Internet connection and satisfies user requests.

Wide Area Information Services (WAIS) Server

A wide area information services server provides multiple indexes to other Internet-attached WAIS servers. These servers also act as search engines based on particular wording or text.

Security Server

A security server is a physically secured server maintaining security-related information such as names and passwords.

Middleware

Middleware is an essential second level of the company's infrastructure between the network and application levels. Middleware is the software layer facilitating the communication and accessing of data between clients and servers. Middleware starts with the client asking for a service and includes the transmission of the request over the network and the resulting response. It is the glue allowing the client to obtain services from the server. In other words, middleware serves as the nervous system of the client/server infrastructure. Middleware also has a distributed system management (DSM) software component as does the client and server. The DSM applications run on every node in the client/server network. Differences or irregularities are reconciled. For example, middleware has to decipher and interpret data coding differences in applications and databases. Middleware does not apply to the software providing the actual service since that is in the server's application domain. It does not include the database. Server-to-server interactions are usually client/server in nature. For instance, servers are clients to other servers, and clients are servers to other clients. Thus, a server can act in both client and server roles. Some server-to-server interactions require specialized server middleware. For example, a two-phase commit protocol may be used to coordinate a transaction executed on multiple servers.

Middleware is needed because of a multitude of applications, and the number of locations. It is necessary that a set of core data and services be moved from their multiple instances into a centralized institutional offering through the use of middleware. Types of middleware include Internet, groupware, systems management, object, and database. Messaging middleware transports data through the network as independent messages. A company should have a flexible middleware architecture so they can integrate new applications with existing systems.

Client and server middleware products bought have to be compatible with the operating system, and must support the interconnected clients and servers. Further, the system must support the protocol required by the middleware. The middleware must also be compatible with the network database management system. Finally middleware must be compatible with all platforms the distributed network has deployed. Middleware's biggest attribute has been to integrate legacy network systems with object-oriented software.

Middleware is software that satisfies the requirements of the networked company to initiate, integrate, and manage large-scale, distributed applications in widely diverse environments. It allows one network system to be symmetrical with another. Middleware is a layer between an application and the network, host operating system, and any resource servers (e.g., data base servers). Middleware enables different platforms appear the same to an application and can save money and time in developing client/server systems. The purpose of middleware is to make the multitude of technology transparent to end-users. Middleware attempts to resolve all the challenges of making diverse technologies compatible with one another.

Middleware structures data and meaning in applications. Middleware systemizes and organizes data flow between integrated resources configuring it into a business process. Middleware acts as a communication channel network of resource messages including request/reply, store and forward, and publish and subscribe. Middleware provides security and queuing functionality to message delivery. It resolves data coding and layout differences.

Data from agents on the network are accumulated by managing the workstation which also presents the data in graphic form. The managing agent can also request its agents to carry out operations. In essence, we have an autonomous network within a network.

Middleware attempts to achieve scalability, openness, and integrity. Scalability assures protocol standards while increasing capabilities. Openness applies to interoperability across unrelated domains. Integrity seeks data security and auditability.

A distributed network uses and manages many different kinds of applications. Middleware servers can be designed to support a wide range of advanced applications including in the areas of electronic commerce, distributed computing, and distributed databases.

Remote procedure call (RPC) lessens the complexity in developing distributed processing by keeping the components of a remote call the same irrespective of whether the server and client are on the same system.

The types of middleware are:

- Message Oriented Middleware (MOM). The provides program-to-program communications by message passing. Since MOM typically supports multiple protocols, it consists of an infrastructure that will support scalable and reliable high performance distributed application networks in diverse and complex environments.

- Transaction processing (TP) monitor technology gives the distributed client/server environment the ability to develop, run, and manage stable transaction applications. TP monitors track client/server traffic. It is an operating system for transaction processing. It conducts database updates and business logic rule computations. It furnishes application services to thousands of clients in a distributed client/server environment. TP monitor technology accomplishes this by multiplexing client transaction requests (by type) onto a controlled number of processing routines that support specified services. Transactions are managed from their initiating point, typically on the client, across servers, and then back to the originating client. TP monitors can run transactions, route them over systems, execute them, and restart them after failure. It is a type of middleware acting as a supervisor of distributed applications that mandate postings to many distributed databases. It also applies to a series of activities taking place in response to a business transaction.

TP monitoring enhances responsiveness and reduces resource overhead. TP monitoring is a very active system including delivery services for network access, data management, order processing, forms and terminal management, authorization, and security. TP is cost-effective for online applications.

In conclusion, a TP monitor is in effect message oriented middleware (MOM) with processing capabilities.

- Object Request Broker (ORB) distributes computing across multiple platforms. ORBs are the messaging middleware by which objects in heterogeneous distributed environments make and receive requests and responses transparently.
- Enterprise JAVA Beans (EJB) are visual components reused across applications. EJB is a cross-platform component architecture to develop and deploy multi-tier, scalable, distributed, and object-oriented JAVA applications.

With an n-tier architectural structure, middleware provides a platform to run server-side components, load balancing, security, high-availability, and managing and controlling transaction reliability and integrity.

Most middleware products run on TCP/IP.

Upperware

Upperware is a type of middleware that furnishes specific services for applications relieving such applications of the need to do it themselves. One type of upperware, business application middleware, is of particular interest to companies. Practical services provided by business application middleware are message handling, application gateways, object resource brokering which obtain business information in distributed environments, and transaction monitoring to conduct audit activities.

Groupware

Groupware is a type of client/server software facilitating collaborative work. It allows for people to communicate and achieve activities and tasks together electronically regardless of location. In consequence, productivity and communication are enhanced regardless of place or time. Client/server groupware builds on certain foundation technologies consisting of data transfer and sharing, workflow, group scheduling desktop conferencing, e-mail, conferencing, document review and management, and multi-media document management. Groupware allows individuals or businesses to collaborate on a job using client/server networks. Groupware is software supporting the creation, flow, updating, and watching over of non-structured information to support collaborative group activity. Groupware relates to managing data, knowledge, and activities. Automated replication is used for continuous updating of copies of a project database placed on local servers throughout different geographic operating areas of the company.

Distributed Computing and Applications

Distribution is the process of distributing interrelated shared information over a network. Distributed transaction processing refers to transactions being processed over multiple distributed linked computers in multiple geographically situated locations. Distributed functions relate to having the management, processing, and error recovery activities in a network scattered over the nodes. Distributed applications are divided applications executed cooperatively in at least two computers whether they be servers or clients. The processes are most likely not running at the same time. Each process conducts part of an application in sequence. This is distinct from having the operations focused at a central place.

Portability in distributed computing relates to the ability of client/server application development in one computer environment and deployed in another one(s).

Two major characteristics of distributed processing are scalability and transparency. Scalability relates to the ability of distributed processing systems to add clients and/or servers without detracting from the system's overall performance. Scalability of a distributed processing system is essential only if application programs can be split between clients and servers. Transparency is the ability of distributed processing systems to merge clients and servers of different operating systems and protocols into a cohesive framework that processes distributed applications. Often, the data used in a distributed processing environment is distributed across platforms. While complete transparency cannot presently be accomplished, limited transparency among a limited number of vendors' equipment, operating systems, and network operating systems is currently achievable. This transparency is achieved via middleware. Middleware is in the middle of the distributed processing system, serving as transparent insulation surrounding the enterprise network over which the client/server communication travels.

With regard to distributed processing, the client part (front-end) runs on the client workstation as required by the user. Its major operations are formatting data processed and received from the server, and user interfacing. The server-part (back-end or engine) runs continuously and stores and retrieves data, performs calculations and application processing, and furnishes security and management operations. The services provided are for multiple users sending requests for processing and data from multiple client workstations.

Distributed objects are objects located anywhere on the network. The objects interact with each other as if they were on the same machine. They are software elements with well-defined interfaces. Distributed object management systems (DOMS) furnish a way of putting various components together and managing their communications. They address the following:

- A method to locate independence for client applications.
- A single interface to manage the difficulties of a heterogeneous environment.
- A uniform framework based on standards and extensibility to build, integrate, and deploy distributed computing applications.

A distributed device manager probes accumulated data from multiple vendor network devices.

Distributed computing integrates database clients to relational database servers and to applications. The distributed computing environment (DCE) network security services have more sophisticated processes relying upon authorization, authentication, and user account management. DCE authenticates and substantiates clients relying upon a security server which retains a depository having information such as names and passwords. This is often a dedicated server. Authorization comes after authentication to determine the access level of a particular client.

Cells are administrative units (domains). The customer defines the domain of the cell. A cell is usually composed on the set of computers used by the group(s) working on associated tasks. At a minimum, a DCE cell must have one security server and one cell directory server.

The distributed application environment (DAE) is comprised of: (1) messaging services, (2) database services, (3) presentation services, and (4) device services to capture data and to control a broad category of controllers, devices, equipment, and sensors.

A distributed database is a collection of tables spread over two or more servers in the company at different locations. For example, a distributed database is one in which data is stored in different computer systems. This process updates databases on multiple servers. A distributed request diagram is represented in Figure 9.5.

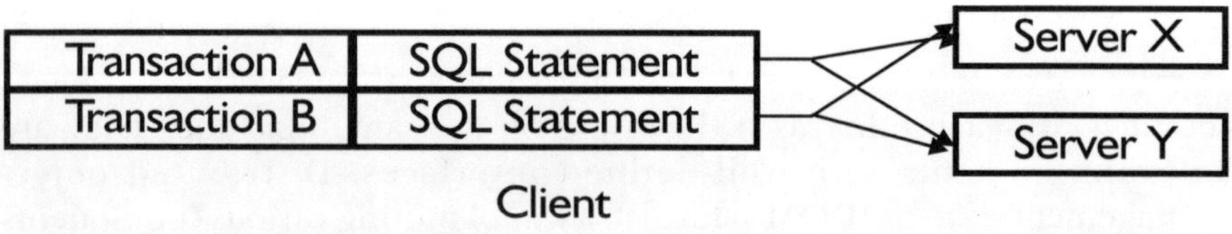

Figure 9.5: Distributed Request Diagram

Transaction Processing

A transaction is either one or a series of entries resulting in automatic events within the network. It is a sequence of predefined steps or operations within an application program to sufficiently record that business transaction. The proper recording of a business transaction usually infers posting to the impacted files or databases. Transaction processing ensures that postings to a relevant business transaction have successfully been achieved. The monitoring of transaction posting is conducted by software called transaction posting monitors. In the case of client/server information systems, multiple geographically dispersed computers are linked so that transactions may be posted across multiple distributed computers.

A transaction server relates to a client asking for services residing on the SQL database engine. The remote procedures on the server execute or group SQL statements for the database. Here, transactions are defined as the group of statements. The network exchange consists of one request/reply message. The SQL statements are a unit. In a transaction server application, there is a code written for the client and server components.

Transaction processing systems (TPS) process voluminous, routine, and repetitive daily corporate transactions in an automated way. Examples are the processing of sales on account and backup transactions. TPSs lower transaction costs, improve service, and enhance efficiency and effectiveness. A TPS having online processing updates data at the time of the transaction.

1 This chapter was coauthored by Robert Fonfeder, Ph.D., CPA, a Professor at the College of Business at Hofstra University.

10

Network and Internet Security

The need to protect computer network information and data from hackers, crackers (malicious hackers who break into network systems), criminals, and other intruders has expanded virtually exponentially as the need for networks has increased and their costs have decreased. The Internet, itself, is nothing more than a network and as connectivity to it increases due to popularity, high speed modems, etc. so do security considerations. As modern technology improves and individuals are able to more easily and quickly access the internet (e.g., through the use of ethernet cards and high speed DSL lines, cable modems and the use of local area networks), the probability that intruders may also access your computer through the same connection increases. Nevertheless, users must have access to computer networks. Therefore, it is imperative network administrators be familiar with security practices and procedures that will ensure that network and internet communication and processing is safe and that the organization's resources are protected.

The Concept of Network Security

The concept of "network security" is relatively elusive in that it cannot ever really be completely achieved by an entity. It may be compared, for example, to halving any number continually. If you then halve the result and then halve that result again and again, you will find that although you can continue this process to perpetuity, you will always be left with some amount that can be halved. That is, you approach zero but you can never reach it. The same holds true for network security. We can never hope to reach absolute network security. It is truly elusive in that there undoubtedly will always be someone who will figure out how to circumvent the most sophisticated of security systems. In addition, as the technological environment changes and new improved ways are invented for data communication and processing, new control procedures are needed to ensure protection. Since the rules are constantly changing, we must devise new security practices and procedures to ensure a secure network environment.

General Strategies for Protection

Although specific control techniques may change as technology improves, the following are some general security parameters that should be followed to ensure good network control. These principles follow:

1. *Commit the entity's security policies to a written form* – Besides the security team, make sure that the security policies are physically distributed to all employees who are a part of the network. Another way of distribution is to establish a network web-site that incorporates all network security policies. The site should be set up so that sections of security information relating to various parts of the network can be easily printed.
2. *Encourage ethical values and integrity among all personnel of the firm* – Standards of behavior should be demonstrated by the CEO and other members of senior management. High standards of ethical behavior should be consistently demonstrated through communication to employees, written policy statements, codes of ethics, etc. Everyone should:
    - Know what is expected of them;
    - Have the responsibility for reporting any violations to a predetermined higher level of management; and
    - Be cognizant that violations will result in penalties.

    In addition, all opportunities for advancement and other company enticements that might be perceived by employees as being unobtainable which might cause them to engage in unethical behavior should be reduced. For example, unrealistic performance results, unachieveable short-term goals, budget targets that are not realistic, and bonus and profit sharing plans with terms which might cause employees to commit fraud in order to earn stock options.
3. *Make competence and the pursuit of excellence a requirement* – In order to insure that the company's network is secure, personnel at every

level must possess the skills and requisite knowledge needed to perform their jobs accurately and effectively. This requires that company employees have the proper training and experience as well as have the requisite ability to perform well. Personnel should screen potential employees carefully and then ensure that they have the ability, education, and training throughout their tenure with the firm to succeed. Entities should strive to hire individuals whose skill and knowledge supercedes the level of those who would attack the company's network. Management should strive to bring the security level to state-of-the-art status so that intrusions by individuals from outside the company (i.e. without knowledge of the specific control procedures of the system and cryptographic techniques used) are minimal.

4. *Design your network in a way that balances restrictiveness and flexibility* – The more restrictive your network is the less flexible and operational it is. However, a restrictive system more easily highlights deviations and reduces the losses associated with them. For example, a highly restrictive system might require major reconfigurations each time a new service is added to the network. However, if the system could be modified so that the security process can be accordingly altered without too much inconvenience, it might well be worth it. Generally, a hacker who does not have insider information will make mistakes which a conservatively designed system will hopefully detect.

5. *Ensure that the defenses that you have built into your systems are adequate and working* – A good network defense system must be tested to insure that the internal security systems are working as they were designed to. In addition, it provides a way to insure that bad procedures are not maintained and are removed and good ones are improved. Test attacks should be frequently run with varying degrees of intensity to see how the system reacts and to ascertain what changes are needed to maintain security. The more you run these compliance tests, the more knowledge you will have regarding how your routers, computers, and other devices will react. (Frequently neither you nor your vendor will completely be able to predict how the system will react under unusual circumstances.)

6. *Make security modifications and improvements an ongoing procedure as part of a general network maintenance program* – Every entity should incorporate a policy of executing regular security maintenance procedures, frequent testing, and periodic auditing of all security procedures. In general, the older a security system, the weaker the defense against hacking. The older the system, the more people are familiar with it and the greater the probability of attack with

new technological know-how. Constant updating and improvement are necessary to insure that the network security is providing the company with full impenetrable protection.

7. *Make sure that employees are ever cognizant of the basic security policies of the entity.* At company meetings and seminars, concepts of network security should be reinforced so that employees understand why security is important and why they have to personally contribute to insure the company's overall protection. Security procedures and concepts should be positively reinforced whenever possible.

About Networks

Networks are complex systems that are highly susceptible to security risk by their very nature. Because networks are conglomerations of connected computers that need to communicate with each other, each computer on the network represents a potential security risk to every other. Obviously, the more machines that are interconnected the greater the vulnerability to security breaches.

Local Area Networks Versus Wide Area Networks

Local area networks (LANs) and wide area networks (WANs) are fundamental structures of a computer network. LANs were developed to enable multiple computers in a room, floor, or building to communicate with each other. As an alternative to having each of the computers in a group be connected to the others to communicate, a LAN may be set up to share a common data source that is instantly disseminated to all the computers connected to the LAN. In a typical office LAN, hundreds of computers, file servers, printers and other equipment are joined together. Two configurations that are used most commonly in LANs are the hub LAN and the cable LAN. In the former, each computer connected to the network sends its information to a central hub where it is then transmitted to the other computers connected to the hub. In the latter configuration (called the cable LAN), each computer sends its messages to a shared cable. More recently, wireless networks are being utilized that function through the use of an infrared spectrum.

The security considerations relating to a LAN primarily relate to opening the LAN to the Internet, as well as to allowing employees and other users to gain access to the network from outside sources. In addition, security procedures must be established to insure that company data and information are restricted and not fully available to all who use the network. In addition, sufficient internal control procedures must be implemented to protect the company data files from inadvertent errors, unauthorized users, malicious intent of some employees, and overall general safety. These considerations will be addressed when network controls are discussed later in the chapter.

A wide area network (WAN) is a network consisting of computers that are located in geographically separated, diverse areas. For example, a company whose executive office is located in New York may have regional offices located in Los Angeles, San Francisco, Oklahoma City, Tampa, and others. These offices may be connected through telephone lines, the Internet, and leased high speed cable connections. The difference between the LAN and WAN is simply the amount of geography separating the users of the network.

The security considerations relating to WANs are similar to those of LANs. Not only are we interested in ensuring that there is restricted access to the network and that only those who have authorization should be able to log on, but also we want to make sure that the internal security of the network is protected and maintained.

Finally, the Internet itself is really nothing more than a huge network made up of a multitude of smaller networks. There are important security considerations related to it such as making sure that unauthorized changes aren't made to the company's website. This and other Internet security considerations will be covered later in the chapter.

Network Security Procedures

Security network administrators are always faced with the possibility that a network will be susceptible to hostile attacks. These may range from direct attacks by insiders of the entity itself to outsiders including hackers, crackers, and partners (customers, vendors, etc.) of the organization. There is a myriad of damage that may occur from a belligerent attack on an entity's network. These might include: precluding authorized users from using network services by consuming CPU time or memory; infecting the system with a virus, back door, or Trojan Horse; destroying data by writing over or deleting the information; and illegally reading or copying confidential information – to name just a few. Clearly, network intrusions must be controlled. A discussion of the network security procedures needed to combat such intrusions follow.

Protection of Network Hardware

Protection of the network and network databases within the network begin with the physical protection of the computer system itself. This encompasses protecting the hardware and plant to insure that there is no loss of equipment, loss of information that may come from such a theft, and insurance that there is not a physical security failure that could result in a compromise of the integrity and accuracy of the company's data. The physical hardware that must be protected in a network environment includes the servers, wiring area, hub area, and buildings and areas containing production facilities, disks, tapes, archives, etc.

The facility should be designed to insure that the effects of the surroundings do not damage the hardware in any way. For example, it should have an electric backup to insure that the air conditioning will work continuously and reliably. This will insure that heat and humidity do not build up and potentially damage the equipment. The air in the facility should be constantly filtered to minimize the possibility of dirt or dust causing problems with the hardware. In addition, a power supply system should be acquired that guarantees continuous power to the network equipment to prevent against power failure, brownouts, etc., all of which could cause severe damage to the equipment as well as destroy or alter precious data. Clearly, the facility should be protected against the damaging effect of water by checking nearby pipes, windows, sprinkler systems (a fire extinguisher should be used instead), and other potential sources of water entry. The facility should have water alert systems wherever water can damage the network installation. These mechanisms activate alarms or pumps at the first sign of flooding. They are helpful in not only protecting equipment such as water-cooled computer cabinets but are also useful in protecting the electrical cables underneath the floor. Preferably, however, the facility should be situated away from any potentially damaging water sources.

Limit Access to the Physical Site

Network computer systems receive, store, process and disseminate sensitive information. Information technologists who are responsible for protecting this information must realize that a major consideration of protection is securing the perimeter of the computer facility and allowing only authorized users to access the company's hardware site. By allowing only those with authorization to gain entry to the network facility, the risk of having data compromised or stolen is greatly reduced. In devising a hardware protection strategy, perimeter security, interior security, and physical protection of the hardware itself, should be considered. Perimeter security consists of preventing entry through such obvious portals as windows and doors. Windows should be protected with easily mountable devices that will sound an alarm when the glass is broken. Access by means of the doors can be controlled with authorization devices that can range from a simple keypad entry device to a more complex card reader system that tracks all employees entering or leaving the network hardware environment. Interior security can be achieved through motion detectors at night and surveillance camera monitoring during the evening and daytime so that if an intruder gained access, he or she could be identified and apprehended. Physical protection of the hardware itself can be achieved through alarm systems that go off if the machines are moved or disturbed. If any hardware is sufficiently small to be carried away, cable kits can be used to unobtrusively control this problem.

Other Physical Security Considerations Relating to Network Access

The following should be considered in strengthening physical security relating to network hardware access:

1) Areas containing the network computer servers and related hardware housing areas should have only a single entry door that is operated by physical controls monitored by a log. The log should work automatically and should enumerate such information as the user's identification number, time of access, and specific activity performed. Software such as data dictionaries provide the capability of generating automated logs of access to database information. In addition, such items as cameras and motion detectors are able to record and monitor sensitive areas and insure that the presence of any unauthorized users will be posted for review.
2) Network access should be assigned different levels of confidentiality and security such as generally accessible information (unrestricted), internal use only, confidential, highly confidential, and top secret. Confidential information should not be displayed on computer screens without the appropriate access codes.
3) Employee compliance should be periodically reviewed to insure that individuals are honoring the controls that have been established. Each database function, such as accounts receivable, accounts payable, payroll, etc. should require its own password so that employee users only have access to the areas that they have been approved and authorized for. In addition, as was previously noted, the database should maintain a log of the identity of the user, activity performed, and the date and time that the database was updated.

4) To further insure control of network and database information, there should be appropriate segregation of duties. For example, programmers who could modify or patch a program should never be given access to the computer database area or library to insure that data manipulation is prevented. In addition, a temporary employee should never be given unrestricted access to a company's system. Their activities should, for example, be limited to a specific project, activity, and time period.

5) For sensitive database information, terminal access controls to the network may be used to prevent against unauthorized use. These controls include:

    a) Time Lock – The network system will deny access to a user during specified hours, typically, for example, after normal business hours.

    b) Automatic Sign-Off – The network system will automatically sign an authorized user off if he or she fails to sign off after accessing and transmitting information to the database.

    c) Call-back – Before access is granted to the network system a phone-call may be made to the terminal site to confirm the user's identity.

Authentication Procedures

In general, the process of authentication involves determining an individual's identity and ascertaining whether he or she is the person they profess to be. Computer network authentication is the process of verifying a network user's identity and determining whether he or she is authorized to gain access to the information that is being requested. That is, does the person have the authorization to gain access through the network's firewall and what level of access is allowed. Firewalls and related levels of access will be discussed in a subsequent section of the chapter. It is imperative that the network has sufficient controls to insure that access remains secure. Some intruders attempt to gain access to an entity's network while others already have valid low level access (internal users) and are trying to achieve a higher level (or unlimited) access. Most network software or communication software packages contain encryption features as well as other security features that will be discussed in this chapter. Virtually all these packages have password controls. Password controls are by far the most popular way of authenticating network users. The following are some important considerations relating to network authentication and passwords.

1) General Guidelines for Choosing a Password include:

- Avoid the use of patterns in the choice of passwords such as: 246810, 10203040, abcdefg, etc.
- Try not to use passwords that are words in any language. American and foreign dictionaries are frequently used to ascertain passwords.
- A company policy of changing passwords regularly should be incorporated. Thus, the company's computer system should be programmed to periodically require new passwords and not allow users to use their same passwords again and again or allow them to choose ones that they have recently discarded.
- A technique that results in passwords that are easy to remember but fairly difficult to guess is to choose a password that can be easily derived by taking the first letter of a phrase that is easily remembered by the user. For example, the phrase, "My favorite food is hamburger," generates the password, MFFIH, which clearly would be difficult to figure out.
- Geographical terminology such as New Jersey or Los Angeles should not be used.
- Numeric and nonalphabetic characters should be used as much as possible in preparing a password. In addition, a combination of upper and lowercase letters should be used.
- If the password characters of the system require alphanumerics, do not choose a word and simply add a numeric to meet the system's requirement: for example, Airplane5. Hackers are familiar with this technique and may easily determine your password.

Symantec, producers of computer security software (www.symantec.com), has a website that provides assistance in choosing a password and determining the quality of the ones chosen. It is important to note, however, that although passwords provide the first line of security against hackers, experienced ones can generally bypass this control. In addition, software programs are now available to assist hackers in ascertaining passwords. Sometimes, a serious flaw is uncovered in an entity's network security system and a hacker is able to create a universal password that satisfies the log-in program without having to actually determine a user's password giving the intruder unauthorized access to the network.

2) Double Identification Protection – Clearly, if a system places importance on insuring that a network be protected because of

highly confidential information, a sophisticated insider may not have difficulty in gaining access if the only protection to the network is an individual user's single password. Security Dynamics Technology, Inc. (www.securitydynamics.com) has developed a mechanism that is more complex, making it extremely difficult for an unwanted user to gain access to a highly confidential network. The protection mechanism is based on a token that requires a dual-determinant identification. The first identification is a secret code that only the user knows, such as a password or PIN (Personal Identification Number). The second identification is more complex, based on a SecurID token that only the proper user will possess. The SecurID token is a fairly small device no larger than a credit card that displays a constantly changing ID code. This mechanism is basically a microprocessor constructed on a handheld device that generates random, unique, access codes every minute. In order for a bona-fide user to gain access to the network, the user's name must first be entered together with his or her individual password and then the specific identification code that is shown by the SecurID token. The security server of the entity's network will then compare this information – especially the SecurID password for that specific time period. A primary advantage of the SecurID is that it does not require a card reader or time consuming response procedure. In addition, it clearly prevents unauthorized access to confidential data sources on restricted databases and can be used at the network, application, or transmission level. It is ideal for remote access devices such as a PC, workstation, laptop, or similar device. It is also possible to send authentication access data for a specific database remotely over public networks. Typical usage for this type of access security, for example, is medical patient clinical information found in hospitals. It is also compatible with popular software security access programs.

3) Smart Card – A smart card is a credit card device that has a microprocessor embedded in it that contains a small amount of memory as well as interface capabilities that allow it to communicate with database workstations or networks. The most popular use of smart cards is the common phone card. But it can also be used for identity access within any network system. With smart cards, the individual's identification number or password are carefully encrypted on the card. If access requires biometric identification to gain network access, it would not even be necessary to use a PIN number. Thus, a smart card can be personalized to the needs of the user.

4) Tokens – A token is a small device no bigger than a credit card that displays a constantly changing identification number. Once the user enters a specific password, the token displays an identification code that is then used to log onto the network. Aladdin Knowledge Systems (www.aks.com), an information security system, has developed a token known as eToken which is a car key sized token. Token technology can be used to protect network database sensitive information by controlling access.

5) Biometrics – Compromising the security of a network as a result of inadvertently divulging one's password or the inconvenience of forgetting, losing, or misplacing it has led to the realization that a methodology was needed that potentially would be a more reliable and secure form of access. Information from the Compaq Corporation recently indicated that up to fifty percent of calls to its corporate help desk were related to forgotten or expiring passwords. Out of this need grew the concept of identifying people based on their unique physical characteristics or behavioral traits such as fingerprints, eyes or other facial features, speech, writing, and ways of doing things. This process of identification is known as biometrics. Although biometric identification has not yet gained wide acceptance, it is predicted that in the future its usage will grow in excess of one third of its current usage and will be able to better handle access applications. While currently biometric technology is relatively expensive without being as effective as other security products. However, as the technology improves and becomes considerably cheaper, it will undoubtedly become a highly reliable and secure form of data access. Some of the commercial biometric identification methodologies that are either currently available or becoming available in the near future include:

- Fingerprints
- Retina prints
- Iris scans
- Voice prints
- Signature and handwriting analysis
- Palm prints and hand geometry
- Hand veins
- Facial recognition
- Facial thermographs and
- Silhouette Identification and Gait Prints

Biometric identification consists of two parts. The first consists of physically scanning the personal characteristics of the individual seeking to access the company's network. These features are then encrypted into numerical data and stored as distinctive templates. In the second part of the process, the potential user attempts to gain access to the network. His or her personal characteristics are then presented for access which is read by a biometric reader. The computer then compares the personal information read and compares it to the template which is stored in the network. If a match is made, the individual is given access to the network. The aforementioned biometric methodologies that were enumerated will now be discussed in more detail.[1]

Fingerprints

The fingerprint of an individual has unique characteristics. It includes whorls, arches, loops, ridge endings and ridge bifurcations. Verification fingerprint readers are able to capture the flat image of a finger and perform a comparative one-to-one validation. For example, Compaq Computer Corporation (www.compaq.com) uses a fingerprint identification technology that is able to significantly simplify the log-in process. Compaq's Fingerprint Identification Technology uses software algorithm technology to convert it to a distinctive map of minutiae points. Minutiae points are the distinctive data positions in the fingerprint that uniquely describe it. The total of these points is then aggregated into an encrypted map which is then stored. The person requesting access to the network only need place his or her finger on the reader for a comparison to be made of the stored fingerprint encrypted map so that access may be either obtained or denied. It is important to note that the actual fingerprint itself is not stored anywhere. This is significant. Because the fingerprint cannot be recreated from the encrypted map of an individual's minutiae points that are stored in the network, the fingerprint itself cannot be recreated. This consideration should not be understated. Garfinkel[2] notes that although the fingerprint access process is highly accurate, even proponents agree that until individuals are made to realize that an actual fingerprint could never be recreated from the data map that is stored, there will always be stiff resistence to its use. Although fingerprints have been used as a means of successfully identifying individuals for over 100 years, the method has been unable to shake off the stain of criminality that is frequently associated with its use.

Retina Prints

This process analyzes the patterns inside a person's eye. Although in the 1980s, retina prints were popular, the minutiae of detail do not remain fixed for a lifetime as it does in fingerprints. For example, when a woman becomes pregnant, the fetal hormones that are secreted cause the new arteries and veins to branch out in the mother's eye. In addition, retina prints require the use of a relatively expensive retinal scanner that is somewhat intrusive in its application and not convenient. Because of this and the lack of general reliability, retina prints are not widely used.

Iris Prints

Unlike retina prints, the patterns in the human iris do remain fixed for life when formed in utero and therefore are considered quite reliable. Each human iris contains more than 266 specific discriminators (compared, for example, to a fingerprint which contains only 35 discriminators). In addition, an iris print can be measured with an inexpensive standard video camera. IriScan is a leader in the development of iris print technology. Its machines are currently being used in prisons, automatic teller machines, and subway stations. Its partner in the venture is currently developing technology that will be able to capture the print of an individual while driving at up to 50 miles per hour. Iris prints have become very popular in the 1990s and millenium decade. Unfortunately, like fingerprinting, much of the public considers iris scan technology to be an intrusion on their privacy because their eye (iris patten) is being photographed and stored.

Voiceprints

Using this biometric methodology, network access is accomplished by first enrolling the individual into the voiceprint database by having him or her repeat a simple pass phrase three times. For example, the individual might state "My name is John Smith." The newly created voiceprint may then be stored either on a security server, smart card or local PC. The voice print of the individual seeking access to the network is then compared to the voiceprint that was previously stored. If the computer deems that a match has occurred, access is given to the individual. VoiceGuardian, software from Keyware Technologies (www.biometrics.org), for example, analyzes an individual's speech patterns by looking for points of inflection and articulation that form a person's unique voice patterns. Generally, people do not consider this method of identification as being an intrusion on their privacy. Unfortunately, to date, experts consider its identification accuracy as being only fair and voiceprint recognition is considered one of the less accurate methods of biometric identification. That is, it is questionable whether computers will ever be able to identify speakers with complete accuracy.

Signature and handwriting analysis

Signature and handwriting analysis is one of the more commonly used biometric methodologies. Personal signatures and handwriting are digitized and electronically compared with stored templates. It is also possible to measure the speed at which the pen moves as well as the pressure exerted against an electronic pad. These three parameters, speed, pressure and position, establish a unique biometric that is very difficult to forge.

Palm prints and hand geometry

This biometric relies on the wrinkles in an individual's hand or the hand's relative length. Sometimes, the ridges of the palm are measured. Clearly, wrinkles, ridges, and hand length are not consistent over an individual's life and therefore are not as reliable as fingerprints.

Hand Veins

In this process, the vein pattern on the back of the individual's hand is scanned by the computer to create a digital template which is then stored in the computer's database. When access to a network is requested, a current scan is performed which is then compared by the computer to the stored template to determine whether a match exists. Given the difficulty of changing one's circulatory system (except perhaps by surgery), hand vein scanning is considered fairly accurate.

Face Recognition

Face recognition identifies people based on their appearance. Current face recognition technology requires that an individual's face completely fill the computer video camera and that the background be somewhat controlled. Future technology will probably be able to recognize a person in a crowd as people do. The process of face recognition consists of converting the video image of the face into a digital template which can later be compared to a currently recorded image. Because of the ease of taking someone's picture with minimal negative perceived consequence, it is assumed that face recognition should become a very popular biometric identification mechanism in the future. Miros Inc. (www.miros.com) utilizes neural network technology to identify an individual's face. Their True-Face Isolator represents an automated neural network locator.

Facial Thermograms

This biometric mechanism predicates identification of people based on the patterns of veins and arteries underneath their skin. Given the virtual impossibility of changing one's circulatory system, as was noted in discussing hand vein scanning (although surgery could effectively produce that effect), facial thermograms, as hand vein scanning, is considered a highly reliable technique.

Silhouette Identification and Gait Prints

These biometric parameters have been suggested and coined by Garfinkel based on peoples' physical size and proportions, the way they walk, and the kind of clothes they wear. Although this technique is in its infancy, Garfinkel believes that since people frequently identify others based on this technique, computers will soon be able to do so as well.

There are many biometric access systems in use today. It is important that a user entity decide whether a particular biometric access control system fits their organization. In making this decision, the Biometrics Consortium (www.biometrics.org) recommends that the following questions and considerations be addressed before a final commitment is made to undertake biometric technology:

(1) Regarding the level of access security that exists in your current control application, the following parameters should be ascertained:
   - Time required (elapsed time) for enrolling an individual
   - Time required (elapsed time) for an individual to utilize the access application
   - Percentage of false rejections
   - Percentage of false acceptances
   - Degree of consistent and uniform results across the population upon which the biometric application will be performed.

(2) What improvements does management expect regarding the biometric application in the future with respect to the aforementioned parameters. Specifically determine target improvements for the application.

(3) Select the specific biometric that is desired for the access application as well as the manufacturer. The Biometric Consortium has available lists of manufacturers who could provide various biometric capabilities.

(4) A compliance test must be performed to determine if the device that has been installed actually performs to the specifications that were required of, and promised by, the manufacturer. The

compliance test should be performed by the company's own employees under actual operating conditions. If needed, an independent consultant may be present to ascertain if the application will operate successfully given the requirements of the company.

(5) Following a successful test of compliance, the biometric should be fully implemented.

During test processing but before final commitment, the following queries should be answered:

- Suitability of the biometric itself relative to the company's needs –
  - Is the biometric that is being installed suitably satisfying the company's needs?
  - Are there any company groups or persons who cannot use the biometric application?
- Satisfaction with the biometric device –
  - Is the cost of the biometric device and its installation in line with the company's budget?
  - Does the size of the biometric device fall within the acceptable parameters of the company or is it too obstrusive or cumbersome?
- Reliability and reputation of the biometric manufacturer –
  - Is the company satisfied with the manufacturer's reputation, reliability, and quality of warranty work based on inquiries made?
  - Who are the manufacturer's major clients? Are they comparable to your company?
- Potential for fraud and failure –
  - Is it easy for someone to gain access to the company's network database surreptitiously by gaining the information that the manufacturer has? That is can forgery easily occur?
  - In a nutshell, will the biometric application significantly prevent the committing of fraud by criminals?

Network Communication Security

Obviously, the communications systems that link data between sites needs to be secure, private and most of all reliable. Network communications systems are affected by a myriad of factors including environmental, software problems, and hardware malfunctions. In general, communications security encompasses protection against all network attacks that does not require physical access. The imperative consideration in a communication network is that the physical links between the computer networks involved are functioning properly all the time. This means that all the types of data that are being transmitted in a typical network including traditional, sound, and video have to be security transferred. Every effort must be made to avoid breakdowns, delays, and disturbances during transmissions. In addition, transmissions must be protected from perpetrators who might tap into and intercept data.

The following represents important guidance relating to communication security:

A) Line Security – All unauthorized access to the communication lines that connect the various parts of the computer system should be restricted.

B) Transmission Security – Any unauthorized interception of communication should be prevented.

C) Digital Signatures – This provides validation to the receiver that the sender is authentic and that the message is not a forgery. A bonafide digital signature represents an electronic method of signing messages that represents a certificate of authentication providing assurance to the receiver that one who is purporting to have sent the message is in fact the one who did so and that the information sent has not been changed or altered in any way. The process of digital signing uses public key encryption to sign electronic documents. The actual digital signature is a code that is appended to an electronically transmitted message. The receiver, in this situation, is guaranteed that the individual who sent the message is in fact that person. Digital signatures so accurately authenticate and insure the integrity of messages that individuals who transact business electronically using these signatures have been unable to deny participation in the exchange. In addition, the receiver knows that there was no alteration during transmission because any changes to the message after it has been electronically signed would invalidate the digital signature. Unfortunately, the reliance on digital signatures is predicated on whether or not you can depend on the assurance that the public key belongs to the individual or entity who is signing the document. Thus, forgery is possible if you cannot be assured, through verification, that the public key belongs to the signer. Therefore, it is important that there be a system of digital authentication. One such system is called public key infrastructure (PKI). PKI is a system of digital certificates, certificate authorities, and other registration groups whose function it is to authenticate each party to an Internet transaction. Various certificate authorities exist to provide this service. Clearly, the users of one certificate authority should be able to communicate securely with users from another certificate authority. A managed PKI is a means of maintaining a relationship of trust between certificated authorities. Other authentication systems exist as well. Clearly, an entity's network will be protected if the Internet transactions entering the network are authenticated with these digital signatures security measures.

D) Cryptographic security – The process of converting plaintext data into an unintelligible format if the transmission is intercepted and correspondingly converting encrypted data into intelligible form (decoding). For increased security double encryption is possible. The technique consists of using algorithms combined with the text of the message with "keys" (a large binary number) to form an encryption. Before the creation of the public/private key method, the technology that was used was called symmetric cryptography. Symmetric cryptography utilizes a secret key to encrypt a message and then makes use of the same key to decrypt the message. Although this technique is still used (e.g., in Web browsers each time a secured transaction is opened) there are several problems with the process. For example, it is very difficult to deliver the secret key to the receiver of the message so that decryption can take place even through another secure transmission channel. In addition, if the secret key is compromised, all transmissions utilizing it are also compromised. Thus, if two or more parties are sharing a secret key, there is no assurance that the key will not be disclosed to others. Also, in a large organization, thousands or millions of secret keys may be necessary resulting in a difficult, cumbersome situation. Moreover, the use of secret keys does not allow the receiver to be assured of the sender. That is, its use does not provide a way of checking on the legitimacy of the sender.

A significant improvement occurred with the invention of public key cryptography. Public key (asymmetric) cryptography uses a public key and private key that are mathematically tied. This method encrypts a message with a public key and then requires the use of a private key to decrypt it. The encrypted message may only be decrypted with a matching private key that is only known to the receiver. This methodology reduces the risk of compromising the encrypted message during its delivery. One of the major uses of public/private key technology is in on-line trading. Security in this area is clearly of great significance to both investors and brokers. The use of encryption in on-line trading helps insure that the broker's customer account network database is free from security threats from outside sources.

Another security based methodology based on the use of public key technology is Kerberos. Kerberos is an authentication system used to achieve authentication of identity over a network that is not secure. Developed by the Massachusetts Institute of Technology, Kerberos enables two parties communicating on an insecure network to exchange private data by assigning a unique key, called a ticket, to each user that logs onto the network. When information is sent to the network from a user, each message has embedded in it the encrypted ticket that identifies the sender to the server. However, Kerberos is designed for encrypted authentication. It does not ascertain whether or not the individual user has the proper authorization to perform specific activities.

A good technique for encrypting messages and protecting them on the Internet is called Pretty Good Privacy (PGP). PGP is based on the public key method and represents an easy, inexpensive (it is free) way to protect the confidentiality of messages on the Internet. In order to successfully encrypt messages using PGP, it is necessary to use a PGP encryption package. Network Associates latest encryption package is called PGP Enterprise Security 3.0. This package is part of Network Associates' network security suite which includes such security functions as multitier virus security, firewall security (see next section), encryption, digital certificate authentication, virtual private networking (see next paragraph), intrusion detection system (see later section of this chapter), etc.

A further way to achieve privacy in communication over a network is through the use of a Virtual Private Network (VPN). VPNs allow secure communications between locations on a public network such as the Internet. It is really a private network connection that requires encryption of data at the sending point and decryption of data at the receiving side. Using the Internet a user may be able to connect, using a VPN, to his or her office LAN, a vendor LAN, or customer LAN if such an integration would be deemed cost beneficial. Commonly, VPNs are built into firewalls (see next section) as a means of connecting two LANs. In each of these cases, a virtual secure "tunnel"[3] is created between the locations. However, it is important to note that a VPN is much more than just a data encryption process. It includes, for example:

- Integrated security procedures – These include: Data encryption to guarantee the security of the connection, granular access control, user authentication, privacy and integrity of data communication.
- Communication controls – To ensure reliability and qualitative performance, VPN systems have acceleration of data encryption and decryption and band/width management.
- Comprehensive security policy – VPNs are integrated into an overall comprehensive security policy that incorporates the security procedures of the entire enterprise that the VPN is a part of.

E) Emission security – Prevention of electromagnetic radiations from electronic devices which can be intercepted and read by unauthorized individuals.

Firewalls

A firewall is a security system or series of systems that insures security level screening between two networks. That is, it enforces the security of transactions flowing between a trusted network and an untrusted one such as the Internet. The security administrator of the company generally determines how the security regulations should be enforced. Firewalls perform many security functions to protect the entity's internal corporate network . For example, it

- establishes a chronological log of activities and transactions transpiring between company networks for auditing purposes creating a trail of documentation when a fraudulent or suspicious event occurs. Thus, any transaction which attempts to change or actually changes the entity's database can be subsequently reviewed for authenticity;
- notifies audit or security individuals when events that appear questionable transpire;
- is in control of all connections both into and out of the company's network;

- validates the user and the application being requested before the entity's databases are actually updated and internal data modified;
- secures network access with a strong security perimeter defense.

High Performance Firewalls

Entities that require high-performance firewalls that are relatively easy to configure generally utilize packet filter firewalls. Packet filter firewalls are configured to check each and every incoming and outgoing packet header against a table consisting of packet rules of security rules and codes. If the address and port data on the header are acceptable, the packet proceeds through the firewall directly and speedily to its destination. If, on the other hand, the data on the packet header is not determined to be acceptable, the packet is not allowed to go through the firewall and is dropped. This type of firewall is generally considered to have a low level of security in that it does not test for potential application weaknesses. That is, once the connection has been approved by the firewall, the source can go directly to its target destination, thus exposing the internal network and its databases to assault by unknown external sources.

An improvement to this level of firewall security is the creation of the application proxy firewall that tests potential connections for full application level awareness. An application proxy firewall examines everything at the highest level of the protocol stack seeing the specific details of each attempted connection. This type of firewall makes it very difficult for external sources to hurt the company's information system and exploit any weaknesses that might exist. The application proxy function will abort any new connection between an entity's system and an unknown source if there is any question regarding the validity of the connection. In this way, it protects the company's network and componental databases.

Another, more elaborate form of firewall is known as the "hybrid" firewall. Foote[4] describes this system as one that combines the best of two worlds combining the speed of the packet filter and the rigorous security features of an application proxy. During high volume activity, the hybrid firewall uses packet filtering. However, the application proxy quickly takes control when any threat to the internal network is perceived.

As more sophisticated forms of threats evolve, firewalls alone have been deemed by some to be inadequate. To satisfy this problem, the security firewall system has evolved into a more fully integrated security-based system of network protection. Such a system may have the following security elements:

(1) Firewalls that protect a system network and its databases against illegal entry from outside sources.

(2) Scanners that continuously scan the internal network for breakdowns or weaknesses in the security system.

(3) Virus and content checking programs that search out any malicious virus programs such as Trojan Horses, Worms, or other hostile codes. (These are discussed in the next section of the chapter.)

(4) Intrusion detection monitors that constantly check the company's internal network for breaches of security and signs of attack. (This topic is discussed more fully later in the chapter.)

(5) A security manager who oversees all the security components to make sure that all areas of the company's information system are being watched and protected.

The system that includes all of these security components is generally known as a distributed firewall system. Network Associates currently has a firewall system known as the Gauntlet Active Firewall which combines such protection features as vulnerability checking scanners, real time security intrusion monitoring, antivirus checking, and other security considerations into a single active system.

Malicious Programs That Can Destroy Or Damage An Entity's Network

There are various programs that have the potential to destroy or damage an entity's network and databases. This occurs most commonly when the entity's business operates from a web site on the Internet. It may even come from an intranet or an extranet connected to the Internet. Obviously, the risk of system or database infection is significantly reduced if your system is stand-alone. However, the risk is not reduced to zero because a virus may not only be downloaded from the Internet, but may also be obtained from a simple floppy disk. Information technologists should be familiar with these viruses and the kind of damage that they may inflict to a system. The following is a brief discussion of these programs:

Trojan Horse

A Trojan Horse is a harmful program that is surruptiously placed in another program. When the program is run, the Trojan Horse program then initiates and executes some predetermined unexpected task. In general, Trojan

Horse programs do not replicate themselves as other viruses do (to be discussed momentarily). However, they have the ability to steal passwords, delete or destroy data, and cause other serious problems.

Worm

A Worm is a program that may operate in a networked computer system or non-networked system. In a networked system, the worm program has the capability of copying itself from its original source across network connections infecting the system's hard disks and databases. In a non-networked system, it can copy itself to different locations on a hard disk or database that may exist.

Bomb

A Bomb is a malevolent program that is schedule based. That is, it generally uses the system's clock to activate at a specific time when some event occurs. For example, a bomb could be programmed to erase every tenth data file on a data base on Lincoln's birthday. It could also cause the erasure of important system programs (producing a crash) on New Year's Eve. Another might be activated after a particular program is executed for some arbitrary nth time effecting malicious damage to the system's network.

Dropper

A Dropper is a program that is designed to specifically not be detected by anti-virus programming software. Droppers are generally sophisticatedly encrypted to avoid detection by anti-virus programs. The primary function of a Dropper is to transport and install viruses (to be discussed next). They operate comparably to a Bomb in that they are generally designed to execute when a specific event occurs. At that time, the virus is then launched to infect the host system.

Salami Techniques

A dishonest individual modifies an entity's computer program to affect very minute changes that will in all liklihood not be discovered. However, the cumulative effect of this modification, if disclosed, would undoubtedly matter and be noticed. For example, in calculating customers' interest in a bank, a program may be directed to compute interest each quarter and truncate the calculation at the second decimal place. The program would then take the remaining fraction of cent and credit this amount to the perpetrator's account. Although each of the credits would be minuscule, for millions of customers, over a relatively long period of time, the cumulative amount would indeed be material.

Back Door or Trap Door

In preparing a network computer program, programers may insert a code to allow them to bypass standard security procedures to enable them to more easily complete and test the program. When the programming process is complete, the back door may have been left, either intentionally or accidentally in the program. Perpetrators who know about this trap door may use this coding to bypass security and commit illegal activity.

Masquerade

A computer program is created that simulates a real program for the express purpose of capturing a valid user's identification and password. When the user attempts to initially log in, the masquerade program runs and asks the user to input his or her identification and password and then displays an error message after the information is successfully input. After the user puts in his ID information a second time the actual program is run allowing the user to successfully log-in to the network. The user never realizes that his access identification has been compromised and captured by the first masquerade impostor log-in.

Scavenging

When a user deletes sensitive information or data from files on the network, that information is not destroyed initially. A scavenger program is able to steal and accumulate the information that the user thought had been deleted but was actually collected and saved.

Data Input Alteration

One of the simplest and most effective ways of modifying data intentionally or committing fraud is to alter the information during (or even before) the input process. Oftentime, data is simply input erroneously unintentionally. Although this does not represent a fraudulent situation, it could still cause many problems and difficulties for a company. Although audit software is available to review transactions for reasonability which provides evidence that the data may have been modified, it is best to try to catch such errors in the bud: that is, at the point of input. A series of computer routines exists to check to see if data input into a network may be incorrect or illogical. These tests are known as validity tests. Some of these include:

- **Check digit control**

  This control performs an arithmetic computation on an employee number or account number and then compares the result to a check digit which is actually part of the identification number. This calculation provides some degree of assurance that the identification that was input was properly done.

- **Missing field test**

  Data fields are checked to insure that they are properly filled out and there is no missing data.

- **Data field test**

  Program tests the characters that were input into an information field to see that they are of the right nature. For example, it might check a name field to make sure that it contained no numerical characters. A social security number field might be checked to make sure that it contained no numerical information.

- **Hash total**

  This is a predetermined nonsensical total derived from a source other than the individual who input information (e.g., the sum of social security digits of a transaction run of individuals who are being identified by their social security numbers.) It is derived for a short period of time such as a day's processing run to insure that all that was supposed to be input was done. If the hash total does not agree with the amount that the system calculates as a result of the input, there is reason to believe that an error was made in the input process.

Virus

A virus is a computer program that attaches itself to an entity's software and then attaches itself to other programs. Viruses can secretly pass many system security defenses and thus are quite dangerous. When the program that is infected is run, the virus then executes some effect that can range from simply displaying a message on a given predetermined date (a relatively harmless outcome) to slowing the program's performance to much more serious problems such as erasing files on a database, destroying data, or outright crashing a system. Viruses can be hidden and then launched in several different ways. For example, they can be concealed through a dropper and then operate as a worm in replicating itself. They can also operate as a

Trojan Horse by either attaching themselves to an existing program in the system, hiding in the program until the program is executed or at some other specific time when the virus will be launched to commit some malicious damage to the system.

It is important to note that new viruses are constantly being designed with unpredictable outcomes. However, some of the different types of viruses currently in use that could damage an entity's network include:

Boot Sector Virus

This virus is resident on a hard disk that is read and executed by the computer at the time of boot. It can infect either the DOS boot sector or the master boot record that is being loaded into memory during the boot of the system.

File Infector

This type of virus attaches itself to executable files and then waits in memory for another program to be run. If the database program is run it will then infect the database files.

Macro Virus

This type of virus affects programs that contain other programs that are built in, such as a database program with several applications. When a document or template containing the virus is opened, the virus then executes and does its damage. It copies itself into other documents and continually spreads.

Multipartite Virus

This type of virus combines file infections with boot sector viruses.

Stealth Virus

This type of virus, as its name indicates, has the ability to fool anti-virus software into believing that the system is not being infected. It obtains information from files that have been infected and provides anti-virus programs with information that indicates that everything is alright. For example, the stealth virus will intercept an anti-virus program searching for altered files by giving the anti-virus program the old, unaltered information regarding files that have in fact been infected.

Polymorphic Virus

This type of virus has the ability to actually change itself when it reproduces so as not to be discovered by anti-virus software. Anti-virus software looks for specific patterns in attempting to recognize the existence of a virus. By altering themselves during replication, they are able to avoid recognition.

The antidote to protecting your computer information system, databases, and network is to use current anti-virus software. Norton Anti-virus software, Mc Afee Software, Network Associates Software (Total Virus Defense) are all very helpful in protecting a network database system against the thousands of viruses that currently exist as well as the hundreds of ones that are being created each month. Clearly, an entity will not be able to prevent against all losses of data. However, by planning for the contingency that some damage may occur in a given period, the entity will not be brought down by the event. Frequently backing up the company's database together with archive logging (although not very easy or efficient) will keep a loss from becoming a catastrophe[5]. More will be said about these activities (system back up and archive logging) later in the chapter.

Server-Side Logic Controls[6] Related to Internet Security

Besides protecting an entity's network by restricting access to the computer plant and equipment and by identity authentication, internal access can be further controlled through what is sometimes termed server-side logic. For example, an individual who is a truly bona-fide network user, having a valid network ID and password, should not have unrestricted run of the network. Network administrators must determine the range of access each and every user should have. Merkow and Breithaupt[7] suggest the following logical access controls:

- Restrict Access to Server Files
- Spell Out the Privileges *of Authorized Users*
- Turn Off the Automatic Directory Listing
- Control User Access by Using Access Control Lists
- Use a Domain Name Service and IP Address or Host Name to Determine What Access aUser can have
- Authenticate All Users
- Take Steps to Secure Intruders Use of Programming and Scripting Features of Servers

In general, if the web servers in use are not properly secure, then hackers will ignore the encryption protecting the data in transmission and attempt to get to the data by exploiting the improperly configured web servers. Network servers in the form of e-mail servers, remote login servers, FTP servers, etc. must be protected from attacks. Although systems are available to insure that transmissions occur securely, the host server must constantly be reevaluated through configuration to insure that security is maintained. Each of the aforementioned logic access controls noted above will now be discussed and analyzed.

Restrict Access to Server Files

Typically, Internet servers contain critical information that must not be compromised by unauthorized individuals. The two classes of files that are resident on Internet servers are termed server root files and document root files. The server root directory contains files that are included in the server's installation as well as other system files (generally accessible by the system administrator). The document root files contain the actual Web pages seen by the user of the site. A hacker could cause havoc to the user's site by modifying these files. It is imperative that these critical server files be protected by the use of file access permission. In addition, server log files should be protected to insure that the server is able to determine: who visited the site, the user's source Internet Protocol (IP) address, and the number of requests made. This data should be recorded.

Spell Out the Privileges of Authorized Users

The system administrator should set forth the rights of the users to access or modify data files on the network servers. The system administrator must distinguish between those who have access to a limited number of data files and those whose authorization is broader based. As part of the initial installation and configuration of a web server, a system administer establishes subordinate processes (called child processes) which can satisfy unrestrictedly various program requests including bringing up the Web pages themselves. A cracker would have little difficulty using these processes to gain control of the system, stealing confidential information, stealing passwords, creating bogus accounts, destroying important customer information, to note just a few. To combat this situation, it is imperative that all servers be configured to allow only those with the authority of system administrators (highest access level possible) to have access to these subordinate processes. If this access is not so restricted, then it is possible that a cracker may be able to execute these effects with only limited authorization.

Turn Off the Automatic Directory Listing

If the automatic directory listing option is turned on when a server boots up on some web server software and an unwanted user happens to be browsing the directory, a listing of all files will be disclosed. If interface files (e.g. CGI scripts) happen to be included in the directory, source code could be copied and scrutinized and means of manipulating the system may take place. The Automatic Directory Listing is generally turned on (default mode) when the server is installed and booted. It is important that the Automatic Directory Listing is turned off to minimize the chance of this exploitation from happening and avoid the aforementioned manipulation from taking place.

Control User Access by Using Access Control Lists

Because of the importance of protecting company files, confidential information and documents, it is important that the system administrator set up a structured plan spelling out which user group has access to what network information. For example, in a UNIX system, the group that a user is part of will define the access that he or she has on the network and the information that can be accessed. Access control lists allow the system administrator to perform this allocation. The list contains the identification, passwords, and specific operations that users are allowed to perform. In addition, web page access (via host name or IP address) or specific inaccessibility is also spelled out. As a matter of general policy, some security specialists suggest that an entity should initially deny access to all then allow access to specifically determined individuals. A new security policy model that is being utilized in new software on the market provides that access to network information and files be divided into secure zones based on the way the network is structured. This allows the resources of the entity to be compartmentalized and protected through the use of specific defined access rules. Partitioning the network in this manner provides better protection than that provided by firewalls alone which protect a network from only a single access point. The use of secure zones together with firewalls, routers, and intrusion detection systems insures that the network is protected from external and internal attacks and protects the entity by limiting access of all individuals who do not have the proper authorization.

Use a Domain Name Service and IP Address or Host Name to Determine What Access a User Can Have

Packet filtering firewalls looks at each of the packets that enter and then examines the basic information that is in the packet. For example, it examines source and destination addresses, protocol, etc. to insure that the information provided satisfies the criteria for admission. To pass through such packet-filtering routers, crackers may impersonate the IP addresses of trusted machines in the case of firewalls that make admission decisions on the basis of the requester's IP address or host name. This intrusion occurrs because the server restricted access to services, documents, or files based solely on the IP address or host name. A more secure alternative, however, is to use a Domain Name Service (DNS) to specifically denote what access is available to the user. The DNS associates IP addresses with machine names to determine the availability of a particular host. Access is blocked if the IP addresses and machine name is not listed and agree with what is indicated in the DNS table predetermined by the system administrator. However, DNS servers are not without their problems. DNS servers operate through recursion. That is, they will resolve DNS names recursively. Thus, when a DNS server obtains a positive outcome to a client request, it in turn becomes a client to the next server. That is, the sequence that follows is predictable. It is possible for a cracker to send a request to the DNS server obtaining a fabricated response which will then be believed and processed by the next server. It is possible, in this manner for DNS servers to be corrupted and be taken advantage of.

Authenticate All Users

A user password file associates a user to a password and thereby grants general limited access to users based on the password database in use. If a cracker gains control of the password file, it may not be difficult to ascertain and identify a user's password. The defense in such a case is to utilize a shadow password file. Password shadowing enhances password security by creating a file from the encrypted password file. In the password file, all the encrypted passwords are replaced by asterisks and the actual encrypted strings are then moved into the shadow password file. Although the password file is word readable the shadow password is not. *UNIX Authentication Tools* includes a shadow program that enables any system to use shadow password files. It includes aging passwords, restrictions on which port root logs in, logs of failed attempts, and other features. It allows up to 16 character passwords. Other shadow password packages include:

Shadow in a Box by Michael Quan. This is a compilation of utilities for managing a company's password.

Password+ by Matt Bishop. This program offers extensive logging of successful and unsuccessful logins.

Anlpasswd by Argonne National Laboratory. This program features rules relating to passwords such as number and spaces, uppercase, lowercase, all numbers, and leading capital letters.

Merkow and Breithaupt also indicate that the system administrator must decide whether to store ACLs centrally or allocate them across the areas that they are supposed to protect. For example, in the former case, the ACL is resident in a single file that is referenced any time a request is made for one of the listed files. Centralizing this list makes access easy and also makes updating simple (by only going to one location) as well. The alternative is to decentralize each of the ACLs by placing it in each of the specific function areas that require security. This enables access to be controlled at the local level while precluding these individuals from gaining access to the network controlling files. Of course, localizing ACLs is more expensive and although more control oriented, they do introduce an increased risk of error in that more individuals are involved in maintaining these files. The system administrator must consider all of these factors and determine which policy is best for his or her entity.

Take Steps to Secure Intruders' Use of Programming and Scripting Features of Servers

Common Gateway Interface (CGI) and application program interfaces (APIs) built into the Web servers software of today provide a truly interactive environment and make the Internet a viable, successful marketplace. However, CGI programs are extremely insecure. Exploitation and security weakness consist of using hidden variables specifying any filename on the system and revealing more about the system than should be revealed. Crackers may be able to manipulate files, execute malicious programs, destroy files, and exploit Internet Web pages through manipulation security lapses in the CGI scripts. Merkow and Breithaupt note some simple control steps that can be taken to protect your system against crackers abusing the programming and scripting characteristics of Web servers. These include:

- Eliminate the sample CGI scripts that are supplied during the software installation from the cgi-bin directory. Many older script versions can be exploited by crackers.

- Store all CGI scripts in a central directory which should not be called the cgi-bin and make sure that access to this directory is highly restricted. In fact, only system administrators should be given access to write and delete these files.
- Protect your program source code by converting your CGI scripts to C or C++ (compiled languages) to prevent hackers from gaining access.
- Review the scripts used in the development of CGI scripts for quality control purposes to determine errors.

With respect to the aforementioned software weakness that make an entity's network susceptible to attacks, Ghosh[8] recommends two important protective policies that go directly to the heart of combating software vulnerabilities. These are:

- Carefully review server programs to assure that there are designs that reduce the possibility of security breaches.
- Programs should be carefully reviewed as part of the quality assurance program to seek out design flaws and other problems that could result in security problems down the line.

Monitoring Security Breach Attempts on Your Network[9]

Although the exhaustiveness of the network security considerations discussed thus far might imply that an entity's network is totally secure from security breaches and attacks, in fact, these procedures really constitute a degree of security rather than absolute security. As our dependency on network systems increases, it is imperative that we increase our vigilance of them. An intrusion detection system (IDS) attempts to detect intrusions into an entity's network. Firewalls generally notify systems administrators of attempts made to gain entry into an entity's network by posting this information to log files. These files let one know of intrusion attempts but do not convey information about the nature of activity on the network. The firewall acts as a form of electronic cage around the network. It does not tell the firewall administrator that the network is being broken into. It simply restricts access to the network. That is, it is not proactive. An IDS, on the other hand, is proactive in that it is able to recognize and report attack against the network or computer host. The IDS is able to act as a monitoring system for suspicious activity and provides an extra layer of protection against attack. In addition, firewalls protect a network against external attack only. Recent studies indicate that 4 out of 5 losses due to crackers are a result of internal attacks. If estimates of the benefits of such a system exceed its costs and resulting reduced network speed, then an IDS system should be implemented. Intrusion detection systems can be viewed as really a combination of three types of monitoring scenarios. These include the following.

Network Intrusion Detection Systems

This system looks for packets on the network wire and checks to see if an attempt to break into the system is taking place. A network intrusion detection system monitors many of the computers on the network.

Log File Monitors

This system looks for recurring patterns on the entity's log files indicating that a hacker or cracker is trying to get into the network. Typically, the log file's monitors will look for intruders (e.g., hackers or crackers) who try well known security holes in the network.

System Integrity Verifiers (SIV)

These programs check the network's systems files to ascertain whether or not a cracker may have changed anything and whether a network user may have acquired access privileges that he or she should not have. For example, "Tripwire" for Servers is a SIV that monitors file changes (regardless of whether they originated inside or outside of the entity), verifies data integrity, and notifies the system of any violations of data on the network servers. In addition, "Tripwire" identifies changes to system attributes including file size, access flags, write time, etc. Although SIVs detect the content changes previously described, they do not alert the network on a real time basis of a break in.

Honeypots and Deception Systems

These are systems that are designed to intentionally allow hackers and crackers to successfully gain access to a network for the purpose of tracking them down and preventing further break-ins. The culprit(s) in this scenario are actually entrapped. Honeypots allow for extensive logging of the intruder's activities so that these individuals may be tracked down and eventually stopped.

How Does IDS Work?

There are two primary models upon which IDS programs have been developed. These are (1) Anomaly Deviation and (2) Pattern Recognition. A discussion of these follow:

Anomaly Deviation (Aberration Testing)

This concept requires that baseline data be established for the purpose of measuring and detecting deviations which could signal an intrusion. This approach may be viewed as auditing by exception. For example, data may be collected in the form of file activity, disk activity, user logins, CPU utilization, workstation activity, etc. After collecting this baseline information regarding normal usage you may find over a period, for example, after the close of business, that there appears to be an unexplained increase in usage after the close of each business day. This aberration clearly requires further investigation. This process is difficult in that, over time, normal patterns must be established for the purpose of making measured comparisons. Audit data must be compiled, compared, and even then investigations may not show any definitive patterns.

Pattern Recognition (Testing for Patterned Attacks)

This technique is predicated on looking for established, well known patterns of intrusion. The manner of detection consists of looking for certain sets of activities that take place. That is, the program is told to examine packets for specific strings of data. Once the cracker's technique is established it is coded for identification so that it may be watched and searched for.

Attributes of a Quality Intrusion Detection System[10]

The characteristics of a quality intrusion detection system are listed below:

- The IDS system should continually run in the background of the system that is being protected and not require human supervision.
- It should impose minimal overhead (e.g. reduction of network speed) to the network.
- It must be compatible to the usage patterns of the network or host it is attempting to protect and should easily adapt to these patterns.

- It should be difficult to undermine.
- The IDS system should be able to adapt to new applications that might be added to the network. Although components of the network might change over time, the IDS should be able to accommodate these modifications.
- It should be able to detect deviations from normal usage and behavior.
- The IDS should be subversion resistant. In addition, it should be able to detect if it has been subverted.
- The IDS should be crashproof. That is the system should be able to survive a system crash without having to rebuild the entire IDS database after each crash.

Where should an IDS system reside on a Network[11]

Some of the places where an IDS system may reside include:

Network Hosts

IDS software may be placed on hosts in the network. In this mode, the software is able to protect the integrity of files by determining, for example, if important files have been modified or whether a particular user has attempted to gain access to files for which he or she is not authorized.

Network Perimeter

Network IDS perimeter software packages monitor the activities of the entire network. The software may be placed outside or inside the firewall or on both sides of it. This software package continually scans the network for suspicious packets and may also reside on hosts that continually feed information to the scanning monitor (regarding patterns of known attacks or other suspicious activities.) It also generates reports on the status of the network and related configuration.

Servers

IDS software packages may also be placed on important servers which do not have very high volume. IDS packages generally cannot keep up with very high volume traffic. For very important servers, sensitive IDS software may be installed that monitor the server's link. In general, application servers which have lower volume usage than a file server are better suited for IDS software.

What Procedures Should be Followed When an Intruder has Successfully Broken into Your Network.

Prior to the identification of a successful intrusion (or the threat of such an attempt), the following measures should be taken:

Set up an incident reaction unit – An interdepartmental group of individuals should be identified and should include representatives from the following groups:

- *Upper Management* – Someone from upper management should have the authority, for example, to shut down processing altogether if the intrusion is serious and has the potential of causing major damage to the network, databases, files, etc. The upper management representative should only be involved if the intrusion is considered serious enough to require a significant reaction.
- *Management Information Technologists* – If an intrusion has or is taking place, security must contact a management information technologist who has direct knowledge and responsibility in the area that has suffered the transgression so that the problem may be directly dealt with and be resolved as soon as possible.
- *Personnel Department Representative* – Since a majority of attacks are from individuals inside the entity (e.g., an employee gaining access to databases and files in excess of the level for which he has authority or illegally making changes to records) a personnel department representative may be helpful in tracking down the perpetrators.
- *External Members* – Based on the severity of the intrusion, it is a good idea to have the names of contact people in the local police department or FBI available so they may be called in if necessary.

Predetermine Your Plan of Response Before the Intrusion

An entity should predetermine its response to network intrusion so that its reaction to a break-in is swift and decisive. For example, when it is apparent that a serious intrusion is taking place, the system may be shut down to minimize damage and loss. On the other hand, the security team in conjunction with the members of technical members of the management information staff after weighing the degree of severity of the intrusion may decide to allow the intrusion to continue so that more evidence may be gathered against the intruder.

Ensure That There is Adequate Logging/Monitoring Capabilities on Your Network

In order to help ascertain the identity of the intruder and determine what happened, make sure that your network has adequate logging procedures so that all his or her actions are recorded.

In Case of an Intrusion Predetermine Who You Want to Notify

Predetermine the guidelines for notifying individuals in the event of an intrusion. Should information about the intrusion flow vertically up to the top of the organization or should it flow horizontally to other units of the entity. Do you also notify vendors, important customers, or others who are involved and connected to the network? Should the police, FBI, or other law enforcement agencies be notified? How should the press be handled?

Make Sure that all Individuals Know Their Responsibilities in the Event of a Break In

Management should insure that the members of the Incident Reaction Unit are cognizant of their responsibilities and know what has to be done in the event of an intrusion. Members should be prepared to execute predetermined response procedures in the event of an attack and should measure their reaction in accordance with the severity of the intrusion. If necessary, drills should be performed to insure that everyone knows what to do.

Using the Internet to Back Up Data[12]

Historically there has always been difficulty with utilizing tape or disk backup to insure the safety of data backup. Mechanically, the data must be removed from the premises to insure that safety is insured in the case of significant or catastrophic loss to the premises. In addition, even if transporting the information is properly and successfully done, excessive moisture and heat in the storage area can still adversely effect the data. An alternative that has recently been offered by data protection service providers is to utilize the Internet to transfer data to a remote data safe. It is now possible to have an entity's data securely and quickly stored or retrieved regardless of their users' location. Utilizing these secured data centers gives entities a viable alternative to current more mechanical ways of data backup.

What should a company look for in choosing a backup provider?

The following considerations should be evaluated when looking for an Internet data backup provider:

- Make sure that the Internet provider's system provides automatic PC data backup.

  To maximize data protection, it is imperative that the Internet provider can insure that automated backups can be originated from any user location and does not require user interaction. Thus, the system should be able to provide demand backup from any location. This consideration is particularly important for mobile users.

- Does the Internet Backup Provider Utilize State-of-the-Art Technology to Reduce Backup Time?

  It is imperative that a company's Internet backup provider employ extensive cutting-edge data reduction technologies to insure that even under the most constrained circumstances backup takes place as efficiently and effectively as possible. For example, one should not attempt to back up a multi-gigabyte harddrive through the use of a slow-speed dial-up connection.

- Make sure that the server the vendor is using for the backup service is reliable.

  Question the vendor to insure that the server that is being used to backup clients is highly reliable as well being capable of processing the maximum number of users simultaneously.

- Can the Vendor Guarantee that the Internet Backup Provider Insure that a Lost or Stolen PC or Hard-Drive Crash Can be Replaced with Identical Data?

  Is the backup service capable of taking a PC (that is identical or nearly identical to the one that was lost or stolen or has experienced a hard-drive crash) and replace all the required files, applications, and settings that were lost so that productivity may be maintained and downtime reduced.

- Will the use of automatic PC data backup reduce the company's Help Desk workload?

  An automatic data backup should significantly reduce any downtime due to reinstallation of data and or programs. This was probably done by the help-desk previously. Thus, by utilizing a backup service, the help desk should be freed up to handle other contingencies and matters of assistance.

- Other general considerations:

  Recently, Data Protection Services (DPS) of New Orleans, Louisiana identified the following features of a comprehensive Internet backup service:

  1. Daily backups.
  2. If the vendor fails to successfully generate a successful backup, the client is contacted. This is known as monitored service.
  3. Virtual backups are performed rather than incremental ones.
  4. Typically, 20 backups are generated each full month.
  5. Triple redundant backups are performed to insure that data is not lost.
  6. Each quarter, entire backup CDs are created and archived for protection.
  7. If needed and requested by the client, backup CDs may be burned and sent to a client at any point in time.
  8. If setup and configuration is important to a client, the Internet backup service should be able to perform a bare metal restore. This consists of being able to restore the following: registry files, address books, program files, and configuration files. This is the kind of restoration that is required when a hard-drive crashes. A bare metal restoration makes sure that your computer's files and desktop look the same after the restoration when compared to before. A quality Internet backup service will insure that no programs lost in the crash will have to be reinstalled and the configuration that is restored is exactly the same as the one that crashed.
  9. The vendor's technical support system should be geared to the client company's needs. That is, there should be readily available access to network engineers and other high-level technical support as needed.

10. The Internet backup service that is chosen should be able to backup all local drives and network drives from one installation.
11. The service should back up more than just client files. It should be viewed as a complete disaster recovery mechanism.

Conclusion

As was noted in the beginning of this chapter, one can only strive for but never really achieve a completely secured network. The process is always ongoing and requires changes and adjustments as the technology of the external environment changes. This chapter has presented some of the practices, procedures and concepts that need to be followed in the process of striving for network and Internet security.

1. Garfinkel, Simon, *Database Nation: The Death of Privacy in the 21st Century*, January 2000 (O'Reilly and Associates Publishing Co) pp.55-59. Part of this section is based on Garfinkel's work.
2. Ibid., pp. 55-59.
3. Norton, Peter, *Network Security Fundamentals*, (SAMS) 2000 p.16.
4. Foote, Steven, "19Inforsecurity Prediction for '99'," www.inforsecuritymag.com/novcover.htm. 1998.
5. Theriault, M., and Heney, W. *Oracle Security*, October 1998 (O'Reilly and Associates Publishing Co.) P.380.
6. Merkow, Mark S., Breithaupt, James, *Internet – The Complete Guide to Security*, 2000, (AMACOM) pp153-169.
7. Ibid.
8. Ibid., pp. 162. (Quoted in Merkow and Breithaupt).
9. This material is based on information found in "FAQ: Network Intrusion Detection System," Http://wwwticm.com/kb/faq/idsfaq.html. Pp. 3,4,12.
10. Op. Cit. Merkow and Breithhaupt, Pp.131.
11. Op. Cit. "FAQ: Network Intrusion System," Pp. 14, 18-19.
12. Salerno, Cheryl, "Spotlight on:Internet Backup Services," *Contingency Planning and Management*, May-June 2000, Pp. 40-41.

11
Wireless Networks[1]

The feverish pace of technological advancement seems to have reached a temporary area of stability. As companies have cut spending over the last few years, less money has been budgeted to finance expenditures related to systems upgrades and new technology. This has created a situation in which many businesses have been content using older equipment. Yet, productivity remains high, as programmers, software companies, and technology people iron out virtual creases in operating systems, programs, and networks. It seems that a sort of equilibrium with technology has been achieved. However, one cannot overlook new technological developments. Managers, business owners, technology officers, etc. must keep a watchful eye out for new technology that will further boost their business's productivity and efficiency. As the economy picks up over the next few years, businesses will have more money to spend on upgrades. Furthermore, as time goes on, hardware and software will inevitably become out of date and will have to be replaced. Therefore, now is a golden opportunity for many businesses, large and small, to get ahead. By upgrading now, a business can give itself the edge against its competition. An attractive area for companies to invest in is the wireless network. The purpose of this chapter is to present how wireless networks may be applied in practice by businesses to improve their overall applications, become more efficient, and improve corporate operations and profitability.

Wireless Technology: Bluetooth and Wi-Fi

One of the most significant technological breakthroughs over the past few years has come in the form of wireless networks. In a wireless network, an adapter is installed in each computer, so there is no need to drill holes and run cable. Wireless networks are easy-to-use, and can integrate a wide range of devices. However, the further the wirelessly networked computer is from a wireless access point, and the more solid objects standing in the way, the slower will be the connection. TIP: To maximize network speed and range, place the wireless access point at least a few feet above the floor and *not* near metal objects. It is probably best to position the access point in the middle of the coverage area. However, another line of reasoning is to position the access point where it will be as close to as many computers as possible. In the case of larger areas, or those with numerous obstructions, one should have multiple access points. CAUTION: Make sure the access points have identical settings. Almost all wireless network adapters support "roaming." In areas having access point coverage overlapping, the adapter will go to the strongest signal.

Performance, reliability, and range in wireless networking depend on the quality of the signal. RECOMMENDATION: Add an *extra* antenna to your network adaptor and access point to significantly improve signal quality.

> TIP: For security purposes, mix numbers and letters in your wireless network's Service Set Identifier, which every wireless network uses to log in.

There are three major wireless technologies in existence. The first is Bluetooth, which has the shortest range. Secondly, there is Wireless Fidelity (Wi-Fi), also known as 802.11b, whose range is longer than that of Bluetooth. However, note that Wireless-G operates at speeds almost 5 times faster than 802.11b. Wireless-G is fully compatible with 802.11b products. Lastly there is cellular, which has the longest range. This chapter discusses all of these technologies.

Two major special interest groups have been formed to drive the development and advancement, while keeping a global standard, for Bluetooth and Wi-Fi. The first is the Bluetooth SIG, which is an industry group created to integrate and ensure compatibility of devices supporting Bluetooth technology. The Bluetooth SIG has over 2000 members worldwide. In a nutshell, Bluetooth is wireless integration and synchronization of devices. With Bluetooth, a cell phone, electronic notebook, laptop, and desktop may be synchronized with one another. From then on, these devices are able to communicate freely and regularly with each other while within a certain proximity. Bluetooth can also be used to create Wireless Personal Area Networks (WPAN). A Wireless Personal Area Network is a system of connected PCs and other devices within the same proximity. The purpose of a WPAN is to create an environment in which these devices can share resources such as Internet connections, printers, files, and disk drives. Traditionally, these types of networks are built using wires and cables (known as Local Area Networks or LAN). Using Bluetooth, a WPAN can be built without using a single wire or cable. For a series of devices to be synchronized and/or networked, each must support Bluetooth technology. This means that the product must carry a label or Bluetooth logo on it. The major drawback to a Bluetooth-built WPAN is the limited range of the network's broadcasted signal. For this reason, Bluetooth is most attractive for single and small office wireless networks, and for the integration and synchronization of a variety of devices.

The other industry group is the Wi-Fi Alliance. The Wi-Fi Alliance is an international nonprofit association. It was formed in 1999, and its purpose is to certify interoperability of Wireless Local Area Network (WLAN) products based on IEEE 802.11 specification (IEEE stands for the Institute of Electrical and Electronic Engineers). This institute develops standards and ratings for hundreds of electrical and electronic devices. Wireless Local Area Networks (WLAN) use Wi-Fi to connect devices. The Wi-Fi Alliance currently has 202 member companies from around the world. The association gives Wi-Fi certification in order to ensure interoperability and enhance users' experiences. Wi-Fi, or Wireless Fidelity, allows you to connect to the Internet without wires. Wi-Fi is a wireless technology similar to cell phones. It enables a computer to send and receive data anywhere within range of a base station. Furthermore, Wi-Fi is fast. It is several times faster than most cable modem connections, and has a longer range than Bluetooth. You may connect to the Internet from anywhere with Wi-Fi access if your computer is configured with a Wi-Fi Certified radio. Wi-Fi certification is administered by the Wi-Fi Alliance. When a device is Wi-Fi certified, it means that a user will be able to connect from anywhere there are other Wi-Fi Certified products. The advantage to Wi-Fi is that it is being installed in many public places such as airports and hotels. Therefore, if a company installs a Wi-Fi WLAN in its office, employees may use their Wi-Fi certified equipment anywhere outside of the office where there is a Wi-Fi base station. Due to its longer range, it is ideal for large and multi-office networks.

Cellular technologies are known as Wireless Wide Area Networks (WWAN). This is the technology that we see in today's cell phones. WWAN enables a cell phone user to send and receive email over long distances. It also enables people to access the Internet from their cell phones. Any cellular phone with Internet access makes use of this technology.

Bluetooth

Bluetooth is a radio frequency specification for short-range data transfer. Any device containing Bluetooth technology, whether it be a handheld PC, cell phone, laptop, or standard PC, receives the signal broadcasted by the network. As aforementioned, the Bluetooth SIG was formed to develop and maintain a global standard for Bluetooth wireless technology. This facilitates interoperability, advancement, and development of the technology. Over 2,000 companies from around the globe are members of this industry group. The Bluetooth SIG consists of leaders in telecommunications and computing including such household names as 3Com, Agere, Ericsson, IBM, Intel, Microsoft, Motorola, Nokia, and Toshiba. The Bluetooth specified standard contains the information necessary to ensure that all devices supporting Bluetooth are able to communicate with each other globally. This means that it will not matter who manufactures a device. As long as it has the Bluetooth logo or label on it, it can be fully synchronized with any other Bluetooth device. This is a huge step toward full integration. Instead of using many independently operating devices to accomplish one task, businesses may utilize these devices as one tool by integrating their performance. Using Bluetooth can also give a business many more options in terms of purchasing new hardware and upgrades in the future. Because all Bluetooth devices are fully interoperable, businesses are not forced to go back to the same manufacturer. They can look for the best prices, as long as the alternative device supports Bluetooth.

The current Bluetooth specified standard calls for the support of several elements. First off, there must be general access among devices. This means that they must be able to link, synchronize, and communicate with each other. Next, cordless telephony must be supported. This means that cell phones with Bluetooth must be able to operate as cordless phones when they are in proximity to their base station. For example, the cell phone would act as a cordless phone and a laptop or desktop the base station. The serial ports on Bluetooth supportive devices will act similarly to wire serial ports. Each device will also be able to receive and transmit voice data as well as send faxes. Dial-up networking between a cell phone and a laptop computer is also part of the Bluetooth standard. As already mentioned, Bluetooth devices will have full WPAN access, with the ability to transfer files.

Bluetooth Piconets, Scatternets, & Security

Bluetooth provides a 10-meter personal bubble. It supports the simultaneous transmissions of information and voice data. A network of Bluetooth devices is called a piconet. Each piconet can support a maximum of eight devices and a minimum of two. Any and all devices containing Bluetooth can be potentially networked with each other. Each piconet has a master unit and slave units. The master unit synchronizes all of the slave units. The slave units are all of the other networked devices, besides the master unit. Piconets may be integrated to form a scatternet by setting up a master device to synchronize several piconet master devices. Therefore, the master device of a piconet can also be a slave in a scatternet. The gross data transfer of Bluetooth devices is 1 megabyte per second, while the actual data rate is 432 kilobytes per second. Bluetooth technology is as secure as a wire with up to 128-bit public/private key authentication. Furthermore, it supports good encryption. As mentioned earlier, Bluetooth transmits its signal for up to 10 meters. That is only when a 0dBm radio is used. Using a +20 dBm radio, the link range can be increased to up to 100 meters.

Interoperability

The most attractive feature of Bluetooth is that all products containing this technology will work together. All manufacturers implementing Bluetooth into their products must get them tested and certified, so as to insure interoperability. Accordingly, this feature is perhaps the most attractive for a business. All of your hardware will be integrated and able to share information. You will be able to synchronize your mobile computer with your desktop simply by placing the mobile computer near the desktop. This will save much of the time wasted plugging and unplugging. This technology is also useful outside of the office. You can leave your cell phone in your pocket and using your laptop, dial up to the Internet, as opposed to lining up the phone's infrared port with that of your mobile computer. The best part is, this synchronization will be unconscious and automatic when the devices are within a certain range of each other.

University of Kansas Begins a Study on Bluetooth

The University of Kansas Information and Telecommunications Technology Center (ITTC) and the Bluetooth Special Interest Group (SIG) announced on December 17, 2002 that they would collaborate on the testing of devices that communicate at short ranges. It is possible that devices containing Bluetooth technology could serve as universal remotes. These remotes could be used for anything from opening garage doors to printing computer documents. Director of the ITTC, Victor Frost, feels that the "postage stamp-sized Bluetooth chip can be inserted into a wide range of electronic devices."

Practical Usages of Bluetooth

Currently, Bluetooth's most practical usages are its ability to wirelessly synchronize devices, and serve as small networks. Its range limits its ability to serve the needs of a large company with multi-office networks. However, for small business or sole practitioners, which require smaller networks, Bluetooth can adequately serve this purpose. The main advantage of Bluetooth, though, is the ability a businessperson has when making use of this technology. A person can coordinate each piece of his or her equipment to create a more powerful and efficient business tool. For more information about Bluetooth technology, please see http://www.bluetooth.com. To learn more about the Bluetooth SIG and its members, please go to http://www.bluetooth.com/sig/about.asp.

Wireless Fidelity (Wi-Fi)

Wireless Fidelity (Wi-Fi) transfers data at 11 megabytes per second. Many large corporations use Wi-Fi wireless devices to extend standard wired networks to areas such as training classrooms and large public spaces. In addition, many companies make use of Wi-Fi to provide wireless networks to their workers. These wireless networks may also be accessed remotely from workers' homes or other offices. Wi-Fi is also used to bridge the information flow between two or more offices in different buildings.

Wi-Fi networks are also found in public places such as hotels and airports. When in an area that supports Wi-Fi, a Wi-Fi certified product is automatically linked to the network when the device is turned on. Wi-Fi is ideal for mid- and large-sized companies. It enables businesses to boost productivity and efficiency through constant real-time information streams, full sharing of data, and constant, uninterrupted communications. People are able to continue to work on projects or prepare for meetings in the airport while they are waiting to board their flight. With wireless technologies such as Wi-Fi, companies can keep their employees, whether they are in or out of the office, up-to-date at all times. Costly omissions of information will be virtually unheard of. As time progresses, Wi-Fi networks will be just about everywhere. Members of the Wi-Fi Alliance plan to bring Wi-Fi to urban areas, central cities, and even some major highways. Now, in reading this, one might be reluctant to believe that such an implementation will take place. Just take a look at some of the members of the Alliance. They are all major players in their respective markets. The site is http://www.weca.net/OpenSection/index.asp.

Another advantage of Wi-Fi is mobility. Having a wireless network gives your company many options. It is easy to add other computers to the network. To add another wireless computer to a Wi-Fi network, just plug in your card or universal serial bus (USB) connection. After doing so, the computer is instantly networked. Computers can just as easily be removed. Furthermore, if your business needs to move, the company will not have to abandon its network infrastructure or hire someone to rewire your new office location. All that needs to be done is simply unplug at your old office and plug in the base station to a power outlet at your new office. Then turn on the other devices. Presto! You have assembled your network.

Wi-Fi Range & Security

When a Wi-Fi computer is close to the base station, it is capable of transferring and receiving data up to 11 Megabytes per second rate. As the Wi-Fi device is moved further away from the base station, the data rate will drop to around 5.5 Megabytes per second. Generally, the lowest data transfer rate at the maximum range from a base station will not drop below one or two Megabytes per second. This is a drastic deterioration from the 11 Megabyte maximum transfer rate. However, transferring information at one Megabyte per second is still an acceptable performance level, as compared to other connection-types. For instance, one Megabyte per second is faster than most DSL and cable connections. Therefore, it is still an above average high-speed transmission. This makes it more than adequate for such tasks as sending and receiving e-mail, browsing the Internet, and entering data from a mobile computer. A standard Wi-Fi antenna broadcasts from 750-1000 feet outdoors and in open spaces. In an office or other work-place setting, which has objects and other factors that could interfere with the signal, the standard antenna broadcasts from 150-350 feet. Keep in mind that dense materials such as certain metals may affect the transmission of radio waves, and therefore affect a Wi-Fi broadcast. If the walls in your office are heavily reinforced with such materials as metal, heavy wood, brick, or stone, Wi-Fi devices may have difficulty transmitting from one room to another.

Wired Equivalent Privacy (WEP) is 802.11's encryption standard. WEP works by encrypting the data transferred between the Wi-Fi computer and the Wi-Fi access point. A Wi-Fi access point (AP) is a hardware device or software that acts as a communication hub for wireless devices to connect to a wired local area network (LAN). They provide wireless security, while extending the service range that a user of wireless can access. When data is sent to the access point, it is done via the public Internet. Therefore, unless it is encrypted by some other means, the data is unprotected during its transmission over the Internet. When the data reaches the access point, it is encrypted and transferred to a Wi-Fi device by WEP. Other wireless technology standards with better security are also in development. IEE 802.1X is a more secure wireless access, which makes use of an Extensible Authentication Protocol (EAP). There have been several security problems with WEP. Therefore, the Wi-Fi Alliance is planning on implementing a new security standard. This new standard is Wi-Fi Protection Access (WPA). Wi-Fi Protected Access implements 802.1X and the EAP to provide better security.

Many corporations use Virtual Private Networks (VPNs) to protect remote-access to their WLAN's. The VPNs also protect their connections. A VPN creates a secure virtual tunnel from the worker's computer all the way to the corporation's server. This means that a secure connection is established from the worker's Wi-Fi hardware, through his or her Wi-Fi access point, over the public Internet, through the corporation's access point, to the corporation's hardware. Existing systems can be modified to make use of VPN, so as to support a Wi-Fi network. There are many different types of VPNs, some of which are very expensive. Microsoft, however, provides a very basic VPN technology that is free. The VPN is provided with Microsoft's advanced server operating systems. For more information, visit http://www.microsoft.com/ntserver/support/faqs/VPNSecFAQ.asp. The site is a Q & A covering the Microsoft VPN. A VPN works through a company's server. It encrypts any data transferred to company computers from outside the office. VPN software also operates on remote computers and laptops. Virtual Personal Networks enable the transfer of data, while protecting it from being viewed by unintended parties. Setting up a VPN gives a company a lot of mobility. It supports communications from areas such as airports and hotels, enabling people to stay abreast of developing situations, which may be crucial to a company. Through VPNs, individuals can access information in company networks from anywhere. With VPNs, companies can expand their networks to the homes of their employees. Therefore, through the use of a WLAN with a VPN, part-time employees can do more from home. In addition, full-time employees can spend more time at home and with their families, due to the fact that more tasks may be performed from their home computers. Another important aspect of Wi-Fi security is its ability to provide different levels of access. For example, a company can provide top-notch security while allowing open access to the Internet and email accounts for guests on the network. This is accomplished by granting network access at different levels. Therefore, a company's visitors can still have unlimited access to the Internet and e-mail, while the VPN access is only given to authorized employees.

In addition to a VPN, wireless networks have a standard technology called Remote Access Dial-Up User Service (RADIUS). RADIUS is used to protect access to wireless networks. Remote Access Dial-Up User Service is security that allows only approved users to access the network. In order for someone to access the network, and the corresponding files contained on the network, he or she must input a user name and password. The RADIUS server then verifies the name and password, and if correct, allows that person to access the network. A key feature of RADIUS is that it may be used to provide different types of access levels. A company may use a RADIUS server to provide access to the Internet, another one for network databases, and another one for access to email. Microsoft provides a basic RADIUS server with its advanced server operating systems for free. Other RADIUS servers may be purchased in the forms of software and hardware, but can be rather pricey.

One of the most formidable lines of system defense is the firewall. Firewalls work by blocking unwanted and unauthorized people from viewing company databases. Many Wi-Fi access points have built-in firewalls. In addition, all Wi-Fi's include a network capability enabling a group of devices to share one Inter Provider (IP) address, regardless of whether the IP is from a dial-up, cable, or digital service line (DSL) connection. This capability is called Network Address Translation (NAT). NAT works by creating new IP addresses for each computer of the group from the original single IP address. Therefore, the new IP addresses are invisible to users of the public Internet. The new IP addresses function separately "under the cover" of the original one.

Public Area Wi-Fi

Wireless networks are not only making their way into today's businesses, they are steadily becoming a reality in many public areas, as well. Companies such as Wayport, a company that provides Wi-Fi to hotels, airports, and other public areas around the country, are teaming up with major companies such as 3Com, Sony, Sharp, and other members of the Wi-Fi Alliance to provide wireless fidelity in public areas. Major hotels such as Four Seasons, Holiday Inn, Hampton Inn, Ramada, and many others are also teaming up to provide wireless fidelity, through Wayport, in their hotel chains. In addition, more and more remote access and wireless network service providers are beginning to pop up. Figure 1 shows a map of areas in the U.S. that support wireless computing. With members of industry groups such as the Bluetooth SIG and Wi-Fi Alliance pushing toward the development of a global wireless infrastructure, more hardware will be built on this standard. If you haven't noticed, this is already happening. Cellular phones and laptops equipped with Bluetooth and Wi-Fi certification are already on the market. These companies are trying to phase out wires. Therefore, eventually, more wireless will exist. With hotels and airports around the world upgrading to wireless, companies that go wireless will be a step ahead. They will have the up to the minute information needed, full integration of technological equipment, full access to their network databases, and the infrastructure to make it possible from anywhere in the world. An interested party's best bet is to begin researching the benefits and detriments of going wireless. The benefits of going wireless are: (1) it is cost effective because you do not incur the costs for wiring and labor to do so (you only have to set up a software protocol); (2) it is a portable network, so you can just move computers around the wireless router (a radio transmitter); and (3) it is peer-to-peer flexible because of the absence of wires, therefore a traditional router is not needed. The detriments of going wireless are: (1) transmission time is slower than the fastest wired networks; (2) controversy exists because radio frequency transmission is believed by some to cause health problems similar to microwaves; and (3) there may be security problems. Of course, with the use of a *bridge* a wireless network may be hooked up to a wired network (e.g. local area network). Therefore, a company's system may have an integration of wired and wireless networks. Beginning to upgrade now may be beneficial in the long run.

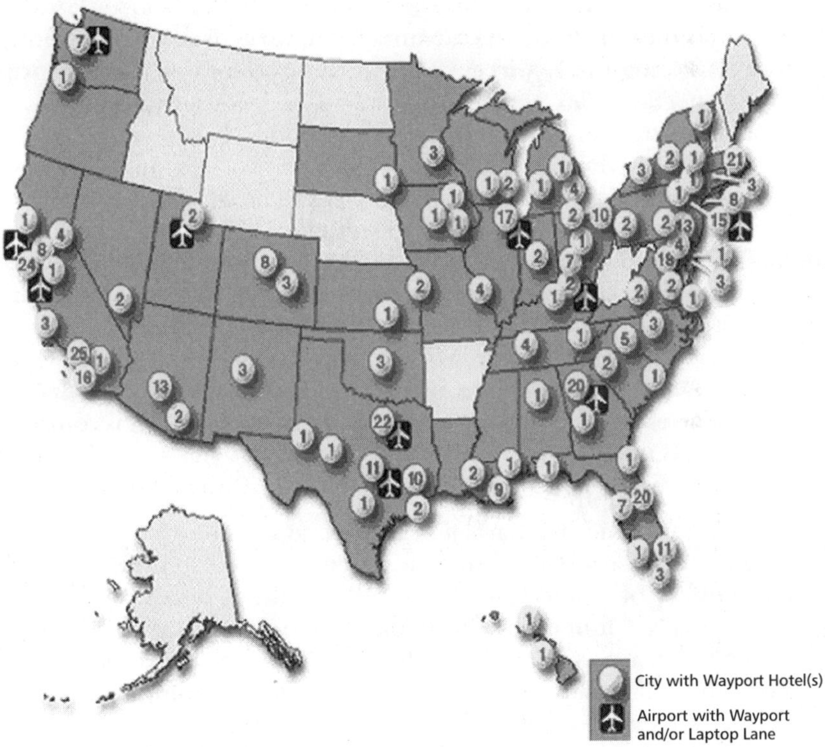

Figure 11.1: Public Areas with WI-Fi in the United States

Map was taken from http://www.wayport.net/locations. Please visit this site. At the site, click a geographic area on the map to view a list of hotels and airports in that area that have Wireless Fidelity Capabilities.

Wireless Gateways

Signal strength, your card's ability to receive data under different conditions, is essential. As you move away from the gateway, there is a degrading of signal strength and performance. The measure of gateway performance is the combination of range and throughput. Range refers to how far you can go from the gateway and still have a good signal. Throughput is a measure of the amount of work performed by a computer system over a given period of time. Note that throughput decreases significantly when several users access the gateway at the same time. The card in the gateway has a fixed bandwidth that all client cards must share.

Wireless vs. Wired

Mobility and efficiency are the advantages of using wireless. Using other network technologies like wired, phone line, or power line based networks require connections through wire or cable. Wireless uses radio waves that are capable of traveling through walls and floors. Furthermore, you can connect from anywhere, regardless of whether you are in or outdoors. Wireless networks do not require phone jacks to be close to the devices for them to be networked with the rest of the system. This completely eliminates the inefficient and expensive task of hiring people to rewire an office to support a wired network. Furthermore, wireless technology simply eradicates the clutter of wires around offices. In addition, power line networks are generally more expensive than Wi-Fi networks. Plus, transformers, large appliances, and surge protectors may interfere with power line networks (a type of wired network). Lastly, no wired network is portable. It is completely fixed when it is intact.

An employee at an office with a wired network cannot pick up his or her laptop and go to another part of the office to continue his or her work, unless there is an available connection. With wireless networks, having a limited number of connections in an office will not be an issue, because everywhere is a connection. Probable situation: you already have an expensive wired network in your office, and would like to add wireless components. This would not be a problem. Wireless components can be added to your existing wired network. Many wireless access points and gateways allow users to connect to wired equipment, as well. Many corporations frequently do this. Companies can extend their wired networks by adding Wi-Fi networks. By connecting wireless access points to an existing wired network, a company can provide wireless access to all areas of its office.

The major problem with wireless technology is its price tag. The components for wired networks are still cheaper than wireless network components. Access points can cost up to $1,000 each. Therefore, the cost of hardware that supports wireless connections can be easily two to four times more expensive than wired network components. The argument, however, is that this higher cost for hardware is offset by the ease of installation and maintenance. Accordingly, long-run costs to maintain the equipment are lower. There will be no expenditures to replace old wires. Furthermore, companies using wireless will never have to hire an electrician to run electrical wiring or install outlets to make a part of the office network-able. If the company is wireless, all that need to be done is power on the new hardware, and it is instantly networked. For a wireless network, the major expenditures will be for new hardware to add to the network.

Access points are a bit pricey, but in most instances the purchases of additional access points will be few and far between. The only foreseeable instance, in which the purchase of a new access point will be necessary, is when a company needs to significantly expand its network. Another point to keep in mind is the fact that wireless technology is still relatively unknown because it has not caught on yet. Therefore, as more and more hardware is produced to support wireless, more and more people will switch to this newer technology. Therefore, as demand grows, the price will decline. Also offsetting the higher costs of wireless are the savings the technology generates. As a result of lower maintenance costs, companies will spend less on information technology (IT). This coupled with greater efficiency and a higher level of productivity (generated through the use of wireless), will mean higher bottom lines.

Expected Growth of Wireless

The following section contains supplemental information. Keep in mind that the independent study discussed in this section is displayed on the Bluetooth website. Please read about it at http://www.bluetooth.com. The information contained in the study is significant when one considers the size and composition of the Bluetooth SIG. The companies that make up this industry group are household names. Furthermore, many of them are also members of the Wi-Fi Alliance.

Promoter Members of the Bluetooth SIG:

(Taken from http://www.bluetooth.com/sig/membership/promoter.asp)

3Com
Agere
Ericsson Technology Licensing AB
IBM Corporation
Intel Corporation
Microsoft Corporation
Motorola, Inc.
Nokia
Toshiba Corporation

* For a complete list of the group's members, please see the above website.

Some Members of the Wi-Fi Alliance:

(Taken from http://www.weca.net/OpenSection/members.asp?TID=2)

- 3Com
- Agere
- Apple
- Epson
- Ericsson Technology Licensing AB
- HP
- Intel Corporation
- Microsoft Corporation
- Motorola, Inc.
- Nokia
- Sharp
- Sprint
- Toshiba Corporation
- U.S. Robotics
- Yamaha

* For a complete list of the group's members, please see the above website.

Recently, the Bluetooth Special Interest Group, contracted with a research firm to project future demand for wireless technology. The group that performed the independent study was the Zelos Group LLC. The Zelos Group LLC is an advisory service firm, which helps assess the nature and timing of the adoption of new products and services by consumers. The group serves technology industries. Their website is http://www.zelosgroup.com. The group predicts that over the next four years, Bluetooth technology will generate around $12 billion. The majority of this revenue will be the result of purchases of Bluetooth-enabled equipment used to link pocket computers with mobile handsets, equipment used to wirelessly view and download personal digital assistant (PDA) files, and other equipment capable of wirelessly accessing email and the public Internet. The United States and Western Europe were covered in the studies for the years 2001 through 2006. The study assessed likely usages of Bluetooth in those regions to be for PDAs, headsets, handsets, notebook PCs, and automobiles. This information is not just beneficial to businesses seeking to go wireless, but also businesses that distribute hardware, software, and the like. Please view the website for more information.

UPS Going Wireless

Shipping giant United Parcel Service has decided to implement a wireless tracking system into its operations. The company has begun to deploy a wireless tracking system that combines Bluetooth with Wireless Fidelity. UPS plans to issue Bluetooth bar code readers to 55,000 of its package handlers. This is the first major wireless implementation of such a system. The project will cost around $120 million.

This large-scale implementation, which combines Bluetooth with Wi-Fi to create one system, may be an indication of future trends. As discussed earlier, Bluetooth's strength seems to be in its capabilities to facilitate communication between devices. On the other hand, Wi-Fi is more adept at providing wireless and remote access to the public Internet and private networks. Combining these technologies may prove to be the most efficient means of building a company's infrastructure. Wi-Fi access points could be used to provide access to a network, and Bluetooth can be used to connect devices. For example, a person would be able to access his or her email from their laptop wirelessly through Wi-Fi. That person could then download it, forward it, or edit it. Suppose the person wanted to print it. Bluetooth would enable the transmission of information from the laptop to a printer, and allow the computer to command the printer to print the document, all without wires. It's kind of like a "one-two wireless punch." There must be a reason for a company the size of UPS, which must keep track of a huge number of shipments daily, to spend $120 million on this new technology. Furthermore, a significant number of blue-chip technology companies are members of both the Bluetooth SIG and the Wi-Fi Alliance. Perhaps they too see the potential relationship between Bluetooth and Wi-Fi.

Wireless Networks and Your Business

Wireless systems have all the aspects needed to alter the way in which people work. Through remote access, more can be done from the home and on the road. Important information can be received in real time, virtually anywhere. The need to spend long hours in the office over weekends is greatly decreased by Wi-Fi. Small and medium sized businesses can now obtain the same (and perhaps better) technology to keep track of its inventory and distribute information to its employees, as their larger-sized competition. The best part is that a small business can install a wireless network itself. It will not have to enlist outside help for installation. Businesses can now provide each of its employees with live streaming, real time data, regardless of where they are. This type of wireless technology is ideal for retail, wholesale, shipping, and manufacturing companies. Figure 11.2 depicts a wireless network.

Conclusion

Already, many major producers are manufacturing and marketing wireless technologies, some of which you may unknowingly be using. Wireless networks have many advantages and should be strongly considered. A completely or partially wireless communication infrastructure may be implemented depending on your company's needs.

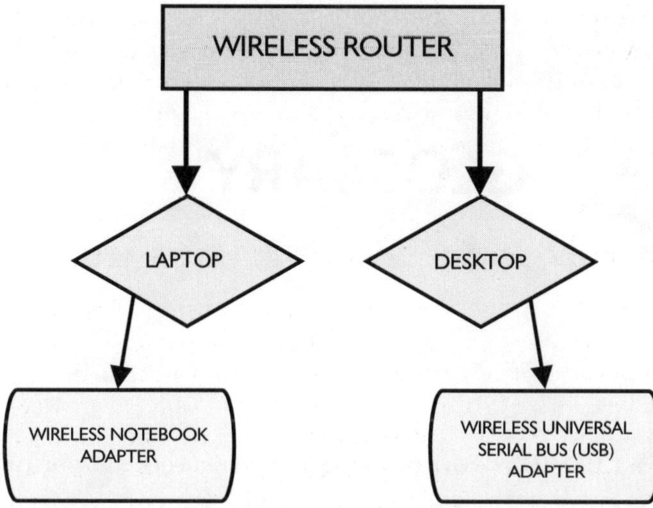

Figure 11.2: Wireless Network

[1] This chapter was coauthored by Matthew Derba of Deloitte & Touche, LLP

GLOSSARY

ACCESS POINT DEVICE device bridging a wired network to/from a wireless network.

ACTIVE MONITOR computer tracking the status of the token ring network.

AMPLITUDE height of a signal relative to a base line.

APPLICATION operational system conducting a function or activity such as database software.

APPLICATION SERVER specialized server furnishing access for client/server applications and information associated with those applications.

APPLICATION SHARING users such as in a workgroup share software and information running on one computer.

ARCHITECTURE designing and constructing a network to properly carry out the company's business objectives.

ASYMMETRIC DIGITAL SUBSCRIBER LINE (ADSL) modem term referring to the conversion of twisted pair telephone lines into access paths for high-speed data and multimedia transmissions.

ASYNCHONOUS TRANSFER MODE (ATM) call switching technology having the ability to integrate multimedia with data over a single type transmission method.

ATTRIBUTE features and characteristics associated with a file.

AUTOBAUD automatically selects the baud rate based on which the destination modem is able to perform.

BACKBONE one cable segment in a bus configuration linking computers in a straight line.

BACK-END server in a client/server arrangement.

BANDWIDTH frequency range of a transmission medium. It can refer to the maximum amount of data a medium can transfer.

BAUD speed of a modem.

BRIDGE device linking communication channels. Bridges can conduct protocol conversions on LANs. Remote bridging refers to bridging geographically remote networks. Bridges can merge multiple LANs into a single virtual network. Thus a bridge is a device used to connect independent networks together, allowing them to act as one. Different types of networks may be connected.

BROADBAND analog transmission method having simultaneous multiple communication channels.

CENTRALIZED COMPUTING processing occurs on a central computer only.

CHANNEL medium over which data is transmitted. A host computer is the physical device in which the data to be transmitted over the channel is located.

CLIENT computer asking for services and resources from a server.

CLIENT/SERVER a local area network (LAN) consisting of at least one server providing information to clients.

COAXIAL CABLE same cable used widely today for cable TV. It is a popular connection for bus networks.

COLLISION occurrence in an Ethernet network when two nodes broadcast simultaneously.

COMMUNICATION SERVER server enabling users to share a modem on the LAN for outgoing data calls. It can house multiple modems allowing access over different lines and speeds.

CONCENTRATOR monitoring, tracking, and managing network information.

CONNECTIONLESS protocol type transmitting data to an addressee without guaranteeing its recepit. This differs from a connection-oriented protocol that does assure data receipt.

CONTIGUOUS SECTOR storage locations on media that are in consecutive locations.

DATA CHANNEL cables and infrastructure of a network.

DATAGRAM data packet or transmission unit in a network layer.

DEDICATED SERVER computer which is almost exclusively operating as a server.

DESIGN planning a network.

DEVICE hardware topology item such as a modem.

DOMAIN

1. computers sharing a common database.
2. collection of computers organized into one security database.

DOMAIN CONTROLLER directory server furnishing access controls over users, groups, computers, and network resources.

DUPLEXITY capability of data flow. In a *simplex* system communication flows in one direction only. In a *half duplex* system communication flows in both directions but only one way at a time. In a *full duplex* system communication flows in both directions.

DYNAMIC DATA INTERCHANGE protocol for transferring information among applications.

ENCRYPTION security coding to safeguard data transmission.

ENTITY data transferred as the payload of a request or response.

EVENT LOGGING recording network activity for evaluation purposes.

FAULT condition that makes a device fail to operate properly.

FAULT APPRAISAL identifying a fault requiring correction. Fault management corrects network errors.

FIBER OPTICS cable using laser light for data transmission. It is the most costly medium of transmission.

FILE SERVER computer storing files, master software package, and databases to be accessed by clients. A dedicated file server is a computer that instructs the activities and applications on the network. A file server can operate multiple programs at the same time for multiple users carrying out different tasks.

FLOW exchange of data among devices such as telecommunications software.

FRAME RELAY data packet-switching network service. It is the means of transmitting data over a wide area network by segregating the data into packets or frames. However, it does not conduct error checking activities because it expects that to be done at either end of a connection typically by the hardware.

FREQUENCY number of cycles a wave makes per second.

FRONT-END client in a client/server arrangement.

GATEWAY device linking heterogeneous (unlike) network segments.

GLOBAL NETWORKING company's network crosses national, commercial and government boundaries. International activities are supported.

GROUPWARE network application allowing more than one user to interact at the same time.

HUB central connection point in a network over which data passes. An intelligent hub can generate information about data packets transmitted through it and is able to identify any difficulties needing correction. An active hub is capable of retransmitting information received to other hubs.

HYBRID NETWORK network having the aspects of both peer-to-peer and client/server.

INTERFACE connection of two devices.

INTERNETWORK network of local area networks.

INTRANET a company's private network.

LOCAL AREA NETWORK (LAN) linking of computers and devices in a limited area such as in company buildings located in close proximity. LANs provide the building blocks for wide area networks.

LOCALLY ATTACHED device that is directly connected to only one computer.

LOCKING a system to prevent more than one user accessing data at the same time.

LOOP twisted-pair copper telephone line.

MANAGEMENT PROTOCOL a network's communication language.

METROPOLITAN AREA NETWORK covers a larger area than a LAN. MANs interconnect network sites on a need basis. MANs usually connect LANs and WANs.

MULTIPLEXING method enabling more than one message to share a communication channel. This can increase the amount of data that can be transmitted through available circuits.

MULTITASKING running multiple applications on the computer simultaneously.

NAME SERVER software running on a host that maps names to addresses.

NETWORK computers linked together to perform some function such as sharing resources and services.

NETWORK (ADAPTER) INTERFACE CARD (NIC) card attaching a computer to a network medium. The NIC translates information moving over the network into standard sized "packets" of data. The NIC checks incoming packets to make sure all of them were received correctly – a function called error checking.

NETWORK MEDIUM connection (e.g., cable) linking network computers.

NETWORK TOPOLOGY arrangement of computers in a network.

NODE ending point for connected communication devices.

PACKET string of bits that has been built for transmission over a network. A packet usually contains a part of a file because files are divided between many packets before transmission and reassembled at the receiving station.

PACKET SWITCHING transmission approach having packets transmitted over a networking medium supporting multiple pathways between sender and receiver.

PACKET TRAILER data added to the end of information being transmitted. Such addition is typically of an error-checking nature.

PATHWAY way to get information moved from one computer to another.

PEER-TO-PEER network where connected computers can be a client and/ or server. Centralized control over shared resources is non-existent.

PERIOD cycle length.

POINT-TO-POINT PROTOCOL standard technique to transport multi-protocol datagrams over point-to-point links. It connects transport packets between two peers.

POLLING channel access technique in which a primary device questions secondary ones if they have information to send.

PROTOCOL transmission paramenter associated with computer linkage. It refers to rules associated with transmission over a network so devices can properly communicate.

PROXY intermediary software functioning as both a server and a client making requests on behalf of other clients. Requests are serviced internally or by passing them on, with possible translation, to other servers.

RADIO WAVES data transmission over radio frequencies with practically no distance restrictions. They may be used to connect far distant LANs. This form of transmission is costly and vulnerable to interference.

REPEATER device repeating a signal to make it stronger and extend its travelling distance. An intelligent repeater can conduct packet filtering.

RESOURCES devices, services, and information available in a network.

ROUTABLE protocol having network addresses.

ROUTER device connecting two or more LANs with different physical media or translating between different architectures. It routes data. In dynamic routing, the routers learn from each other which paths are available. Routers can determine the most efficient path to use to send each network message.

SECTOR smallest logical storage unit.

SEQUENTIAL DIGITAL NETWORK one whose outputs depend not only on the current inputs but also on the sequence of previous inputs.

SERVER computer furnishing resources or services to client computers.

SERVER SESSION connection of a server to a node.

STANDALONE computer or device not connected to the network.

SWITCHING directing data communications between sender and intended destination.

TERMINAL SERVER implements multiple dumb terminals concentrated into a single device.

TOKEN RING NETWORK an electronic token is passed from station to station. The station that obtains this token is allowed to transmit while the others that do not have this token cannot transmit. If the unit with the token does not have anything to transmit it just passes the token on.

TOPOLOGY configuration (physical layout) of a network so data may be transmitted among its components. In an active topology, the computers themselves send information along the network. In a passive topology, the computers listen to the data signals being transmitted but do not participate in the communications.

TRANSMISSION CONTROL PROTOCOL (TCP) application providing the ability to conduct remote logons to remote hosts. TELNET operations using a client and server. TCP/IP was developed to allow dissimilar machines to talk to one another.

TUNNEL intermediary software functioning as a blind relay between two connections.

WIDE AREA NETWORK group of interconnected LANs located distant apart including globally.

WIRELESS NETWORK a LAN in which transmission is not over wires such as using infrared technology or radio frequencies to connect computers.

Index

A

Address resolution protocol (ARP) 58
Agent event manager 205
Application layer protocols 66
Asynchronous Transfer Mode (ATM) 30
Authentication procedures 237
face recognition 243
facial thermograms 243
fingerprints 241
gait prints 244
hand
geometry 243
veins 243
handwriting analysis 243
iris prints 242
palm prints 243
retina prints 241
signature analysis 243
silhouette identification 244
voiceprints 242

B

Backbone configuration 2
Backup 167, 217
data compression 172
facilities 168
identifying critical data records 168
Internet
choosing a provider 269
using 269

lossless compression 172
lossy compression 172
network data management protocol 171
policy 167
recovering data 171
remote 171
techniques 170
testing 171
Bandwidth 3
Bluetooth 270, 272
members of SIG 282
practical usages 274
Bridges 29
Bus 14

C

Cable
ARCNet 90
cabling and interfaces 85
cabling considerations 92
coaxial 74, 87
types 88
fiber optic 77, 90
modem 80
plenum cabling 90
shielded
twisted pair 75
twisted-pair cabling (STP) 87
thicknet 89
coaxial 74
thinnet 88
coaxial 75
topology 72
twisted-pair cable 85
types 74, 85
comparisons 92
unshielded twisted pair 76
cabling 85
Capacity design and management 36–48
Cisco's LocalDirector 133
Client 211
workstation, configuring 112
Client server
architecture 199
agent event manager 205

configuration 201
connecting 207
groupware 226
middleware 223
network operating system 205
standards 157
technologies 197–229
upperware 226
Clustering 216
Connections
analog 79
dedicated remote 78
dial-up 78
digital subscriber line (DSL) 80
integrated services digital network (ISDN) 79
switched multimegabit data service (SMDS) 82
T1 and T3 81
wireless networking 82

D

Data
integrity 138
packet collision 42
return network infrastructure 146
storage and management 162–172
transmission
block mode 52
compressed mode 53
stream mode 53
understanding storage 162
Databases 207
hierarchic 209
Digital subscriber line (DSL) 80
Directory services 4
Distributed computing 226
Domain name system 210
Dynamic Data Exchange (DDE) 32

E

E-mail 54
Enterprise network 203
Ethernet 24, 41
Extranet 7

F

File transfer protocol (FTP) 34, 51
Firewalls 98, 133, 249
high performance 250
Fragmentation 216
FTP, *See* ***File transfer protocol***

G

Gateway 30
Groupware 226

H

Hubs 27, 115

I

IBM e-business Solutions 128
Integrated services digital network (ISDN) 79
Internet
Protocol Suite Stack 65
security 230–269
server-side logic controls 256
using to back up data 269
Internetworking 10
IP Addressing 57

J

Java
applets 173–196
applications 173–196
arrays 181
class library 183
creating
simple applet 176
application 175
data variables and calculations 179
development kit 174
logic and loops 184
naming
conventions 180
variables 179
object oriented programming 178
program structure 175

running an application 174
strings 181
variable types 180

L

LAN
configuration 11
switching 147–161
technology 147
Topologies 13
Lightweight directory access protocol (LDAP) 32
Linear-bus topology 72
Linux 12, 102
Local Area Network (LAN) 10

M

MAC layer 58
Malicious programs 251
back door 253
bomb 252
data input alteration 253
dropper 252
masquerade 253
salami techniques 252
scavenging 253
trap door 253
trojan horse 251
worm 252
Media 3
Memory 163
MESH configuration 22
Middleware 223
Modem, cable 80
Multistation access unit (MAU) 21

N

NetWare 11
Network 1
access device 5
administrators 48
advantages 9
arrangement 40
availability levels 123
backbone 131

basic input/output system (NetBios) 33
binding 2
client/server standards 157
data management protocol (NDMP) 171
design 44
device interface specification (NDIS) 94
enterprise 203
fault tolerant 119–146
hardware 1
high availability
system clustering 136
design 132
hybrid 7
interface card (NIC) 2
introduction to high availability 119
intrusion detection systems 262, 267
deception systems 267
Honeypot 267
location 269
log file monitors 267
models 268
system integrity verifiers (SIV) 267
local area network (LAN) 10
networking standards 157
operating system
NOS 9
OS 9
peer-to-peer 6, 94
performance 68
and troubleshooting 8
physical
cabling needs 71
design 1, 70
planning 7
remote
access 56
users 57
securities authentication procedures 237
security 66, 97, 230–269
communication 245
general strategies 231
physical considerations 236
procedures 234
protection of hardware 234
server based 95

application servers 97
configuring 97
operating systems 101
speed 100
stand-alone system 1
subnets 62
troubleshooting 115
using to back up data 269
wireless 82, 270–285
wiring 70
Networking 4
node 4

O

Open Datalink Interface (ODI) 94
OSI Stack 64

P

Packet 3
Protocols 31

R

Radio waves 91
RAID (Redundant Array of Inexpensive Disks) 107
0: striping 140
1: shadowing/mirroring/duplexing 141
3: striping and parity 142
5: striping and parity 142
introduction 139
levels advantages and disadvantages 143
peer-to-peer 1: mirroring over networks 141
software v hardware 144
Repeaters 30
Ring 19
configuration 21
topology 73
Router 28
Routing
fundamentals 60
graphic 61

S

Security 66
Sequenced Packet Exchange/Internetwork Packet Exch 32

Server 39, 212
application 97
server 220
availability 135
backup 217
client/server standards 157
communication server 221
connecting clients and servers 207
connection issues 110
database server 220
dedicated v nondedicated 215
directory service servers 221
file server 218
gateway server 220
groupware server 222
hard disk mirroring 107
hard drives 106
hardware requirements 105
memory requirements 47
message server 221
needs 47
print server 221
proxy server 222
redundant array of inexpensive disks (RAID) 107
remote node server 222
security server 223
side logic controls 256
software 111
tape backup devices 108
transaction server 219
types of 218
uninterruptible power supply (UPS) 110
web application server 222
wide area information services (WAIS) server 222
Simple
Mail Transfer Protocol (SMTP) 54
network 37
Star
bus topology 23
configuration 16
ring 24
shaped topology 73
Storage
assessing requirements 162
backup 167

facilities 168
network data management Protocol (NDMP) 171
remote 171
techniques 170
devices and media 163
memory 163
compact disc 167
drive arrays 166
EPROM 164
hard drives 165
magneto-optical or floptical drive 166
RAID 166
RAM 164
removable hard drive technology 166
understanding data storage 162
Subnets 62
Switched multimegabit data service (SMDS) 82
Switching 30
System clustering 136
Systems network architecture 41

T

T1 81
T3 81
TCP/IP (Transaction Control Protocol /Internet Protocol) 34
basics 49–69
file transfer 51
use of 49
Token ring 43
topology 73
Topology
BUS 15
cable 72
hierarchical 13
linear-bus 72
ring 73
star-shaped 73
token ring 73
Transaction processing 228
Transport layer protocols 63
Tree configuration 17

U

Uninterruptible power supply (UPS) 110
Unix 11

Upperware 226

V

Virtual local area network (VLAN) 12
Virus 254
  boot sector 255
  file infector 255
  macro virus 255
  multipartite virus 255
  polymorphic virus 256
  stealth virus 255

W

WAN technology 147
Wi-FI range & security 276
Wi-Fi (Wireless Fidelity) 270, 275
  Alliance members 282
  public area 278
Wide Area Network (WAN) 25
Wireless
  network 91
    internetworking 83
  networking 82
  technologies 91
    infrared transmissions 92
    microwaves 91
    radio waves 91
Wireless networks 270–285
  compared to wired 280
  gateways 280
  interoperability 274
  technology 270
    Bluetooth 270
    Wi-Fi 270
Wiring networks 70